IMPACT OF COMPUTATIONAL MECHANICS ON ENGINEERING PROBLEMS

PROCEEDINGS OF THE SEMINAR ON IMPACT OF COMPUTATIONAL MECHANICS ON ENGINEERING PROBLEMS / SYDNEY / AUSTRALIA/ 2 AUGUST 1993

Impact of Computational Mechanics on Engineering Problems

Edited by

V.A. PULMANO & V. MURTI

University of New South Wales, Australia

A.A. BALKEMA / ROTTERDAM / BROOKFIELD / 1993

The texts of the various papers in this volume were set individually by typists under the supervision of each of the authors concerned.

Published by
A.A. Balkema, P.O. Box 1675, 3000 BR Rotterdam, Netherlands
A.A. Balkema Publishers, Old Post Road, Brookfield, VT 05036, USA

ISBN 90 5410 324 8

Printed in the Netherlands

Impact of Computational Mechanics on Engineering Problems, Pulmano & Murti (eds)
© 1993 Balkema, Rotterdam, ISBN 90 5410 324 8

Table of contents

Impact of Computational Mechanics on Engineering Problems, Pulmano & Murti (eds)
© 1993 Balkema, Rotterdam, ISBN 90 5410 324 8

Preface

This Seminar was convened to mark the 60th birthday of Val and was initiated by the second editor, his former Ph.D. student. The papers in this volume are dedicated by the authors to Val, and are divided into two parts, namely: (a) personal notes and (b) technical papers. We thank all those who took part and contributed to this Seminar in any way. We are specially grateful to Professor O.C. Zienkiewicz, a long standing friend of Val and a former mentor at Swansea, for his personal note which gives an account of his close association with Val and his family, and his academic accomplishments.

To Val, we wish many more years of productive work, and please accept, on behalf of all contributors, this volume on the occasion of your 60th birthday.

V.A. Pulmano
V. Murti
August 1993

Personal notes

Impact of Computational Mechanics on Engineering Problems, Pulmano & Murti (eds)
© 1993 Balkema, Rotterdam, ISBN 90 5410 324 8

S. Valliappan: An eminent civil engineering scientist

O.C.Zienkiewicz
University of Wales, Swansea, UK

1 INTRODUCTION

The 60th birthday is an important event in anybody's life. For a man who has accomplished so much as Professor Valliappan in his civil engineering and academic career this is even more significant. It marks for him the moment when he can look back, hopefully in a relaxed and philosophical manner, on the strains and efforts of the past and contemplate the future directions of his activity. This activity I am sure will be continued with energy for many years to come and reach even greater heights.

Happy Birthday, Val, and best wishes of health and productivity for the future.

Of course Val is the name by which his friends know Professor Valliappan and by which I have called him since 1965 when we first met. Indeed, I seem to recall that shortly afterwards, when we were publishing a joint paper, he had to invent an initial to meet the usual Western practice. Here he chose the first letters of his father's name: Somasundaram.

This long period of our accquaintance of almost thirty years makes me, of course, an "authority" on writing the present, brief, biographical appreciation and I shall try to do justice to his past and to his achievements.

2 THE YOUTH (1933-1965)

This is a period of which I know little. He was born in Devakottai (India) of father Somasundaram and mother Meena on the 15th of June, 1933. Then he finished his primary and secondary education, joining at the age of 16 the American College of Madurai for two years and completed his bachelor's degree in Annamalai University in the state of Madras in 1955.

The American influence was evident. Despite being selected due to the excellence of his degree for an academic post, and reaching the level of Readership in 1960, he took the opportunity of higher study in the USA when it became available. At the Northeastern University (Boston) he completed his M.S. in 1963 and it was because of this experience that I recruited him to join the Swansea Finite Element team in 1965.

3 THE SWANSEA INTERLUDE (1965-69)

The mid-sixties were probably the most exciting period of finite element development. At last the basis of the method was reasonably established and the search of new areas of application was on. The group of my collaborators at Swansea was expanding and included such names as Cheung and Irons, both of whom became eminent in this field. Many visitors and research workers participated and Val joined as one of these with the support of the Civil Engineering Research Association. The objectives were the extension of the finite element procedure into the areas of rock and soil mechanics and here, for the first time, nonlinear problems had to be handled.

Through the work of Professor Valliappan at

that time several serious breakthrough developments were made. One of these was the introduction of the so-called "no tension" material model to describe the behavior of such materials as rock and concrete. Another introduced the "initial stress" approach for elasto-plastic analysis. Although the second is now known by the name of the modified Newton method, both developments are still used extensively in practice!

This and other work assured for Val an excellent doctorate and, immediately after that, a position with the University of New South Wales where his subsequent career developed and where he is still now!

However, more memories of the Swansea period come to mind, many of these evoked by Val himself who, a couple of years ago, wrote some reminisences on the occasion of my own 70th birthday. Through my mind come not only the complicated computing procedures of the early days. These invariably involved delegation of the researcher for one or two nights to London where a more powerful machine was available to us. At the time in Swansea we only had an old IBM 1620 with the power somewhat less than that of a contemporary PC.

Many times Val would attempt (without luck on initial tries) the solution of a problem there while Helen, my wife, consoled Kamala assuring her that the London trips were a strict, scientific necessity. In the end of course it all came right and the computer worked as well as the programs. But Kamala herself contributed much to the work.

At that time we were frequently engaged in consulting activity and one of the problems with which we were concerned was the analysis of the Churchill Falls hydroelectric underground power station for which many finite element analyses were required. Time flies, and today such things as automatic mesh generation are taken for granted. Then a mesh had to be decided on by experience (with both of us participating), then laboriously drawn (by Val) and finally "digitizing" was carried out (faultlessly) by Kamala! It is such a long time ago that most of us here have forgotten the early computer encounters and difficulties.

4 AUSTRALIA NEW SOUTH WALES (1969-)

The start of this period tested the independence of Val severely and here he passed those tests with flying colors. Being the unique finite element expert at that University at the time, he had first to put the teaching of the subject on the map and organize his own team of students and assistants. Further, he had to develop his own approach to problems and decide on research directions not being now supported by a team of colleagues. The publication list attached to this article is a measure of success he has achieved in that period. Three books, six summary survey chapters in books and more than 120 papers are the achievement few can boast about.

The subjects of his research were wide ranging. The natural extension of his interest in plasticity and nonlinear material behavior continued with a first numerical formulation of plastic anisotropy. However, completely new areas such as biomechanics (and the analysis of the hip prosthesis), earthquake dynamics and wave propagation, porous media behavior in soils and electromagnetism testify to his imagination and breadth. In all he has made substantial and widely cited contributions.

Today his interests go even further, and an area in which he now treads is that of "fuzzy logic". This name is not to be confused with poor thinking. It is a field of mathematics permitting the use of uncertain and variable data in the prediction of realistic performance and Val was one of the first to realize its possible potential in real engineering. He is now an expert here, called upon to give keynote presentations and survey articles. This recent work is, I am sure, the foretaste of things he will yet contribute.

Though evidently wide ranging his work has at all times been connected with Civil Engineering in its broad sense and definition of this work. Though much of the profession today seems to be concerned with work schedules, contract prices and other business activity, he has chosen to deal with the sharp end of the work where the analysis and design are undertaken and "one off" problems dealt with. It is this side of the work that has always had the main attraction to myself, and perhaps my influence has remained there. It is, after all, most important to ensure that the construction is safe and well designed than merely to deal with accountancy!

This Civil Engineering interest is manifested in the multitude of his consultancies and practical projects in which he has been involved. These range from road design, tunnel assessment, to the design of important dams in Australia, China, and elsewhere.

In all of this, his teaching activity must not remain forgotten. Both at NSW and elsewhere, he has introduced many special courses and the appreciation of his students is widespread in many parts of the world, and in Australia in particular.

Indeed today Val must be a fully certified "Aussie" and will soon celebrate his 25th anniversary of his arrival on that "island". It gives me much pleasure to observe, however, that he has not forgotten his country of birth and its habits. Our paths crossed in India in 1980 and I recall with much regard that when his father died at the time he still was sufficiently local to offer personally his father's ashes to the Ganges. These new and old roots make the whole man!

5 VAL TODAY

His large professional research and teaching efforts and achievements have been outlined. But at a certain stage of development, public commitment is required. Much of this has been evident throughout his time in NSW through editorial boards, reviews, etc., but today his enthusiasm is being directed to the activities of the International Association of Computational Mechanics. He is the main organizer of the 2nd Asian Pacific Conference on Computational Mechanics that will take place shortly after this "birthday party". Further, he is now the founding president of the Australian section of this Association. Congratulations, Val, and best wishes for this work and also for the happy life with Kamala and your family in the years to come.

PUBLICATIONS

A. Books and Chapters in Books:

1. Continuum Mechanics - Fundamentals, A. A. Balkema Publications, Netherlands, 1981.
2. Finite Element Analysis in Geomechanics, Unisearch Ltd., Kensington, 1979.
3. Soil Mechanics - Recent Developments, Ed. Valliappan, S., Hain, S. and Lee, I. K., Unisearch Ltd., Kensington, 1975.
4. Chapter on Analysis of Soil Settlement, Soil Mechanics - New Horizons, Ed. Lee, I.K. (Chapter 5), Butterworths,London, 1974.
5. Chapter on Application of Finite Element Method to Soil Deformation, Soil Mechanics, Recent Developments, Ed. Valliappan et al, 1975.
6. Chapter on Application of Finite Element Method to Soil Dynamics, Soil Mechanics, Recent Developments, Ed. Valliappan et al, 1975.
7. Chapter on Constitutive Relationships for Composite Material through Micromechanics, Mech. of Engineering Materials, Ed. C.S. Desai and R.H. Gallagher, John Wiley and Sons Ltd., 1984, with I.J. Curiskis.
8. Chapter on Nonlinear Analysis of Contact Problems, Numerical Methods in Coupled Systems, Ed. Lewis, Bettess and Hinton, John Wiley and Sons Ltd., 1984, with I.K. Lee and P. Boonlualohr.
9. Chapter on Application of Fuzzy Sets in the Finite Element Analysis of Geotechnical Problems, Computer Applications in Civil and Building Engineering, Kozo Systems Publishers, Tokyo, 1991, with T.D. Pham and V. Murti.

B. Papers:

[1] Zienkiewicz, O. C., Valliappan, S. and King, I. P., Stress Analysis of Rock as a 'No-Tension' Material, Geotechnique, Vol. 18, 1968, 56-66.

[2] Zienkiewicz, O. C., Valliappan, S. and King, I. P., Elasto-Plastic Solutions of Engineering Problems 'Initial Stress' Finite Element Approach, Int. Jnl. of Numerical Methods in Engineering, Vol. 1, 1969, 75-100.

[3] Zienkiewicz, O. C. and Valliappan, S., Analysis of Real Structures for Creep, Plasticity and Other Complex Constitutive Laws, Proc. of Southampton 1969 Civil Engg. Material Conf., Wiley Int. Sci., 1969, 27-48.

[4] Valliappan, S. and Nath, P., Tensile Crack Propagation in R. C. Beams-Finite Element Technique, Int. Conf. Shear Torsion and Bond in Reinforced and Prestressed Concrete, India, 1969.

[5] Valliappan, S. and Doolan, T. F., Nonlinear Stress Analysis of Reinforced Concrete, J. of Struc., Div., ASCE, Vol. 98, ST4, 1972, 885-898.

[6] Valliappan, S., Elasto-Plastic Analysis of Anisotropic Work-Hardening Materials, Archivum Mechaniki Stosawanej, Vol. 24, No. 3, 1972, 465-281.

[7] Wood, R., Valliappan, S. and Svensson, N. L., Stress Analysis of Human Femur, Theory and Practice in Finite Element Structural Analysis, University of Tokyo Press, 1973, 461-478.

[8] Valliappan, S., Lee, I. K. and Boonlualohr, P., Finite Element Analysis of Consolidation Problem, Proc. of Int. Symposium on F. E. Methods in Flow Problems, Swansea, 1974, 741-747.

[9] Valliappan, S., Lee, I. K. and Boonlualohr, P., Settlement Analysis of Piles in Layered Soil, 7th Conf. of Australian Road Research Board, 1974, Paper No. A 31, 1-10.

[10] Boonlualohr, P., Valliappan, S. and Lee, I. K., Elastic-Plastic Analysis of Shallow Foundations, Proc. of Int. Conf. on Finite Element Methods in Engineering, Sydney, 1974, 479-492.

[11] Valliappan, S. and Germanis, E., Thermal Stress Analysis of Gravity Dams, Int. Con. on Finite Element Methods in Engineering, Coimbatore, India, 1974, 79-92.

[12] Valliappan, S. and Pulmano, V. A., Torsion of Nonhomogeneous Anisotropic Bars, J. of Struc., Div., ASCE, Vol. 100, ST. 1, 1974, 286-295.

[13] Valliappan, S., Wilson, M., Crane, P. F. and Donaghy, A. T., Use of computers in Welding Technology, Proc. of Conf. on Computers in Engineering, Sydney, 1974, 188-192.

[14] Valliappan, S. and Boonlualohr, P., Analysis for Plastic Flow of Anisotropic Materials, Proc. of 1st Aust. Conf. on Engineering Materials, Sydney, 1974, 701-721.

[15] Haque, M. N., Valliappan, S. and Cook, D. J., Tensile Creep Analysis of Concrete Structures, Proc. of Int. Conf. on Finite Element Methods in Engineering, Sydney, 1974, 349-364.

[16] Haren, R. and Valliappan, S., Finite Element Analysis of Electro-Magnetic Field Problems, Proc. Int. Conf. on Finite Element Methods in Flow Problems, Swansea, 1974, 739-741.

[17] Valliappan, S. and Lee, I. K., Consolidation of Non-Homogeneous Anisotropic Layered Soil Media, Proc. of 2nd Australia-New Zealand Conf. on Geomechanics, Brisbane, 1975, 67-72.

[18] Germanis, E. and Valliappan, S., Mining Subsidence at the Graving Dock Site, Newcastle, Proc. of Symp. on Recent Developments in the Analysis of Soil Behaviour and their Application to Geotechnical Structures, Sydney, 1975, 173-187.

[19] Valliappan, S., Application of Finite Element Method to Soil Deformation, Soil Mechanics, Recent Developments, Ed. Valliappan et al, 1975, 113-143.

[20] Valliappan, S., Application of Finite Element Method to Soil Dynamics, Soil Mechanics, Recent Developments, Ed. Valliappan et al, 1975, 327-355.

[21] Hain, S., Valliappan, S. and Lee, I. K., Analysis of Rafts on Nonhomogeneous Nonlinear Soil, Proc. Int. Conf. on Finite Element Methods, Adelaide, 1976, Paper No.28.

[22] Valliappan, S., Boonlualohr, P. and Lee, I. K., Nonlinear Analysis for Anisotropic Materials, Int. Jnl. Num. Methods in Engineering, Vol. 10(3), 1976, 597-606.

[23] Valliappan, S., White, W. and Lee, I. K., Energy Absorbing Boundary for Anisotropic Materials, Numerical Methods in Geomechanics, Ed. C.S. Desai, ASCE, Vol. II, 1976, 1013-1025.

[24] Boonlualohr, P. and Valliappan, S., Elasto-Plastic Torsion of Anisotropic Bars, Jnl. of Eng. Mech. Div., ASCE, Vol. 102, EM. 6, 1976, 995-1008.

[25] Goel, V. and Valliappan, S., Stress Analysis of First Molar, Proc. Int. Conf. on Finite Element Method, Adelaide, 1976, Paper No. 8.

[26] Svensson, N. L., Valliappan, S. and Wood, R. D., Stress Analysis of Human Femur with Implanted Charnley Prosthesis, Jnl. of Biomechanics, Vol. 10, 1977, 581-588.

[27] Valliappan, S., Svensson, N. L. and Wood, R.D., Three Dimensional Stress Analysis of the Human Femur, Computers in Biology and Medicine, Vol. 7, No. 4, 1977, 253-264.

[28] Valliappan, S., Matzusaki, K. and Raja Sekhar, H. L., Nonlinear Stress Analysis of Buried Pipes, Proc. Symposium on Soil-Structure Interaction, Roorkee, India, January 1977, 1-9.

[29] Valliappan, S., Favalaro, J. J. and White, W., Dynamic Analysis of Embedded Footings, Jnl. of Geotech., Div., ASCE, Vol. 103, No. GT2, February 1977, 129-133.

[30] White, W., Valliappan, S. and Lee, I. K., Unified Boundary for Finite Dynamic Models, Jnl. of Eng. Mech., Div., ASCE, Vol. 103, No. EM5, 1977, 949-964.

[31] Valliappan, S., Lee, I. K., Chandrasekaran, V. and Matzusaki, K., Evaluation of Reinforced Earth as Composite Material, Proc. Int. Conf. on Materials of Construction for Developing Countries, Bangkok, 1978, 897-910.

[32] Lee, I. K. and Valliappan, S., Analysis of Retaining Structures, Keynote Address, Int. Conf. on Materials of Construction for Developing Countries, Bangkok, 1978.

[33] Valliappan, S. and Raja Sekhar, H. L., Interaction between Tunnel Openings, Proc, 3rd Australian Tunnelling Conf., Sydney, 1978, 17-22.
[34] Curiskis, J. and Valliappan, S., Solution Algorithms for Linear Constraint Equations in Finite Element Analysis, Jnl. of Computers and Structures, Vol. 8, No. 1, 1978, 117-124.
[35] Geol, V. K., Valliappan, S. and Svensson, N. L., Stresses in the Normal Pelvis, Jnl. for Computers in Biology and Medicine, Vol. 8, 1978, 91-104.
[36] Valliappan, S., Chandrasekaran, V. and Lee, I. K., Interaction between Tunnel Openings due to Vibration Effects, Numerical Methods in Geomechanics, Aachen, Vol. 2, 1979, 685-696.
[37] Valliappan, S., Numerical Techniques in Geomechanics, State-of the Art Paper, Int. Conf. on Computer Applications in Civil Engineering, Roorkee, India, 1979.
[38] White, W., Valliappan, S. and Lee, I. K., Finite Element Mesh Constraints for Wave Propagation Problems, Proc. 3rd Int. Conf. on Finite Element Methods, Sydney, 1979, 531-539.
[39] Valliappan, S., White, W., Germanis, E. and Jagger, A. C., Analysis of Earthquake Effects on the Intake Tower of Mangrove Creek Dam, Proc. Int. Conf. on Engineering for Protection from Natural Disasters, Bangkok, 1980, 85-98.
[40] Valliappan, S. and Evans, R., Finite Element Analysis Of Slopes at Illawarra Escarpment, Proc. 3rd ANZ Geomechanics Conf., Wellington, 1980, Vol. 2, 241-247.
[41] Valliappan, S., Kjellberg, S. and Svensson, N. L., Finite Element Analysis of Total Hip Prosthesis, Proc. Int. Conf. on Finite Elements in Biomechanics, Arizona, 1980, 527-548.
[42] Svensson, N. L., Valliappan, S. and McMahon, R., Finite Element Analysis of Simulated Total Hip Replacement, Engineering in Medicine, Vol. 9, No. 3, 1980, 143-146.
[43] Svensson, N. L. and Valliappan, S., Analysis of Total Hip Prosthesis, Presented at the 21st Conf. on Physical Sciences and Engineering in Medicine and Biology, Melbourne, 1981.
[44] Evans, R., Valliappan, S., Raja Sekhar, H.L. and McGuckin, D., Stability Analysis of a Rock Slope Against Toppling Failure, Weak Rock: Soft Fractured and Weathered Rock, Vol. 2, 1981, A. A. Balkema, 665-671.
[45] Lee, I. K., Valliappan, S. and McGuckin, D., Nonlinear Consolidation Analysis of Clay Layers, Proc. of 4th Int. Conf. on Numerical Methods in Geomechanics, Canada, Vol. 1, 1982, 319-327.
[46] Valliappan, S., Lee, I. K. and Chandrasekaran, V., Composite Analysis of Pile Foundations, Proc. of Int. Conf. on Finite Element Methods, China, Science Press, Okine, 1982, 442-448.
[47] Valliappan, S., Svensson, N. L. and Soliman, F. I., Thermal Modelling of Femur-Acrylic Grout Prosthesis System by Finite Element Technique, Finite Element Flow Analysis, University of Tokyo Press, 1982, 877-889.
[48] Valliappan, S., Lee, I. K., Murti, V., Ang, K. K. and Ross, A. H., Numerical Modelling of Rock Fragmentation, Proc. of the 1st Int. Symposium of Rock Fragmentation by Blasting, Sweden, 1983, 375-391.
[49] Valliappan, S. and Raja Sekhar, H. L., Stability of Buried Pipes Due to adjacent Excavations, Proc. 3rd Int. Conf. on Ground Movements and Structures, Cardiff, 1984.
[50] Valliappan, S., Lee, I. K. and Murti, V., Rock Fracturing Due to Blasting, Proc. of 9th Australian Conf. on the Mechanics of Structures and Materials, Sydney, 1984, 75-79.
[51] Valliappan, S., Lee, I. K. and Rao, Y. V. A., Numerical Modelling of Rock Fracture by Explosives, Proc. of 4th Int. Conf. on Applied Numerical Modelling, Taiwan, 1984, 201-207.

[52] Paramasivam, P., Curiskis, J. I. and Valliappan, S., Micromechanics Analysis of Fibre Reinforced Cement Composites. Jnl. of Fibre Science and Technology, Vol. 20, 1984, 99-120.

[53] Valliappan, S. and Curiskis, J. I., Constitutive Relationships for Composite Materials through Micromechanics, Mech. of Engineering Materials, Ed. C. S. Desai and R. H. Gallagher, John Wiley and Sons Ltd., 1984, 611-632.

[54] Valliappan, S., Lee, I. K. and Boonlualohr, P., Nonlinear Analysis of Contact Problems, Numerical Methods in Coupled Systems, John Wiley and Sons Ltd., Ed. Lewis, Bettess and Hinton, 1984, 231-253.

[55] Curiskis, J. I. and Valliappan, S., Micromechanics Analysis of Unidirectional Fibrous Reinforced Composite Materials, Proc. 9th Australian Conf. on Mechanics of Structures and Materials, Sydney, 1984, 349-354.

[56] Valliappan, S. and Murti. V., Automatic Remeshing Technique in Quasi-Static and Dynamic Crack Propagation, Proceedings of Conf. on Numerical Methods in Engineering : Theory and Applications, Swansea, 1985, 107-116.

[57] Valliappan, S. and Ang K. K., Dynamic Analysis Applied to Rock Mechanics Problems, Invited Paper (Main Lecture), 5th Int. Conf. on Numerical Methods in Geomechanics, Nagoya, 1985, 119-132.

[58] Valliappan, S. and Murti, V., Numerical Modelling of Wave Propagation Problems, Invited Paper (Main Lecture), 2nd Int. Conf. on Computer Aided Analysis and Design in Civil Engineering, India, 1985, G44-49.

[59] Chandrasekaran, V., Lee, I. K., Valliappan, S., White, W. and Coulter, G., Liquefaction Potential of Ash Dam Foundations, Proc. 5th Int. Conf. on Numerical Methods in Geomechanics, Nagoya, 1985, 1359-1366.

[60] Zhao, X., Cao, M., Lee, I. K. and Valliappan, S., Soil-Structure Interaction Analysis and its Application to Foundation Design in Shanghai, Proc. 5th Int. Conf. on Numerical Methods in Geomechanics, Nagoya, 1985, 805-812.

[61] Valliappan, S., Basis of Nonlinear FEM Analysis. Invited Lecture, Macau Short Course on Finite Element Methods, Macau, 1985.

[62] Murti, V., Valliappan, S. and Lee, I. K., Stress Intensity Factor Using Quarter Point Element, J. Engg. Mech. Div., ASCE, Vol. 111, No. 2, 1985, 203-217.

[63] Murti, V. and Valliappan, S., A Universal Optimum Quarter Point Element, Engineering Fracture Mechanics, Vol. 25, No. 2, 1986, 237-258.

[64] Murti, V. and Valliappan, S., The Use of Quarter Point Element in Dynamic Crack Analysis, Engineering Fracture Mechanics, Vol. 25, No.3, 1986, 585-614.

[65] Ang, K. K. and Valliappan, S. Mesh Grading Technique Using Modified Isoparametric Shape Functions and Its Application to Wave Propagation Problems, International Journal for Numerical Methods in Engineering, Vol. 23, 1986, 331-348.

[66] Murti, V. and Valliappan, S., Numerical Inverse Isoparametric Mapping in Remeshing and Nodal Quantity Contouring, Computers and Structures, Vol. 22, 1986, 1011-1021.

[67] Murti, V. and Valliappan, S., A Dynamic Finite Element Preanalyzer, International Journal for Numerical Methods in Engineering, Vol. 23, 1986, 1367-1386.

[68] Valliappan, S. and Murti, V., Energy Dissipation in Finite Element Modelling of Dynamic Mixed Mode Crack Propagation, Proceedings of International Conference on Computational Mechanics, Tokyo, Vol. 1, 1986, 1367-1386.

[69] Valliappan, S., Bell, A. and Deb, P., Stability Analysis of Lake Parramatta Dam, Proceedings of International Symposium on Engineering in Complex Rock Formations, Beijing, 1986, 951-958.

[70] Valliappan, S., Load-Settlement Behaviour of Pile Groups - Composite Approach, Proceedings of the Second International Conference on Constitutive

Laws for Engineering Materials, Arizona, Vol. 1, 1987, 1331-1338.

[71] Valliappan, S. and Murti, V. Partial Energy Absorbing Boundary for Dynamic Crack Propagation, Invited Lecture, Proceedings of NUMETA Conference, Paper No. T6, Vol. 2, 1987.

[72] Murti, V., Valliappan, S. and Ang, K. K., Automatic Preprocessor for Dynamic Finite Element Analysis, Proceedings of Second International Conference on EPMEC, Guangzhou, China, 1987, Vol. 1, 42-45.

[73] Valliappan, S. and Ang, K. K., Numerical Modelling of Tunnel Blasting, Main Lecture, Proceedings of Sixth ICONMIG, Innsbruck, Austria, 1988, Vol. 1, 145-154.

[74] Valliappan, S., Dynamic Analysis of Rock Blasting using Fracture Mechanics Approach, Invited Lecture, Proceedings of International Conference on Computational Engineering Science, Atlanta, 1988, Vol. 1, 9.3.1-9.3.4.

[75] Ang, K. K. and Valliappan, S., Pseudo-Plane Strain Analysis of Wave Propagation Problems Arising from Detonations of Explosives in Cylindrical Boreholes, International Journal for Numerical and Analytical Methods in Geomechanics, 1988, Vol. 12, No. 3, 301-322.

[76] Valliappan, S. and Ang, K. K., Finite Element Analysis of Vibrations Induced by Propagating Waves Generated by Tunnel Blasting, Rock Mechanics and Rock Engineering, 1988, Vol. 21, 58-78.

[77] Valliappan, S., Flow Through Fractured Media, Computer Methods in Flow Analysis, Okayama University Press, 1988, Vol. 1, 627-634.

[78] Valliappan, S., Numerical Integration for Finite Element Dynamic Analysis, Proceedings of 13th International Symposium on Mathematical Programming, Tokyo, 1988.

[79] Valliappan, S., Murti, V., Jayaram, B. and Cao Ming Bao, Rock Anchoring of Dosing Tank for Sewerage Disposal, Theory and Practice of Earth Reinforcement, A. A. Balkema, Netherlands, 1988, 357-362.

[80] Koga, K., Valliappan, S. and Aramaki, G., Finite Element Analysis of Grid Reinforcement, Theory and Practice of Earth Reinforcement, A. A. Balkema, Netherlands, 1988, 407-412.

[81] Murti, V., Wang, Y. and Valliappan, S., Numerical Inverse Isoparametric Mapping in 3D FEM, Computers and Structures, 1988, Vol. 29, No. 4, 611-622.

[82] Miura, N., Sakai, A., Valliappan, S., Bergada, D. T. and Kuribayashi, E., Reinforced Foundation for Box Culvert on Soft and Subsiding Ground, Proceedings of 23rd Japan National Conference on Soil Mechanics and Foundation Engineering, 1988.

[83] Murti, V., Valliappan, S. and Khalili-Naghadeh, N., Time Step Constraints in Finite Element Analysis of Poisson Type Equation, Computers and Structures, 1989, Vol.31, No. 2, 269-273.

[84] Valliappan, S. and Ang, K. K., Method of Numerical Integration, International Journal of Computational Mechanics, 1989, Vol.5, 321-336.

[85] Valliappan, S., Khalili-Naghadeh, N. and Murti, V., Thin Flow Element and the Problem of Ill-conditioning, International Journal for Numerical and Analytical Methods in Geomechanics, 1989, Vol.13, No.6, 611-628.

[86] Valliappan, S. and Wang, Y. C., Software Development for an Integrated Approach to Stability Analysis of Large Dams, Proc. Int. Conf. on Engineering Software, India, 1989, 345-358.

[87] Valliappan, S., Murti, V. and Zhang Wohua, Finite Element Analysis of Anisotropic Damage Mechanics Problems, Engineering Fracture Mechanics, 1990, Vol.35, No.6, 1061-1071.

[88] Valliappan, S. and Khalili-Naghadeh, N., Flow Through Fissured Porous Media with Deformable Matrix, Int. J. Num. Meth. in Engg., 1990, Vol.29, 1079-1094.

[89] Valliappan, S. and Zhang Wohua, Anisotropic Damage Mechanics Applied

to Rock Mass, Invited Lecture, Proc. NUMETA 90, 1990, Vol.II, 745-754.

[90] Cai, W. M., Murti, V. and Valliappan, S., Slope stability analysis using fracture mechanics approach, Theoretical and Applied Fracture Mechanics, 1990, Vol.12, 261-281.

[91] Zhang Wohua and Valliappan, S., Analysis of Random Anisotropic Damage Mechanics Problems of Rock Mass-Probabilistic Simulation, Rock Mechanics and Rock Engineering, 1990, Vol. 23, 91-112.

[92] Nguyen, V.U. and Valliappan, S., Expert Systems in Geotechnical Engineering for Education Using Small Computers, Proceedings of EPMESC, Macau, 1990, Vol.5, 1149-1156.

[93] Zhang Wohua and Valliappan, S., Analysis of Random Anisotropic Damage Mechanics Problems of Rock Mass-Statistical Estimation, Rock Mechanics and Rock Engineering, 1991., Vol. 23, 241-259.

[94] Zhao, Chongbin, Valliappan, S. and Wang, Y.C., An Approximate Method for Simulating Infinite Media in Time Domain Dynamic Analysis, Proc. 3rd East Asia-Pacific Conf. on Struct. Engg. and Construction, Shanghai, 1991.

[95] Valliappan, S., Zhao, Chongbin and Wang, Y.C., Numerical Simulation of Water-Gravity Dam-Foundation System under Earthquake Loading, Proc. Int. Conf. on Computer Methods and Advances in Geomechanics, Cairns, 1991, 887-892.

[96] Murti, V., Wang, Y.C. and Valliappan, S., Numerical Characterization of Material Damping in Wave Propagation Problems, Proc. Int. Conf. on Computer Methods and Advances in Geomechanics, Cairns, 1991, 845-850.

[97] Valliappan, S. and Khalili-Naghadeh, N., Flow Through Fractured Media, Invited Lecture, Proc. Int. Conf. on Compute Methods and Advances in Geomechanics, Cairns, 1991, 1667-1672.

[98] Valliappan, S. and Zhao, Chongbin, Numerical Simulation of Dynamic Soil-Structure Interaction Problems, Proc. 6th Int. Conf. in Australia on FEM, Sydney, 1991, 494-500.

[99] Murti, V. Wang, Y.C. and Valliappan, S., Discrete Dispersion and Dissipation in Transient Dynamic Finite Element Analysis, Proc. 6th Int. Conf. in Australia on FEM, Sydney, 1991, 41-47.

[100] Kanoh, M., Khalili-Naghadeh, N., Murti, V. and Valliappan, S., Finite Element Analysis for Salt Water Wedge Diffusion in Groundwater, Proc. 6th Int. Conf. in Australia. on FEM, Sydney, 1991, 215-219.

[101] Valliappan, S., Parameter Identification in Geotechnical Engineering, Invited Lecture, Proc. 4th Int. Conf. on Computing in Civil and Building Engg., Tokyo, 1991, pp275.

[102] Murti, V., Wang, Y.C. and Valliappan, S., Modelling of Transient Wave Propagation in Infinite Media, Proc. 4th Int. Conf. on Computing in Civil and Building Engg., Tokyo, 1991, pp282.

[103] Valliappan, S. Zhao, Chongbin and Wang, Y.C., Effect of Reservoir Bottom Conditions on Hydrodynamic Response of Gravity Dams, Proc. 4th Int. Conf. on Computing in Civil and Building Engg., Tokyo, 1991, pp283.

[104] Valliappan, S., Analysis of Anisotropic Damage Mechanics Problems, Invited Lecture, Proc. Int. Conf. on Computational Engg. Science, Melbourne, 1991, 1142-1147.

[105] Zhao, Chongbin and Valliappan, S., An Approximate Method for Simulating Infinite Media, Computers and Structures, 1991, Vol. 41, 1041-1049.

[106] Khalili-Naghadeh, N. and Valliappan, S., Flow Through Fissured Porous Media with Deformable Matrix, Implicit Formulation, Water Resources Research, 1991, Vol.27, 1703-1709.

[107] Valliappan, S. and Zhao, Chongbin, An Alternative Wave Input Procedure in Coupled Finite and Infinite Model, Proc. of Asian-Pacific Conference on Computational Mechanics, Hongkong, 1991, 553-558.

[108] Zhao Chongbin and Valliappan, S., Vibration of Three-dimensional Rigid

Raft Foundation on Viscoelastic Medium, Int. J. Earthquake Engg. and Struct. Dynamics, 1991, Vol. 20, 1159-1177.

[109] Murti, V. Wang, Y.C. and Valliappan, S., Wave Propagation in Nonlinear Soils, Proc. of Asian-Pacific Conference on Computational Mechanics, Hongkong, 1991, 851-856.

[110] Murti, V., Zhang, W. and Valliappan, S., Stress Invariants in an Orthotropic Damage Space, Engg. Fracture Mechanics, 1991, Vol. 40, No. 6, 985-990.

[111] Valliappan, S and Zhao Chongbin, Dynamic Response of Concrete Gravity Dams Including Dam-Water-Foundation Interaction, Int. J. for Numerical and Analytical Methods in Geomechanics, 1992, Vol. 16, 79-99.

[112] Zhao Chongbin and Valliappan, S., A Simplified Method for Evaluating Amplification Effect of Mountains on Incident Earthquake Waves, Proc. Int. Conf. EPMESC, Dalian, 1992, 600-605.

[113] Pham, T.D. and Valliappan, S., Finite Element Mesh Generation Using Fuzzy Logic, Proc. Int. Conf. EPMESC, Dalian, 1992, 1139-1144.

[114] Wang, Y.C., Murti, V. and Valliappan, S., A Transient Finite Element Analysis of Linear Viscoelastic Material Model, Int. Journal for Numerical and Analytical Methods in Geomechanics, 1992, Vol.16, No.4, 265-294.

[115] Zhao Chongbin, Valliappan, S. and Wang, Y.C., A Numerical Model for Wave Scattering Problems in Infinite Media due to P and SV Wave Incidences, Int. Journal for Numerical Methods in Engineering, 1992, Vol.33, 1661-1682.

[116] Zhao, C., Valliappan, S. and Wang, Y.C., Mechanism of Solid-water Interaction Problems during Earthquakes, Computers & Structures, 1992, Vol.43, No.3. 397-409.

[117] Valliappan, S. and Zhao Chongbin, Numerical Simulation of Contaminant Transport in Fractured Porous Media, Proc. Seminar on Problems of Lowland Technology, Saga, Japan, 1992, 107-119.

[118] Wang, Y.C., Murti, V. and Valliappan, S., Assessment of the Accuracy of Newmark Method in Transient Analysis of Wave Propagation Problems, To be published in 1992, Int. Journal of Earthquake Engineering and Structural Dynamics.

[119] Valliappan, S., Pham. T.D., Murti, V. and Franklin, J., Fuzzy Logic for Automatic Finite Element Mesh Generation, Microcomputers in Civil Engg., 1992, Vol.8, 1-8.

[120] Valliappan, S. and Pham, T.D., Constructing A Membership Function with Objective and Subjective Information, Microcomputers in Civil Engg., 1992, Vol.8, 75-82.

Impact of Computational Mechanics on Engineering Problems, Pulmano & Murti (eds)
© 1993 Balkema, Rotterdam, ISBN 90 5410 324 8

Professor S. Valliappan: A personal view and comments on unified modelling

C.S.Desai
University of Arizona, Ariz., USA

ABSTRACT: I would like to offer my respects to Professor S. Valliappan (Val) through (1) a personal view, and (2) as Val's contributions have spanned such a wide range of areas, through some philosophical comments on unified modelling applied to behavior of materials

1 PERSONAL

I came to know of Val while I was a student around 1967 at the University of Texas at Austin, working on my doctoral research in the area of finite element method (FEM) and its applications. Subsequently, while working on 'Introduction to the Finite Element Method' with John Abel, we came across the important contributions Val had made in this evolving technique and its applications. One of the classical contributions Val has made, working with Olek Zienkiewicz, is the stress transfer approach in dealing with materials that experience fracture and tensile stresses. This contribution and its wide implications have provided basis for applications of the FEM to many problems by so many researchers. Personally, I have benefited greatly from Val's contributions.

Val's contributions cover a wide range of problems in mechanics, structural mechanics, geomechanics, fracture mechanics and environmental geotechnology. I have already mentioned about his classical contribution on the stress transfer approach. Val's research on such other topics as dynamic analysis for blast loading and wave propagation, fracture and damage is also highly significant and involves many original elements. The significance of Val's work is evident from the international acclaim he has received, the large number of conferences where he has been an invited speaker, and his membership of editorial boards of so many prestigious journals. His book on 'Continuum Mechanics' is used widely in research and in teaching.

2 ON UNIFIED MODELLING CONCEPTS AND THE DISTURBED STATE CONCEPT

Guided by Val's many contributions to a wide range of areas that take advantage of the generality and unity of the finite element method and material models, I thought it is appropriate to make a few comments on unified modelling approaches in engineering.

It is my belief that even though, at this time, it may be difficult to develop most general and unified models, it is possible to develop unified models for classes of materials. The past has seen separate models for different aspects of a class of materials, and very often, it is evident that such models are essentially the same for apparently different phenomena. Recognition of such unity and synthesis of various approaches can lead to unified models for those phenomena, thereby imparting enhanced fundamental understanding and simplifications.

On a personal level, with a number of my students and coworkers, I have been involved in the development of a unified modelling approach called the disturbed state concept (DSC). It permits characterization of both solid engineering materials and interfaces or joints with the same mathematical framework.

In the DSC, it is assumed that applied forces

(mechanical, thermal, environmental) cause disturbances in the material's microstructure. As a consequence, the material experiences transformation of its internal microstructure, during which the initially intact material transforms gradually due to the disturbances, through a process of conscious or natural self-adjustment in the microstructure. At any stage during deformation, the initially intact material experiences transformations and a part of it reaches the fully adjusted state (FAS), at randomly distributed locations in the material. The observed, or average, response of the 'mixture' is expressed in terms of its responses under the intact and FAS.

I will present brief details of the deviations of the incremental stress and strain equations in which relative motions (diffusion) between the two phases can occur, analysis of the approach and typical applications.

3 CONCLUDING COMMENTS

In the course of the last two decades or so, I have had many opportunities to meet, interact and work with Val. In addition to the professional relationship, I can claim a bond of friendship with Val and his family. When I meet Val, we not only talk of technical topics, but also of personal experiences and adventures!

I think that there is absolute unanimity on the fact that Val is a prominent and outstanding international figure.

I am very happy to be a part of this celebration to honor Val on his 60th birthday. From his activities, his drive and his youthful look, Val is still very young! I wish Val and his family the best of luck, success and happiness in the coming years.

Impact of Computational Mechanics on Engineering Problems, Pulmano & Murti (eds)
© 1993 Balkema, Rotterdam, ISBN 90 5410 324 8

A short lecture on Prof. Valliappan's Kan-reki, Koki and Haku-jyu

H.Tano
Nihon University, Japan

I would like to congratulate Mrs. and Professor Valliappan on his 60th birthday. It may be not so well-known in western countries that the 60th birthday is a special one in Japan we call it as "Kan-reki". Prof. Valliappan has often visited Japan and has many Japanese friends. He stayed at my house for a few days in Koriyama and so my family knows well him. I suppose that he has been made aware of the Kan-reki celebration through communication from his friends.

In this note, I will introduce to him a short history of birthday celebrations in Japan including Kan-reki.

1 Fundamental terms for memorial birthdays in Japan

1.1 Eto(干支):Ji-kkan(十干) and Jyu-ni-shi (十二支)

Ji-kan and Jyu-ni-shi was brought from China to Japan long ago and the former consists of 10 Chinese(Japanese) characters and the latter of 12 characters. These are listed in Table 1. Before the current calender came into use, there were no "months" and "weeks" in ancient China and "60 days" were used as one cycle. They probably didn't have numerical characters at that time.

How did they count and recognize "a year and a month etc."? They made a calender using Ji-Kan and Jyu-ni-shi as follows:

Let us solve the following problem. Now there are two measures, A and B as shown in Fig. 1. Measure A is 10 cm in length and measure B 12 cm. They are trued up at their edges to same the direction as shown in Fig. 1. It is obvious that the difference between each edge is 2 cm. How many measures do we need if we want to put this difference to zero? Chinese people recognized that the number of measure A is 6 and B is 5. The total length of both measures is 60 cm. If we give one character per cm on these measures as shown in Table 1, 60 combinations of these characters are given by such operation. To know the number of measures, ancient Chinese people probably tried to do the above operation using such measures and they could make a kind of calender.

On the other hand, now we can easily solve this problem as to find the least common multiple as follows:

```
      A    B
2 )  10   12
      5    6
```

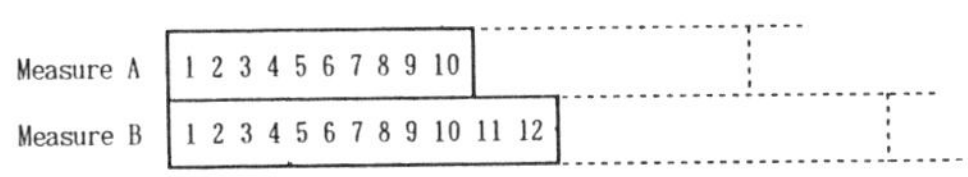

Figure 1 Measure A(Ji-kkan) and B(Jyu-ni-shi)

Table 1 The least Common multiple in Eto: Ji-kkan(十干) and Ju-ni-shi(十二支)

干 (Eto)	支		干 (Eto)	支		干 (Eto)	支	
十干 ji-kkan	十二支 jyu-ni-shi	Arabic numerals	十干 ji-kkan	十二支 jyu-ni-shi	Arabic numerals	十干 ji-kkan	十二支 jyu-ni-shi	Arabic numerals
甲	子	1	戊	子	25	壬	子	49
乙	丑	2	己	丑	26	癸	丑	50
丙	寅	3	庚	寅	27	甲	虎	51
丁	卯	4	辛	卯	28	乙	卯	52
戊	辰	5	壬	辰	29	丙	辰	53
己	巳	6	癸	巳	30	丁	巳	54
庚	午	7	甲	午	31	戊	午	55
辛	未	8	乙	未	32	己	未	56
壬	申	9	丙	申	33	庚	申	57
癸	酉	10	丁	酉	34	辛	酉	58
甲	戌	11	戊	戌	35	壬	戌	59
乙	亥	12	己	亥	36	癸	亥	60
丙	子	13	庚	子	37	甲	子	61
丁	丑	14	辛	丑	38	·	·	·
戊	寅	15	壬	寅	39	·	·	·
己	卯	16	癸	卯	40			
庚	辰	17	甲	辰	41			
辛	巳	18	乙	巳	42			
壬	午	19	丙	午	43			
癸	未	20	丁	未	44			
甲	申	21	戊	申	45			
乙	酉	22	己	酉	46			
丙	戌	23	庚	戌	47			
丁	亥	24	辛	亥	48			

∴ $2 \times 5 \times 6 = 60$

1.2 The relation between Jyu-ni-shi(十二支) and animals

Jyu-ni-shi were used to represent directions or hours in ancient China and Japan. "Jyu" or "Ji(十)" means ten as shown in Table 2. In addition, each Chinese character in Jyu-ni-shi is given an animal as listed in Table 3 and such a custom continues till today in Japan. When Japanese people usually say "Eto", it often means such animals as correspond with Jyu-ni-shi.
Exactly, "Eto(干支)" means the word represented by "Ji-kkan(十干)" and "Jyu-ni-shi(十二支)". These are shown in Table 1.

The word "Jyu(寿)" is often used in celebration of a wedding, birthday and any other happy event.

2 Kan-reki(60th birthday)

As mentioned above Chinese pepole knew that the calender using the combination with Ji-kkan and Jyu-ni-shi was reset every 60 years. That is, 60 years is one cycle in Eto as shown in Table 1. Therefore the 60th birthday has been called the Kan-reki celebration in Japan. Because Kan(還) means "reset" and Reki(歴) means "history or calender" in Japanese, when a man has his 60th birthday(Kan-reki), it is usual that he will retire from his posts or turn over his business to his son. After that he may be able to enjoy his hobby or sports and many other things.

However there are comparatively more elder people today in Japan because recent medical science keeps more alive longer and the average span of life increases gradually. Therefore, it is said that the ability and capacity of a man 60 years old today may be nearly equivalent to that of a man 50 years old in the middle ages.

3 The meaning of Ko-ki celebration

Ko-ki is a celebration for 70th birthday as shown in Table 4. It is said that this custom may be derived from a famouse Chinese poem written by Toho. He said "The Human being can celebrate own 70th birthday may be rare since ancient time". Ko(古) means "since ancient time" and Ki(希) means "rare" in Japanese. Consequently Ko-ki has come to be the celebration for the 70th birthday.

4 Ki-jyu(喜寿), Bei-jyu(米寿) and Haku-jyu (白寿)

Ki-jyu, Bei-jyu and Haku-jyu are the celebrations for the 77th, 88th and 99th birth-days respectively. It seems that thier histories differ from Kan-reki for the 60th and Ko-ki for the 70th birthday. Because the origin of the given name for each celebration is related to the cursive glass style in the Japanese character and its analogy as shown in Table 4.

It is said that these customs for birthday celebration were formed in the middle ages in Japan.

There are other memorial birthday celebrations in Japan. If I can take another chance like this, I would like also to introduce the history continued after Haku-jyu for Mrs. and Prof. Valliappan.

Table 2 Japanese numerals

Japanese numerals	一	二	三	四	五	六	七	八	九	十
Arabic numerals	1	2	3	4	5	6	7	8	9	10

Table 3 Jyu-Ni-shi(十二支) and the animals

十二支 Jyu-ni-shi	animals	example years of birth	
1 子 (ne)	mouse	1924	1936
2 丑 (ushi)	cow(bull)	1925	1937
3 寅 (tora)	tiger	1926	1938
4 卯 (u)	rabbit(hare)	1927	1939
5 辰 (tatsu)	dragon	1928	1940
6 巳 (mi)	snake	1929	1941
7 午 (uma)	horse	1930	1942
8 未 (hitsuji)	sheep	1931	1943
9 申 (saru)	monkey	1932	1944
10 酉 (tori)	hen(cock)	1933	1945
11 戌 (inu)	dog	1934	1946
12 亥 (i)	wild boar	1935	1947

Table 4 Main birthday celebration in Japan

Birthday celebration			Chinese character	meaning of the Chinese character	cursive grass tyle	deformation and analogy
age	Chinese character	pronunation				
60	還 暦	Kan-reki	"Kan(還)" means "reset". "Reki(暦)" means a "calender".			
70	古 希	Ko-ki	"Ko(古)" means "ancient time". "Ki(希)" means "rare"			
77	喜 寿	Ki-jyu	喜	pleasure	七 十 七七 七七	七十 七 70 and 7 →77
88	米 寿	Bei-jyu	米	rice		米 → 八 十 八 → 八+八 80 and 8→88
90	卒 寿	Sotsu-jyu	卒		卆	九 十 → 九 + → 九+ 90 →90
99	白 寿	Haku-jyu	白			百 → 一(one) 白=百 - (一) 100 100 - 1 = 99

Impact of Computational Mechanics on Engineering Problems, Pulmano & Murti (eds)
© 1993 Balkema, Rotterdam, ISBN 90 5410 324 8

Prof. Valliappan in Saga University, Japan

G. Aramaki
Saga University, Japan

1 Invitation of Foreign Professors

For a long time, Japanese government had been prohibited to employ a foreign professor in the national university. In 1985, the ministry of education established a new system to invite a foreign researcher to a university using the legally fixed number of officials. We were looking forward to using the new system to encourage our students to have contact with foreign technology and culture.

We investigated the details of this invitation system for two years; for example, salary and bonus, insurance system, passage fee, an allowance to rent a house, period of stay, etc.

It is very famous about Japanese people's poor ability of speaking and listening English. "Can our students listen to their English lectures?", "Can they follow the foreign professor's special lectures?", "Can they write their graduation theses and present in English?" How it may turn out is anybody's guess? We decided to invite a foreign professor because our students will be able to have a good chance to have contact with different cultures.

Prof. Miura wrote many letters to his friends to recommend a young researcher as an associate professor in the department of civil engineering. We decided to invite Dr Dennis Bergado who was an associate professor at the Asian Institute of Technology. After 2 or 3 weeks, we received a letter from Prof. Valliappan saying that he was able to stay for a year in our university using a sabbatical leave. Prof. Val is a big professor who had many experiences to educate foreign students. We decided to invite two professors in our department.

2 Graduate Theses

We hoped that the two professors would give special lectures for our students and supervise their graduate theses. "Do our students decide to enter foreign professor's laboratory?" There was no necessity for feeling anxiety. Six students applied for Val's laboratory and four students were selected by a lottery. They were all most cheerful and brave and excellent in scholarship in their class. We may possibly change our viewpoint for our students.

Val's special lecture for them had started in his office. When I enter his office, the students raise their eyes to me requiring my help. But I could not join their lectures every time. I rejected their request. Val was only laughing. Their war against hearing, speaking, reading and writing in English had continued for one year.

They were able to communicate with Val in English six months later. But the last struggle is that they should write their theses and present their subject in English. They needed about ten minutes to write only one sentence. I heard their deep breath many times. When a student showed his draft to Val, Val said, "Good, but these two pages need not be included in your paper." His effort for two days had disappeared in a moment.

They accomplished to complete their graduate theses.

Masao Yoshida: "Finite Element Analysis of Spread Footing"
Kitoshi Kaneko: "Stress Distribution Around Tunnel Opening"
Masahito Yahiro: "Settlement Analysis of Pile Foundation"
Yasunori Eguchi: "Finite Element Analysis of Embankment and Dams"

When their presentation started in English, the third year students listening in the room expressed their surprise. Eight students who belonged to Val's and Bergado's laboratory presented in English. Their presentations in English must be a first occurrence in the undergraduate course in a Japanese university.

I explained our system and their presentations in English to many professors of famous universities in Japan. They could not trust me. "Really? Can students of Saga University write their theses and present in English?" Perhaps they wanted to say "I don't think that our undergraduate students can write in English. Can the students of Saga University do this?"

Prof. Val and four of his students shook hands with each other after their presentations.

Prof. Valliappan and Dr Bergado encouraged not only their students but the staff of civil engineering.

3 Terrible Word "Why?"

How many times did I hear Val's terrible word "Why?"
Val asked me as soon as he began to give a lecture to our students. "Gunji, why don't Japanese students reply even to my simple questions? Why do they smile for my questions?"
"Why don't they concentrate in my lecture?"
"Gunji, why aren't there books in the library? Where are the books of civil engineering?"
"Gunji, Japanese computer systems are very useful and cheap. Many countries import those systems. Why don't you have a reference system in your library?"
"Gunji, why do you go to so many meetings?"
I tried to explain for those questions in my poor English. But I had to answer three or four times for every new "Why" during my reply. Val's questions were always much to the point, friendly and sometimes cynical.

I wanted to escape from Val's terrible questions "Why?"

Val proposed to us to have a symposium titled "Finite Element Method - Past, Present and Future". Val invited Prof. Kawai, Prof, Kawamoto, Prof. Kawahara, Prof. Adachi and Prof. Sakurai to join as lecturers. They are all big professors not only in Japan but also in the world. We were in earnest agreement with his proposal. The symposium was a great success.

After two or three weeks, Val proposed a new symposium titled "Constitutive Laws in Engineering Materials". We wanted to refuse it, because the theme was not familiar to the engineers in Kyushu. If we could not gather participants enough to pay travel expenses for the lecturers, we had to cover the deficit ourselves. I did not have the confidence to gather so many people in Kyushu. I wanted to say "No". But I was afraid of the terrible question "Gunji, why?". I could not explain my situation without more "why". I accepted his proposal only because I didn't want to answer for his "why".

Invited lecture in Kumamoto University

That symposium was also a great success. Many participants except Kyushu Island joined our symposium.

Just before his leaving, he said to me, "Gunji, I have understood why Japanese students don't study eagerly. It is very famous that Japanese students of junior and senior high school are the most intelligent in the world. Japanese businessmen and engineers work very hard for companies. If they don't take a rest in the university, they would die earlier." It must be his bitter irony.

According to Prof. Val's many suggestions, I am reconstructing a new educational system at Saga University to encourage students to work harder. Val may ask "Why do you go to many meetings?"

Impact of Computational Mechanics on Engineering Problems, Pulmano & Murti (eds)
© 1993 Balkema, Rotterdam, ISBN 90 5410 324 8

Introductory course on computational mechanics

K.Onishi
Science University of Tokyo, Japan

It was in the city Tainan of Taiwan (ROC), autumn 1981, if I remember correctly, when I first met Professor Valliappan at the International Conference on Mathematical Modelling. Six years later after our long absense, we happened to meet each other in spring 1987 in an administration office of Saga University western Japan, where Dr. Valliappan was invited as guest professor for one academic year long by Professor Gunji Aramaki of the University. In that office I introduced myself to Valliappan with a lot of respect. I had forgotten at the time that I had met him already in Taiwan. He answered to me very gently and friendly with his attractive smile, *"I know you, not because Gunji told me much about you, not because you are part time instructor to 'my' University Saga, just because we met, maybe in Taiwan."* Since then, I have kept close contact with Professor Valliappan in academic as well as in my personal life.

Today I will give you a brief introduction, how a first course on computational mechanics is going in the Department of Mathematics II, Science University of Tokyo, where I started my new life two years ago. Students of my class have had a preliminary course on numerical mathematics. They have learned calculus, linear algebra, classical geometry, topology, and differential equations. They have some trainings in computer science. They are very good students in mathematical courses, but they are quite poor in physics and engineering problems. When a mathematical problem is posed to them, they can show good answers. However, when they are asked to formulate a problem based on the observation of physical or engineering practice in mathematical form, then they may become upset.

Lecture in my class begins with the talk on various source of errors encountered in elementary numerical processes of finite difference, finite element and boundary element methods. Topics are carefully chosen in the field of steady-state heat conduction in solids, steady irrotational flow of incompressible inviscid fluid, and electro-static problem. Through these problems that can be described by the boundary value problems of two-dimensional Laplace equation, students are led to consider errors committed in input operations, in the approximation, in the modeling. They are requested to extend their discussions on how one can treat the singularity of the solution, how the adaptive remeshing is applied, and how extraordinarily high accuray can be obtained. If students want to study computational mechanics further in this introductory course, they can move to the topics in two-dimensional flow problems of viscous fluid, plane elasticity problems, and scattering problems.

Computer programmes coded in FORTRAN and BASIC on a floppy disket are given to students, if they want to have, which run on PC. Students implement the programmes on machines available according to their own circumstances. They must choose a sample problem, solve it, and write a report on the subject with a worked numerical example.

Impact of Computational Mechanics on Engineering Problems, Pulmano & Murti (eds)
© 1993 Balkema, Rotterdam, ISBN 90 5410 324 8

Val: Friend and co-researcher

H. Raja Sekar
Sydney, N.S.W., Australia

I have known Professor Valliappan since 1970. At that time, I was a post graduate student in structural engineering and he was a Lecturer in the School of Civil Engineering, University of New South Wales. It was the time when the finite element method was gaining tremendous attention and popularity among researchers and practising engineers in Australia. The teaching of the method was essentially concentrated at the post-graduate level and Val was deservedly known as 'the man who knew too much' about Finite Element Techniques. It was under those circumstances that my introduction to the analysis of structures using FE Methods and my association with Professor Valliappan began.

Professor Valliappan had a ringside seat at the beginning of the Finite Element revolution, when he developed a valuable 'bounding' solution which employed iterative 'stress transfer' method under the guidance of Professor Zienkiewicz at Swansea in 1965. The publication of his classical paper on stress analysis of rock as a 'no tension' material in 1968 showed how to study many engineering problems which were thought to be insolvable before.

The author of more than one hundred and fifty technical papers and a member of several editorial bodies, Val is privileged to be a part of one of the great lineages of finite element analysis. Over the years he has provided guidance to many of his research students so that they can accept such responsibilities as they are capable of. Nevertheless, he assisted and made sure the research they conducted met high standards. He has been instrumental in helping many young engineers learn and develop a rigorous and disciplined approach to their studies via understanding and using mathematical modelling methods, and many of his students have subsequently launched themselves successfully into entirely different directions all over the world.

Val has never allowed himself to fall into the trap of playing the part of boss or administrator. This is simply not his way, and would have constituted nothing but a hindrance to his scientific goals and work. He has always tried to preserve his physical independence in his professional and private interaction with people. It is this uniqueness which makes him so beloved to those who know him well.

Over the years I have known Val, he has been a good friend, a colleague and a mentor as the case may be. During his numerous visits overseas, especially to Japan, I had the pleasure of travelling with him in 1981 and present a paper as the coauthor at the International Symposium on Weak Rock held in Tokyo. One of the highlights of this visit was the opportunity I had to meet a number of his distinguished academic friends including Professors T. Yamanouchi, T. Kawamoto, S. Sakurai and S. Kobayashi. Among other experiences in Japan, Val and I had an authentic Japanese dinner (Sukiyaki), served with meticulous attention to Japanese tradition, as the guests of Professor Kawai.

In his private life he enjoys good food and wine and has a passion for European cars. We have played tennis together and still do occasionally despite the fact that my game is inferior to his - he is a very good sportsman.

Val and I have one other thing in common; we each have three beautiful daughters, and hopefully we can both look forward to many delightful years in their company in our old age!

I take this occasion to wish Val good health and happiness for years to come.

Impact of Computational Mechanics on Engineering Problems, Pulmano & Murti (eds)
© 1993 Balkema, Rotterdam, ISBN 90 5410 324 8

The King and I

B.V. Rangan
Curtin University of Technology, Australia

I have borrowed the famous movie title for my personal note. The reasons for this choice may become clear at the end of this note.

Reading Tamil poetry is my hobby. As some of you will know, Tamil (the language Val and I speak at home) is about 5,000 years old. In the Tamil literature, FRIENDSHIP is always given the highest regard. A classical example of FRIENDSHIP is the love that existed between a great Tamil King and a humble poet. A Tamil verse written some two thousand years ago describes their friendship and love. The King and the poet were great friends long before they met.

My friendship with KING VAL is analogous to the above. I had heard a great deal about Val through his classmate Professor R S Srinivasan when I was studying for my PhD in mid 1960's. I have had a great desire to meet him ever since. The opportunity came in 1970 when I arrived in Sydney to join UNSW as a lecturer. We arrived on a Saturday morning. Within an hour of our arrival, I found his telephone number in the directory and called him. At that time his two lovely girls, Vimala and Asha, were suffering from measles, along with his wife, Kamala. So the KING expressed his disappointment that we could`not meet immediately as the disease may infect my two little children. However, he contacted one Dr Sunder Das (who became our closest friend later) to assist us to settle down. Two weeks later we met. Thus began my FRIENDSHIP WITH KING VAL.

Our friendship grew stronger and stronger as the years rolled on. It is not possible to fully describe a lifelong friendship in a short note such as this. However, I wish to highlight some moments in my life with KING VAL.

As you may know, my mathematical ability is rather poor. During the early 1970's, I attempted very hard to learn the Finite Element Method and the associated mathematics from the great master KING VAL. My learning took the form of several sessions in which the master explained. In March 1973, Val became seriously ill; it was a great shock to me. When he was lying in bed in the Prince of Wales Hospital I used to visit him every day. Contrary to the doctor's advice, he continued his private lectures to me on the Method. Although the teacher was excellent, the student was hopeless.!

Whenever I was in trouble or had a problem or needed sound advice, I always rushed to KING VAL. I have missed this privilege in the last three years. He was there to comfort me, to offer solutions to my problems, and to advise me. There were numerous occasions; those which stand out in my mind were the time my father passed away (1975), the period after my return from Canada when my house was in a mess (1978), and especially the long period during 1985-1989 when my family problems were monumental.

"People who need people are the luckiest people in the world", so goes the song from the Broadway musical 'Funny Girl'. I have been lucky to acquire the friendship of KING VAL. I needed him more than he realised. Although we are now many miles apart, our FRIENDSHIP is strong and green as ever. My congratulations and best wishes to him on his 60th birthday.

Technical papers

Impact of Computational Mechanics on Engineering Problems, Pulmano & Murti (eds)
© 1993 Balkema, Rotterdam, ISBN 90 5410 324 8

Solution difficulties in multi-wedge stability analyses

I. B. Donald
Monash University, Australia

ABSTRACT: In both conventional vertical slice methods and multi-wedge methods for soil stability analysis solution instabilities may arise for failure surfaces containing geometrically poor features. This paper demonstrates the precautions which are necessary to ensure convergence to the correct value of F when using multi-wedge limit equilibrium and upper-bound solutions. An algorithm is described which automatically chooses the appropriate analysis according to current circumstances. Multiple values of F may be obtained from solutions of the governing non-linear equations, but all except the highest value are connected with negative forces across the basal failure surface.

1 INTRODUCTION

Multi-wedge stability analyses are becoming increasingly popular, as they can model internal shearing in the sliding mass more realistically than conventional vertical slices limit equilibrium methods. As a direct consequence, only a small number of wedges is usually required, but careful attention to the details of the solution routine is essential if the correct solution is to be arrived at. Vertical slice methods are subject to similar problems and in this paper the nature of solution instabilities is discussed and reliable methods for overcoming them described.

2 SOLUTION INSTABILITIES

Several authors have referred to the multiple solutions which are possible in methods-of-slices analyses of slope stability problems (Chugh (1981), Nguyen (1985), Sarma (1987) and Chowdhury and Zhang (1990)). The cause is frequently false convergence of the solution resulting from poor geometry of the failure surface, but provided the highest possible root of the governing equations has been obtained, the remainder may be rejected on the grounds of violation of physical principles, e.g. negative normal forces over sections of the failure surface. Fredlund (1984) explained that "the multiplicity of solutions was not due to the insufficiency of assumptions but the non-linearity of the equations (involving F)".

This statement is also true of multiple wedge solutions, where the governing equations are equally non-linear. When optimization routines are used to select the critical wedge geometry leading to the lowest correct value of safety factor, F, it is inevitable that many geometrically poor failure surfaces will be generated, possibly leading to solution instabilities during the calculation of F. Methods are described below for ensuring that the calculated F will be the highest possible and hence the correct one for the surface under investigation.

3 WEDGE METHODS

The two wedge methods covered by this paper are:-

(a) a limit equilibrium solution, by force and moment equilibrium, of multi-wedge kinematically admissible failure mechanisms; computer program GWEDGEM (Donald and Giam, (1989)).

(b) an upper bound plasticity solution, based on fully associated flow of similar mechanisms; program EMU (Giam and Donald, (1989(a))).

3.1 GWEDGEM Solution

The free body diagram for wedge No. m in an n-wedge mechanism is shown in Figure 1. Before the forces may be calculated it is necessary to guess a trial value of F, so that values may be given to the parameters:-

$$c_{dev} = \frac{c}{F} \; ; \; \phi_{dev} = \tan^{-1} \left(\frac{\tan \phi}{F}\right) \qquad (1)$$

On the base of the wedge the S_j force is given by $c_{dev} \times$ base length, the pore water force is U_j and Q_j is the resultant of the normal effective force on the base and the developed frictional resistance (normal force $\times \tan \phi_{dev}$).

On the inclined sides of the wedge the equivalent forces are X_i, X_k, U_i, U_k and R_i, R_k. Working from the first (toe region) wedge to the $(n-1)^{th}$ wedge, the general equilibrium matrix equation may be written as follows:-

$$\begin{bmatrix} \cos A & -\sin B \\ \sin A & -\cos B \end{bmatrix} \begin{Bmatrix} Q_i \\ R_k \end{Bmatrix} = \begin{Bmatrix} C_1 \\ C_2 \end{Bmatrix} \qquad (2)$$

where

$$A = \phi_{j\,dev} - \alpha_j$$

$$B = \phi_{k\,dev} - \delta_k$$

$$C_1 = W_m + X_k \cos \delta_k - U_k \sin \delta_k - S_j \sin \alpha_j - U_j \cos \alpha_j$$
$$- X_i \cos \delta_i + U_i \sin \delta_i - R_i \sin (\phi_{i\,dev} - \delta_i)$$
$$+ \text{(other external vertical forces)}$$

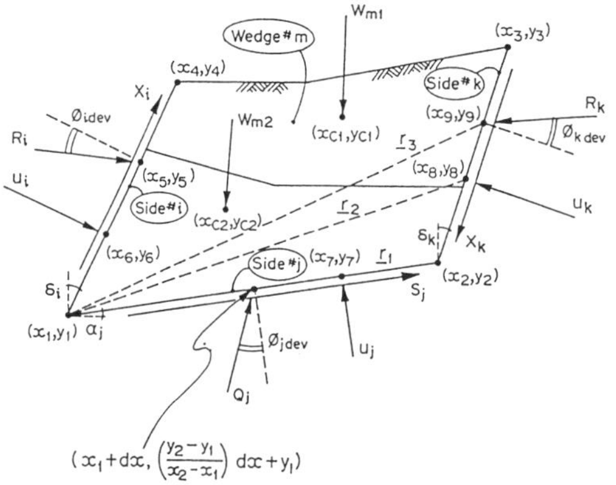

Fig. 1 Forces acting on wedge No. m

$$C_2 = X_k \sin \delta_k + U_k \cos \delta_k - S_j \cos \alpha_j + U_j \sin \alpha_j$$
$$- X_i \sin \delta_i - U_i \cos \delta_i - R_i \cos (\phi_{i\,dev} - \delta_i)$$
$$+ \text{(other external horizontal forces)}$$

Solving for Q_j and R_k,

$$Q_j = (C_1 \cos B - C_2 \sin B) \,/\, \cos(A + B) \quad (3)$$

$$R_k = (C_1 \sin A - C_2 \cos A) \,/\, \cos(A + B) \quad (4)$$

For the nth (crest region) wedge equation (2) is applicable but S_j is calculated as R_k does not exist for this wedge.

$$\begin{bmatrix} \cos A & \sin B \\ \sin A & \cos B \end{bmatrix} \begin{Bmatrix} Q_j \\ S_j \end{Bmatrix} = \begin{Bmatrix} C_1 \\ C_2 \end{Bmatrix} \qquad (5)$$

where

$$A = \phi_{j\,dev} - \alpha_j$$

$$B = \alpha_j$$

$$C_1 = W_m - U_j \cos \alpha_j - X_i \cos \delta_i + U_i \sin \delta_i$$
$$- R_i \sin (\phi_{i\,dev} - \delta_i)$$
$$+ \text{(other external vertical forces)}$$

$$C_2 = + U_j \sin \alpha_j - X_i \sin \delta_i - U_i \cos \delta_i$$
$$- R_i \cos (\phi_{i\,dev} - \delta_i)$$
$$+ \text{(other external horizontal forces)}$$

Solving for Q_j and S_j,

$$Q_j = (C_1 \cos B - C_2 \sin B) \,/\, \cos(A + B) \quad (6)$$

$$S_j = (C_2 \cos A - C_1 \sin A) \,/\, \cos(A + B) \quad (7)$$

Equating S_j to mobilised shear force, to calculate F,

$$S_j = \frac{(\bar{c}\,L)_j}{F} \qquad (8)$$

If the last interface is entirely located in a cohesionless soil, substitute $S_j = 0$ in equation (7) which will then give:

$$\tan(\phi_{j\,dev} - \alpha_j) = \frac{C_2}{C_1} \qquad (9)$$

but

$$\tan \phi_{j\,dev} = \frac{\tan \bar{\phi}_j}{F} \qquad (10)$$

Therefore from equations (9) and (10) F can be calculated.

If the value of F for the nth wedge differs by more than a specified amount from that assumed for the other wedges, an iterative procedure is employed until satisfactory agreement has been reached.

3.2 EMU Solution

For simplicity, a three wedge mechanism is shown in Figure 2, with the directions of relative sliding between wedges given by the developed friction angles, ϕ_{dev}, obtained through the choice of an initial assumed value of F. Applying the upper bound theorem of classical plasticity and equating internal energy dissipated in shearing to the work done by body weights and external forces, the following equation is obtained:-

$$F = \frac{\Sigma_{i=1}^{2n-1} c_i\, l_i\, V_i \cos \phi_{i\,dev}}{\Sigma_{j=1}^{n} (-W_j) V_{2j-1} \sin \theta_{2j-1} + \Sigma_{i=1}^{2n-1} U_i\, V_i \sin \phi_{i\,dev}} \qquad (11)$$

If $\phi = 0$ a value of F may be obtained explicitly from eqn. (11), but for $\phi > 0$ F appears implicitly in the ϕ_{dev} values on the right hand side of the equation and an iterative solution is again necessary.

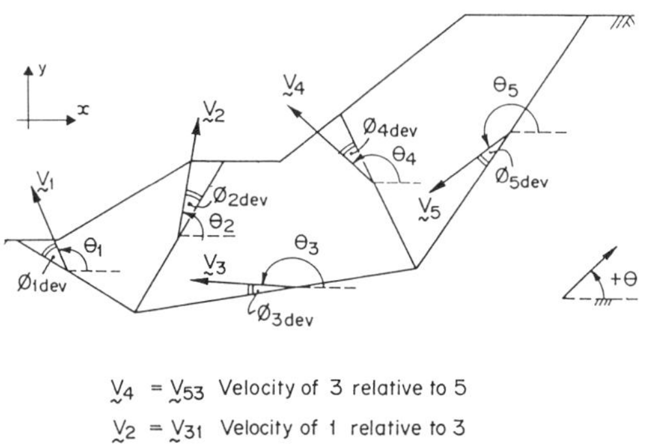

Fig. 2 Relative wedge velocities for EMU solution

4 SOLUTION OF EQUATIONS

For simplified stability analyses such as Bishop's and Janbu's methods the fixed point iteration technique is adequate, though the Newton-Raphson method is often used. For more rigorous vertical slice analyses, such as the Morgenstern Price and Spencer methods Newton-Raphson is required. Many other successful techniques exist, including best-fit regression and rapid-solver (Fredlund, (1984)), modified Newton's method (Chen and Morgenstern, (1983)) and compound iteration (Li and White, (1987)). However none of these techniques proved suitable for either GWEDGEM or EMU.

4.1 Modified Secant Method

A modified secant procedure proved to be an efficient iteration scheme for most realistic failure surfaces. The modification lies in plotting $1/F_{obtain}$ against F_{guess}, in preference to F_{obtain} vs F_{guess}, as the former yields a more well behaved curve. A diagram illustrating the procedure is shown as Figure 3 for an EMU 3-wedge analysis. The method is equally applicable when both F_1 and F_2 are to one side of the required root.

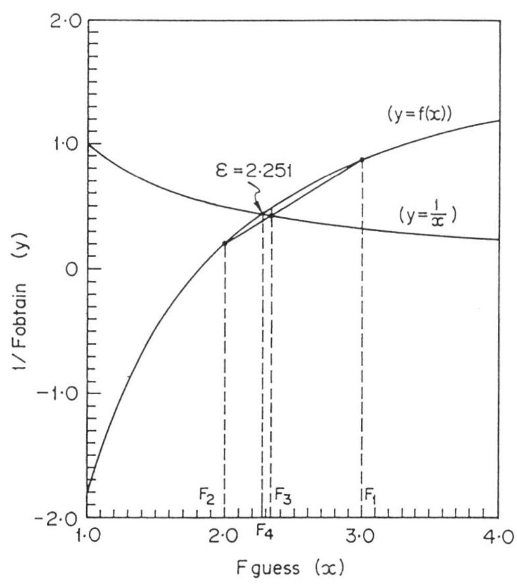

Fig. 3 Modified secant method

4.2 Solution Instabilities

During geometrical optimization of the failure mechanism (Giam and Donald, (1989 (b))) poor failure surfaces which may arise include

1. kinematically inadmissible surfaces
2. non-feasible surfaces
3. highly irregular multi-linear surfaces
4. poorly arranged wedge mechanisms.

Such surfaces tend to be associated with mathematical instabilities when solving for F and the modified secant method cannot be guaranteed to settle on the correct value.

Figure 4 shows a realistic failure surface for which (F_{guess} - F_{obtain}) has been plotted against F_{guess}, for a 6-wedge analysis of a homogeneous simple slope. Several thousand analyses were performed to produce Fig. 4 to define clearly the three apparent roots of the governing equations, of which only the value of 1.00 is physically acceptable. The discontinuities occur when the denominator in eqn. (11) approaches zero. Similar behaviour was observed in a number of examples investigated but in all cases the curve to the right of the highest (correct) root was smooth and continuous.

The modified secant method converges for the problem of Fig. 4 but not for a poor failure mechanism such as is shown in Figure 5 for a 6-wedge analysis in a 3-layered soil profile. The basic problem with this mechanism is the small size of wedge No. 6 in relation to the other wedges as wedge 6 is the critical one for which F_{obtain} is calculated. In geometrical optimization routines it frequently happens that one or more wedges are optimized almost completely out of existence. When this happens, as may be seen clearly in Fig. 5, the discontinuities are very sudden and the first one also defines the correct root accurately. Locating this root then requires a more sophisticated approach than the modified secant method.

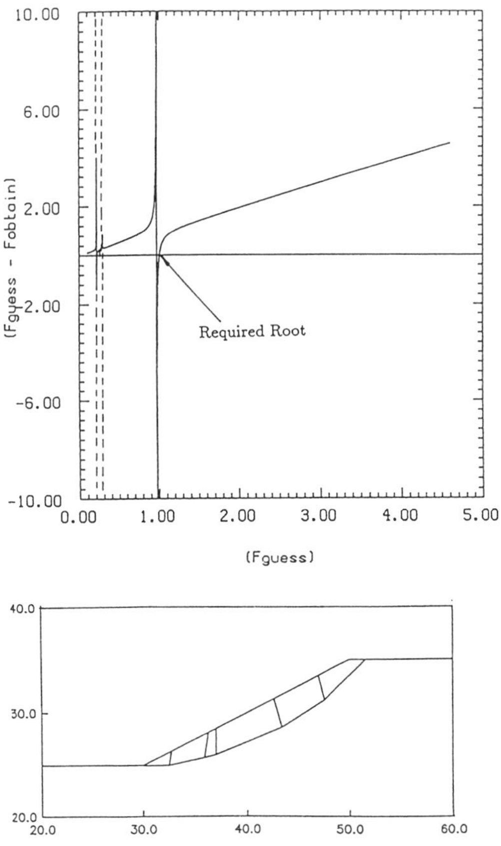

Fig. 4 Solution instabilities - simple slope

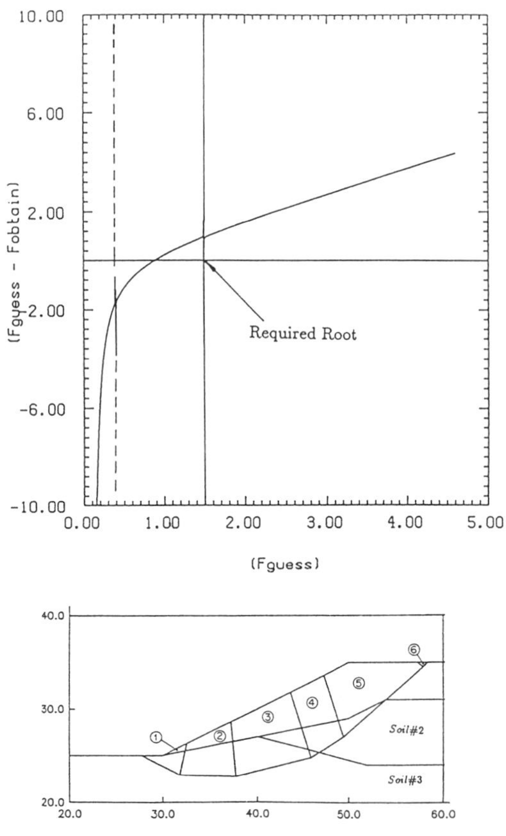

Fig. 5 Solution instabilities - complex slope

4.3 Decision - iteration scheme

General purpose iteration algorithms for the solution of non-linear equations include:-

1. Fixed-point method
2. Bisection method
3. Newton's method
4. Secant method
5. Regula falsi method
6. Müller's (1956) method

Müller's technique is best suited to polynomial-type functions and all the others, with the exception of the bisection method, are unsuitable for GWEDGEM and EMU analyses. The bisection method will theoretically work if a small enough interval is adopted; that is, starting with a high F_1 (to cater for all possible failure surfaces), F is decreased gradually by a small interval until $H(F_i).H(F_{i-1}) \leq 0$, where function $H(F_i)$ denotes the value of (F_{guess} - F_{obtain}) for the ith estimate of F_{guess}, i.e. the graph has crossed the F axis. A very small incremental interval is required in the bisection method to guarantee the $H(F_i).H(F_{i-1}) \leq 0$ condition in view of the asymptotic root shown in Figure 5. However, using a small interval would mean a substantial number of iterations and therefore the bisection scheme is unsuitable because of the slow rate of convergence.

A "decision-iteration" scheme is adopted in GWEDGEM and EMU. From a detailed study of many poor failure surfaces it was found that the final segment of the graph (F_{guess} - F_{obtain}) vs. (F_{guess}) is a gradual monotonically increasing function both before and after the required root. As lower values of root are unimportant, initial iteration therefore begins from a high F_{guess} aimed at capturing the highest root. Tracing down the curve at an interval of (ΔF = 0.2 - 0.4) it is possible to locate the first sudden discontinuity/asymptotic near which the required root must lie (see Figure 6). Since it is not critical to know the exact value of the root, the objective is thus to reduce the interval containing the root until it is smaller than a convergence criterion (ε = 0.001). After first obtaining the interval containing the discontinuity, it is subdivided into halves. A decision is then made as to which half will contain the root (see Figure 7) and the selected half is then further subdivided. This process is repeated until the interval is smaller than ε, giving F accurate to at least two decimal places.

Once the root is obtained, the normal forces acting on the outer failure surface are checked to make sure that they are all compressive. This check is also valuable just in case a root lower than the required root has been captured (note that lower roots are associated with negative normal forces). For such instances, the starting interval of search, ΔF, is reduced and the iteration process is repeated from the start. Reducing the interval of search improves the chances of capturing the discontinuity. For most problems, the number of iterations seldom exceeds 30. Figure 8 outlines the logical sequence involved in the iteration procedure. The scheme is termed "decision-iteration" since at each subdivision of an interval into halves, a decision is made as to which half will contain the discontinuity/root.

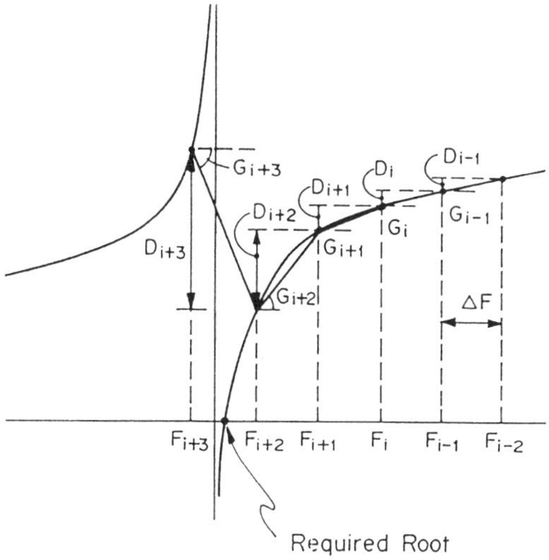

Capturing The Discontinuity

Let $\Delta F = F_{i-1} - F_i$

$D_i = (H(F_{i-1} - H)(F_i))$

$G_i = \frac{D_i}{\Delta F}$

Discontinuity occurs either when:

(1) $D_i > 20.D_{i-1}$; or
(2) $D_i < 0$; or
(3) $G_i < 0$

As in figure, $F_L = F_{i+3}$, $F_R = F_{i+2}$

Fig. 6 Location of first discontinuity

5 DISCUSSION

The method outlined above had been applied to

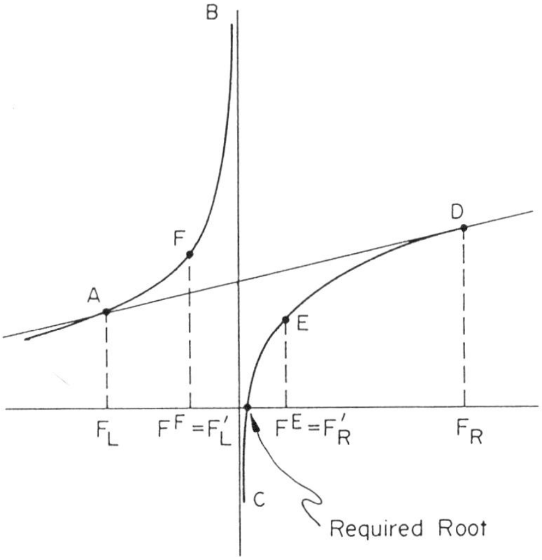

Decision-Iteration Technique

• portion of curve AB is above line AD (left-hand limit)

• portion of curve CD is below line AD (right-hand limit)

F_R = right-hand limit of interval of search.

F_L = left-hand limit of interval of search.

Iteration: $F_{new} = \frac{1}{2}(F_R + F_L) = F^E$

Decision: Since E is below line AD, $F^E = F'_R$

Iteration: $F_{new} = \frac{1}{2}(F'_R + F_L) = F^F$

Decision: Since F is above line AD, $F^F = F'_L$

Repeat *"decision-iteration"* until $(F_R - F_L) < \varepsilon$

• Very often during the "decision-iteration" process, $H(F_R).H(F_L) \leq 0$. When such condition is met, switch scheme to Regula Falsi technique.

Fig. 7 Decision-iteration technique

many problems, with widely differing geometries, soil properties and pore water pressure distributions (Giam (1989)). In all cases it was found to give convergence to the correct (upper) root, usually well within the limit of 40 iterations. For poor geometries the nature of the discontinuities in the solution can be extremely abrupt, but where this is not the case the solution may proceed accurately and efficiently using the modified secant method. The flow chart in Fig. 8 indicates how the analysis proceeds after the initial discontinuity has been located. The decision-iteration technique is used initially, but if the root is significantly higher than the value of F at the discontinuity, the analysis is soon switched to the modified secant technique. If a non-feasible solution arises with $Q_i < 0$, i.e. tensile forces occur across wedge bases, the analysis is recommenced with a reduced search interval.

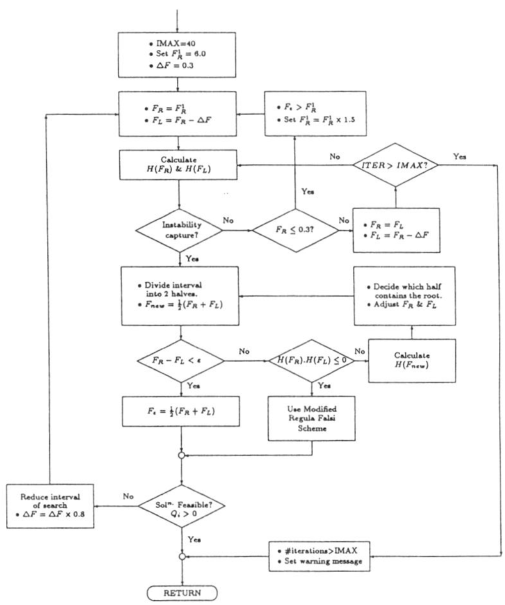

F_R = right hand limit of interval of search

F_L = left hand limit of interval of search

F_R^1 = starting value of F

F_s = required root where $(F_{guess} - F_{obtain}) \leq \epsilon$; ($\epsilon$ = 0.001)

$H(F_i)$ = value of $(F_{guess} - F_{obtain})$ for the ith estimate of F_{guess}

Fig. 8 Decision-iteration flow chart

6 CONCLUSION

A method has been described for solving the non-linear equations for Safety Factor, F, which arise in multi-wedge analyses of soil stability problems. The method is robust and efficient and can cope readily with the instabilities which frequently occur in the solution of non-linear equations.

7 ACKNOWLEDGEMENT

This paper has relied heavily on the work of the late Peter Giam and is a small tribute to his outstanding contributions to soil stability analysis.

8 REFERENCES

Chen, Z and N.R. Morgenstern 1983. Extensions to the generalised method of slices for stability analysis. *Can. Geotech. J.*, 20: 104-119.

Chowdhury, R.N. and S. Zhang 1990. Convergence aspects of limit equilibrium methods for slopes. *Canadian Geotechnical Journal*, 27: 145-151.

Chugh, A.K. 1981. Multiplicity of numerical solutions for slope stability problems. *Int. J. Num and Analyt. Methods in Geomech.*, 5: 313-322.

Donald, I.B. and S.K. Giam 1989. Improved comprehensive limit equilibrium stability analysis. *Civil Engineering Research Report*, 1/1989, Monash University

Fredlund, D.G. 1984. Analytical methods for slope stability analysis. *Proc. 4th Int. Sym. on Landslides*, Toronto, 229-250.

Giam, P.S.K. 1989. *Improved methods and computational approaches to geotechnical stability analysis.* Ph.D. Thesis, Monash University, Melbourne, Australia.

Giam, P.S.K. and I.B. Donald 1989(a). Soil stability analysis by an upper bound plasticity method. *Civil Engineering Research Report*, 2/1989, Monash University.

Giam, P.S.K. and I.B. Donald 1989(b). Appropriate optimization techniques for failure surface determination in geotechnical stability analysis. *Civil Engineering Research Report*, 3/1989, Monash University.

Li, K.S. and W. White 1987. Rapid evaluation of the critical slip surface in slope stability problems. *Int. J. Num. and Analyt. Methods in Geomech.*, 11: 449-493.

Müller, D.E. 1956. A method of solving algebraic equations using an automatic computer. *Math. Tables and Other Aids to Comp. (MTAC)*, 10: 208-215.

Nguyen, V.U. 1985. New application of classical numerical techniques in slope analysis. *5th Int. Conf. Num. Methods in Geomech.*, Nagoya, 949-957.

Sarma, S.K. 1987. A note on the stability analysis of slopes. *Geotechnique*, 37: 107-111.

Impact of Computational Mechanics on Engineering Problems, Pulmano & Murti (eds)
© 1993 Balkema, Rotterdam, ISBN 90 5410 324 8

Finite element analysis of the artificial freezing of sand

G.Gioda
Politecnico di Milano, Italy

ABSTRACT: The main characteristics are illustrated of a finite element approach for the analysis of the artificial freezing process of saturated soils. The effects of the water/ice phase transition are introduced in the calculations through a technique that views the latent heat as a nodal "load" which is activated when the melting/freezing temperature is reached. The results of a laboratory freezing test are illustrated and compared with the numerical ones obtained with the developed finite element program.

1 INTRODUCTION

The artificial freezing of ground is a techniques adopted for improving the mechanical characteristics of soils when dealing with the excavation of tunnels or shafts below the water table [1]. The complex, non linear, nature of the freezing process reduces tha cases in which closed form solutions can be directly applied to actual engineering problems, and in many practical situations the use of numerical methods seems mandatory.

Here the main characteristics are illustrated of a numerical approach for heat transfer analyses with phase change, customarily referred to as Stefan problem, developed for analyzing the freezing process of sandy soils. Among the numerical techniques used for the solution of the Stefan problem [2,3], the finite element method was chosen in the present contest. In fact it permits, with respect to other method, an easier treatment of the non linear aspects of the heat transfer within the soil mass and a more straightforward subsequent extension towards its stress analysis aspects.

The main assumptions which will be introduced in the numerical solution procedure can be summarized as follows:
1) the soil is perfectly saturated;
2) the thermal properties of soil are not influenced by the stress and strain variation caused by freezing;
3) the effects of the possible seepage flow of water and of the heat convection within it can be ignored;
4) the migration of water towards the phase change zone is neglected.

The first and second assumptions seem reasonable and should not appreciably limit the range of application of the numerical technique. Also the third assumption can be accepted since the presence of a significant seepage flow should be avoided in order to permit an effective freezing treatment. Note that this assumption rules out the movement of fluid from the problem at hand and permits to use the same governing equation for the heat transfer both above and below the freezing temperature.

The validity of the fourth hypothesis strongly depends on the characteristics of soil, and it can be accepted mainly for coarse textured soils, such as sands.

2. GOVERNING EQUATION AND BOUNDARY CONDITIONS

The equation of continuity, governing the two dimensional flow of heat through an isotropic solid, can be written in the following standard form [4]

$$\frac{\partial}{\partial x}(K \frac{\partial T}{\partial x}) + \frac{\partial}{\partial y}(K \frac{\partial T}{\partial y}) + \bar{Q} = \rho c \frac{\partial T}{\partial t} \quad , \quad (1)$$

where x and y are the Cartesian axes; T and t denote, respectively, temperature and time; K is the coefficient of thermal conductivity; ρ and c are the mass density and the thermal capacity, or specific heat, and $\bar{Q}$ is the rate of internal flow generation per unit volume.

As to the boundary conditions, let consider first the situation in which no phase change occurs. In this case four main types of conditions can be imposed on different portions of the boundary Γ of the flow domain Ω.

The temperature is known on Γ_1, whilst the heat flow component in the direction normal to the boundary is prescribed on Γ_2

$$T(x,y,t) - \bar{T} = 0 \quad \text{on } \Gamma_1 \quad , \quad (2a)$$

$$K \left(i_x \frac{\partial T}{\partial x} + i_y \frac{\partial T}{\partial y}\right) - \bar{q}_n = 0 \quad \text{on } \Gamma_2 \quad . \quad (2b)$$

In eqs.(2) $\bar{q}_n$ and $\bar{T}$ represent, respectively, the imposed rate of heat flow per unit area and the imposed temperature; i_x and i_y are the direction cosines of the unit outward vector normal to the boundary.

The convection and radiation boundary conditions are not discussed here since they will not be considered in the subsequent parts of the paper.

Note that the material parameters and the coefficients characterizing the boundary conditions depend, in general, on the temperature.

In the case of phase transition a surface $F(x,y,t)=0$ exists, having an a priori unknown shape, that separates liquid and solid phases. Similarly to other moving boundary problems [5], two different conditions, related respectively to the free variable (the temperature, in the present contest) and to its space derivatives, characterize this surface. The first one imposes that the temperatures of the frozen T_f and unfrozen T_u phases facing the moving surface are equal to the melting/ freezing temperature T_o of the material

$$T_f(x,y,t) = T_u(x,y,t) = T_o \quad . \quad (3a)$$

The second one represents the relationship between the net rate of heat flow in the direction normal to the moving surface, the latent heat per unit volume and the velocity v_n of the surface itself

$$K_f \left(i_x \frac{\partial T_f}{\partial x} + i_y \frac{\partial T_f}{\partial y}\right) + \\ - K_u\left(i_x \frac{\partial T_u}{\partial x} + i_y \frac{\partial T_u}{\partial y}\right) = L \cdot \rho \cdot v_n \quad . \quad (3b)$$

In eq.(3b) K_f and K_u are the coefficients of thermal conductivity of the frozen and unfrozen phases, respectively; L is the latent heat per unit mass and ρ is the mass density.

3. FINITE ELEMENT FORMULATION

The finite element formulation for heat transfer problems can be derived by "discretizing" the governing equations, and the relevant boundary conditions, rewritten in a suitable integral form. The case without phase change will be first considered.

For sake of simplicity let rewrite eqs. (1) and (2) in a compact form

$$\underline{\nabla}^T \underline{K} \underline{\nabla} T + \bar{Q} - \rho c \frac{\partial T}{\partial t} = 0 \quad \text{on } \Omega \quad , \quad (4a)$$

$$T - \bar{T} = 0 \quad \text{on } \Gamma_1 \quad , \quad (4b)$$

$$\underline{i}^T \underline{K} \underline{\nabla} T - \bar{q}_n = 0 \quad \text{on } \Gamma_2 \quad . \quad (4c)$$

Here $\underline{\nabla}^T = \{\partial/\partial x \ , \ \partial/\partial y\}$ is the linear differential operator, $\underline{K}$ is the 2x2 conductivity matrix and $\underline{i}$ is the vector of the direction cosines i_x and i_y.

It can be readily recognized that the solution satisfying the governing equation (4a) and the boundary conditions (4b,c), satisfies also the following integral relationship

$$\int_\Omega w \cdot \left(\underline{\nabla}^T \underline{K} \underline{\nabla} T + \bar{Q} - \rho c \frac{\partial T}{\partial t}\right) d\Omega + \\ + \int_{\Gamma_1} w_1 \cdot (T - \bar{T}) \, d\Gamma + \\ + \int_{\Gamma_2} w_2 \cdot (\underline{i}^T \underline{K} \underline{\nabla} T - \bar{q}_n) \, d\Gamma = 0 \quad , \quad (5)$$

where w, w_1 and w_2 are arbitrary "test" or weighting functions, square integrable over the flow domain Ω and over Γ_1 and Γ_2, respectively.

Let now impose a restriction on function w, assuming that its first order spatial derivatives exist. This permits the integration by parts of the first term in eq. (5) leading to the following "weak" form

$$-\int_\Omega (\underline{\nabla} w)^T \underline{K} (\underline{\nabla} T) \, d\Omega + \int_\Gamma w \cdot (\underline{i}^T \underline{K} \underline{\nabla} T) \, d\Gamma + \\ + \int_\Omega w \cdot \bar{Q} \, d\Omega - \int_\Omega w \cdot \rho c \frac{\partial T}{\partial t} \, d\Omega + \\ + \int_{\Gamma_1} w_1 \cdot (T - \bar{T}) \, d\Gamma + \\ + \int_{\Gamma_2} w_2 \cdot (\underline{i}^T \underline{K} \underline{\nabla} T - \bar{q}_n) \, d\Gamma = 0 \quad , \quad (6)$$

where Γ denotes the entire boundary of the flow domain. Note that the integral form (5) requires that the temperature T be twice differenciable, while for the weak form (6) it is necessary only that the first derivative of T exists.

In order to derive the finite element formulation from eq.(6), let subdivide the flow domain Ω into elements within which the temperature is expressed in terms of the relevant nodal values, collected in vector $\underline{T}_e$ for the e-th element, through known interpolation functions $\underline{b}_e$

$$T_e(x,y,t) = \underline{b}_e^T(x,y)\cdot\underline{T}_e(t) \quad . \tag{7}$$

By substituting eq.(7) into eq.(6), the following integral relationship is obtained for the e-th finite element

$$-\int_{\Omega_e} (\underline{\nabla} w)^T \, \underline{K}_e \, \underline{\nabla}(\underline{b}_e^T \underline{T}_e) \, d\Omega \, +$$
$$+ \int_{(\Gamma_{e1}+\Gamma_{e2}+\Gamma_{ei})} w \, \underline{i}_e^T \, \underline{K}_e \, \underline{\nabla}(\underline{b}_e^T \underline{T}_e) \, d\Gamma \, +$$
$$+ \int_{\Omega_e} w\cdot\overline{Q}_e \, d\Omega - \int_{\Omega_e} w\cdot(\rho c)_e \, \underline{b}_e^T \frac{\partial \underline{T}_e}{\partial t} \, d\Omega \, +$$
$$+ \int_{\Gamma_{e1}} w_1\cdot(\underline{b}_e^T \underline{T}_e - \overline{T}_e) \, d\Gamma \, +$$
$$+ \int_{\Gamma_{e2}} w_2\cdot[\underline{i}_e^T \, \underline{K}_e \, \underline{\nabla}(\underline{b}_e^T \underline{T}_e) - \overline{q}_{en}] \, d\Gamma = 0 \tag{8}$$

Note that the boundary Γ_e of the element is subdivided into portions Γ_{e1} and Γ_{e2}, belonging to Γ_1 and Γ_2 respectively, and Γ_{ei} adjacent to other elements.

Now, let choose a test function w that coincides with a "virtual" variation δT_e of temperature

$$w = \delta T_e = \underline{b}_e^T \, \delta\underline{T}_e = \delta\underline{T}_e^T \, \underline{b}_e \quad , \tag{9a}$$

i.e. an arbitrary temperature variation fulfilling the essential boundary conditions

$$w = \delta T_e = 0 \quad \text{on } \Gamma_1 \quad . \tag{9b}$$

In addition let assume that

$$w_1 = -w_2 = \delta T_e \quad . \tag{9c}$$

Substitution of eqs.(9) into eq. (8) leads to the following relationship, where the integrals over Γ_{e1} have been eliminated because of the assumption (9b),

$$-\int_{\Omega_e} [\underline{\nabla}(\underline{b}_e^T \delta\underline{T}_e)]^T \, \underline{K}_e \, \underline{\nabla}(\underline{b}_e^T \underline{T}_e) \, d\Omega \, +$$
$$+ \int_{\Gamma_{ei}} (\delta\underline{T}_e^T \underline{b}_e)\cdot[\underline{i}_e^T \, \underline{K}_e \, \underline{\nabla}(\underline{b}_e^T \underline{T}_e)] \, d\Gamma \, +$$
$$+ \int_{\Omega_e} (\delta\underline{T}_e^T \underline{b}_e) \, \overline{Q}_e \, d\Omega \, +$$
$$- \int_{\Omega_e} (\delta\underline{T}_e^T \underline{b}_e) \, (\rho c)_e \, \underline{b}_e^T \frac{\partial \underline{T}_e}{\partial t} \, d\Omega \, +$$
$$+ \int_{\Gamma_{e2}} (\delta\underline{T}_e^T \underline{b}_e) \, \overline{q}_{en} \, d\Gamma = 0 \quad . \tag{10}$$

After eliminating the virtual variation of nodal temperature $\delta\underline{T}_e$, eq.(10) can be written in the following compact form,

$$\underline{M}_e \, \underline{T}_e + \underline{N}_e \, \dot{\underline{T}}_e = \overline{\underline{f}}_{e\Omega} + \overline{\underline{f}}_{e\Gamma} + \underline{f}_{ei} \quad , \tag{11}$$

where

$$\underline{M}_e = \int_{\Omega_e} (\underline{\nabla b}_e^T)^T \, \underline{K}_e \, (\underline{\nabla b}_e^T) \, d\Omega \quad , \tag{12a}$$

$$\underline{N}_e = \int_{\Omega_e} \underline{b}_e\cdot(\rho c)_e \, \underline{b}_e^T \, d\Omega \quad , \tag{12b}$$

$$\overline{\underline{f}}_{e\Omega} = \int_{\Omega_e} \underline{b}_e \, \overline{Q}_e \, d\Omega \quad , \tag{12c}$$

$$\overline{\underline{f}}_{e\Gamma} = \int_{\Gamma_{e2}} \underline{b}_e \, \overline{q}_{en} \, d\Gamma \quad , \tag{12d}$$

$$\underline{f}_{ei} = \left(\int_{\Gamma_{ei}} \underline{b}_e\cdot(\underline{i}_e^T \, \underline{K}_e \, \underline{\nabla b}_e^T) \, d\Gamma \right)\cdot\underline{T}_e \quad . \tag{12e}$$

In eq.(11), $\overline{\underline{f}}_{e\Omega}$ and $\overline{\underline{f}}_{e\Gamma}$ are the nodal heat flows due to the internal heat generation per unit volume and to the imposed rate of heat flow on Γ_{e2}, respectively. Vector $\underline{f}_{ei}$ represents the nodal flow equivalent to the heat flow across the element sides facing the adjacent elements.

The global system of equations, governing the behaviour of the assembled mesh, is obtained by suitably overlapping eq. (11) written for all elements

$$\underline{M} \, \underline{T} + \underline{N} \, \dot{\underline{T}} = \overline{\underline{f}}_\Omega + \overline{\underline{f}}_\Gamma \quad . \tag{13}$$

Note that the global continuity of the heat flow imposed through the assemblage of the finite element mesh requires that the assembled vector of the interface fluxes $\underline{f}_{ei}$ vanishes. Before solving the equation system (13) the essential boundary conditions should be imposed on Γ_1.

4. TIME INTEGRATION PROCEDURE

The step-by-step time integration of eq. (13) can be based on the assumption that the time dependent variables have a constant rate during a time increment $\Delta t>0$. This leads to the following relationships expressing the values of a time dependent variable, denoted here by x, and of its time derivative $\dot{x}$ at a time ξ within the interval t, t+Δt

$$\dot{x}_\xi = \frac{1}{\Delta t}\cdot[x_{t+\Delta t} - x_t] \quad , \tag{14}$$

$$x_\xi = (1-\vartheta)\cdot x_t + \vartheta\cdot x_{t+\Delta t} \quad , \tag{15}$$

where

$$0 \le \vartheta \le 1 \quad \text{and} \quad \xi = t + \vartheta\cdot\Delta t \quad .$$

Expressing all matrices and vectors in eq. (13) through eqs.(14) and (15) (note in fact that matrices $\underline{M}$ and $\underline{N}$ are in general functions of temperature and, hence, of time), and assuming ϑ=1/2, the following iterative equation is reached that leads to the nodal temperatures at time t+Δt

$$[\frac{\Delta t}{2} (\underline{M}^{i-1}_{t+\Delta t} + \underline{M}_t) + (\underline{N}^{i-1}_{t+\Delta t} + \underline{N}_t)] \; \underline{T}^{i}_{t+\Delta t} =$$

$$= \Delta t \; (\bar{\underline{f}}^{i-1}_{t+\Delta t} + \bar{\underline{f}}_t) + [-\frac{\Delta t}{2} (\underline{M}^{i-1}_{t+\Delta t} + \underline{M}_t) +$$

$$+ (\underline{N}^{i-1}_{t+\Delta t} + \underline{N}_t)] \; \underline{T}_t \quad , \tag{16}$$

where

$$\bar{\underline{f}} = \bar{\underline{f}}_\Omega + \bar{\underline{f}}_\Gamma \quad .$$

The suffix i in eq.(16) denotes the iteration number.

At the beginning of iterations for a given time increment the temperature and all relevant quantities at time t are known, and the temperature at time t+Δt is extrapolated on the basis of its values computed at the end of the preceding time steps. The iterative process for the current time increment, based on eq.(16), ends when the difference between the temperature at time t+Δt evaluated at two subsequent iterations decreases below a given value.

5. PHASE TRANSITION

Within the finite element method, two main approaches have been presented in the literature for the analysis of the Stefan problem. The first one, the so called "variable mesh" approach [2,6], modifies the geometry of the grid during the iterative solution process until a part of its boundary approximates with the desired accuracy the shape of the moving surface at the chosen time. Approaches of this type become exceedingly complex if several moving surfaces exist which tend to join with each other when freezing propagates. Since this situation is likely to take place when dealing with the artificial freezing of ground, an alternative class of approaches, the so called "fixed mesh" techniques [7], has been considered here.

They operate on grid of constant geometry, hence the nodal coordinates remain unchanged during the solution process, and allow the moving boundary to pass through the elements. Usually these procedures are less accurate than the preceding ones, but offer the non negligible advantage of handling several simultaneous moving surfaces without requiring particular provisions.

The technique adopted in this study viewes the latent heat as a positive or negative source of heat, uniformly distributed on the volume, that becomes active when the freezing/melting temperature is reached. The effect of this source is accounted for in the finite element analysis in integral terms, as customary when dealing with this method for numerical solution.

First, the nodal vector of total latent heat is evaluated by the following equation, where the summation runs over the element of the mesh

$$\underline{q}_L = \Sigma_e \int_{\Omega_e} \underline{b}_e \; L \; \rho \; d\Omega \quad . \tag{17}$$

When the temperature at a node reaches the phase transition value, the node is constrained so that no further variation of its temperature may occur. The corresponding increment of nodal heat Δq_i is evaluated for each time step, until the accumulated heat reaches the previously evaluated nodal latent heat,

$$\Sigma_i \; \Delta\underline{q}_i = \underline{q}_L \quad . \tag{18}$$

When the condition expressed by eq.(18) is fulfilled, where the summations runs over the time increments, the node is released and its temperature is again allowed to vary with time. If during a time step the accumulated nodal heat overcomes the corresponding latent heat, the node is released and a heat increment is imposed to it, for that time step, equal to the difference between the previously accumulated heat and the total latent heat.

6. NUMERICAL ANALYSIS OF A FREEZING TEST

Some laboratory freezing tests have been performed in order to calibrate the finite element computer program and to choose a suitable procedure for determining the thermal parameters for the soil. The experimental details of this study have been discussed elsewere [8], here only its main characteristics will be summarized.

The experimental set up consists of a thermally insulated cylinder, having internal diameter and height of 1 m, that before the test is filled with sand with constant density. A steel pipe, installed at the centre of the container, is used for the circulation of liquid nitrogen during the test.

The cylindrical geometry of the container and its insulation (consisting of polyurethane foam) lead to axisymmetric and planar heat flow conditions during the tests. This permits an easier interpretation of the experimental results by means of a one dimensional finite element model.

Before the deposition of sand about 10 cm of water were placed in the container. Then the deposition process was carried out by spreading the sand out from a sieve which was kept at a constant level above the water surface. Additional water was introduced in the container during this process, so that the water level was always about 10 cm above the level of the deposed sand. This simple technique lead to a saturated soil having a fairly constant density throughout the container.

A small gap of about 2 cm was left between the sand and the upper part of the container in order to avoid possible damages of the container itself, and of the insulating material, due to the volume increase of the water during freezing. The drainage of water was allowed by three small diameter pipes inserted in the wall of the container at the level of the mentioned gap.

Nine thermal transducers were placed within the sand during its deposition. They were situated at the level of the mid plane of the container in order to reduce the influence on the readings of its upper and lower bases. The transducers were located along three radii, at an angle of 120° from each other, and at a distance of 150 mm, 270 mm and 390 mm from the cylinder axis.

Three additional transducers were used to measure the temperatures of the outer part of the insulation, of the nitrogen at its outlet from the steel pipe and the room temperature.

The transducers were connected to a personal computer for the real time acquisition of the temperatures, and their graphical representation.

Various laboratory tests have been performed using freezing pipes with different external diameters. In order to facilitate the subsequent interpretation of the experimental data, the same temperature (-183°C) was imposed at the outlet of the nitrogen for all tests, by a suitable regulation of its flux.

The variation with time of the temperature measured by the thermal transducers for one of these tests is shown by dashed lines in fig.1. The diagrams a, b and c, corresponding to the temperatures within the sand, were obtained by averaging the readings of the transducers located at the same distance from the freezing pipe. The maximum difference between the temperatures measured at the same distance from the pipe, on different radii, was less than 1.5 °C. This indicates that the heat flow within the container was reasonably close to the sought axisymmetric conditions.

The flux of liquid nitrogen was stopped when the temperature of the farthest transducers from the axis reached 0°C. This was done in order to avoid that the volume increase due to freezing could damage the container, since some liters of water have already left the cylinder through the drainage pipes a before reaching this temperature.

The recorded data show that after a some hours the temperature of all transducers reached the same value equal to about -1°C which remained practically constant for several days, until the equipment was dismounted. This fact indicates that the polyurethane provided an adequate thermal insulation of the container.

The results of the finite element simulation of the test are represented in fig. 1 by solid lines. The details of the determination of the thermal properties of the soil introduced in the finite element calculations have been discused in [8].

The comparison between experimental and numerical results indicate that the finite element calculation is able to provide an acceptable estimation, from an engineering standpoint, of the heat flow within the sand and that the progress of the frozen boundary can be evaluated with reasonable accuracy.

Note that the stepwise pattern of the numerically evaluated diagrams, in the vicinity of the phase change temperature, depends on the previously dexribed procedure adopted for taking into account the latent heat effects.

7. CONCLUSIONS

A study of the non linear conduction of heat within saturated sands has been presented, with particular reference to the finite element analysis of the artificial freezing of soils.

The experimental data of a laboratory freezing test have been compared with the corresponding numerical results, in terms of the progress of the frozen boundary with time. The reasonable agreement between the two set of results indicates that the developed numerical procedure is able to solve heat transfer problems related to ground freezing with an acceptable accuracy from an engineering view point.

Possible estensions of this study seem possible concerning, in particular, the "optimization" of some of the geometrical and thermal parameters characterizing the artificial freezing treatments, such as the distance and diameter of the freezing pipes and the temperature of the freezing fluid.

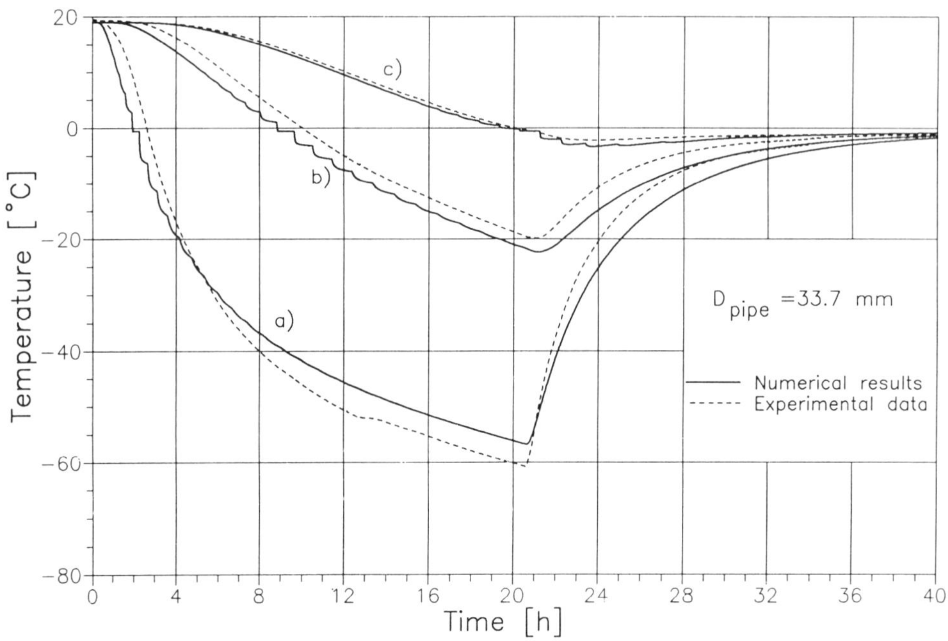

Fig.1 Comparison between experimental data from a laboratory freezing test and results of a finite element analysis.

ACKNOWLEDGEMENTS

This paper is part of a research project supported by RODIO corporation.

REFERENCES

[1] Andersland O.B., Gallavresi F., Goto S., Saareleinen S., Slunga E. (1989) "General report of the specialty session on ground freezing", Proc. XII ICSMFE, Rio de Janeiro.

[2] Zabaras N., Y. Ruan (1990) "Moving and deforming finite element simulation of Stefan problem", Communications in Applied Numer.Methods, Vol.6, pp.495-506

[3] Zabaras N., S. Mukherjee (1987) "An analysis of solidification problems by the boundary element method", Int.J. Numer.Methods Engineering, Vol. 24, pp.1879-1900

[4] Carslaw H.S., Jaeger J.C. (1959) "Conduction of heat in solids", 2nd Ed., Oxford University Press, Oxford, U.K.

[5] Crank J. (1984) "Free and moving boundary problems", Oxford University Press, Oxford, U.K

[6] Zabaras N., Y. Ruan (1989) "A deforming finite element method analysis of inverse Stefan problem", Int.J.Numer. Methods Engin., Vol.28, pp. 295-313

[7] Voller V.R., C.R. Swaminathan, B.G. Thomas (1990) "Fixed grid techniques for phase change problems: a review", Int.J.Numer.Methods Engineering, Vol. 30, pp.875-898

[8] Gioda G., Locatelli L., Gallavresi F. (1993) "A numerical and experimental study of the artificial freezing of sand", to appear.

Impact of Computational Mechanics on Engineering Problems, Pulmano & Murti (eds)
© 1993 Balkema, Rotterdam, ISBN 90 5410 324 8

A generalized material parameter identification procedure in geotechnical engineering

Y. Ichikawa
Nagoya University, Japan

ABSTRACT: Survey plays an essential role in geotechnical engineering, since the materials treated are natural. This implies that the mechanics models are inevitably inhomogeneous. However, the material properties obtained by *in situ* tests such as the plate loading test, bore hole pressure test, Lugeon test, *et al.* have ever been determined on the basis of analytical solutions, which cannot take the inhomogeniety into account. We have proposed a generalized parameter identification method based on a dual boundary control concept, and it is applied to the linear elastic problem (Ichikawa and Ohkami 1992), the viscoelastic problem (Ohkami and Ichikawa 1993), the wave propagation problem (Ichikawa, Kyoya, Ohkami and Wu 1992), and the damage mechanics (Tsuchiyama, Ohkami and Ichikawa 1993). A new horizon is here presented for applying the above concept in geotechnical engineering.

1. PARAMETER IDENTIFICATION IN GEOTECHNICAL ENGINEERING

In the geotechnical engineering it is highly important to know the state of underground precisely. There are a lot of methods for surveying the geological state. The surveying is a procedure to interpret the data which are observed under some external excitation. The geological survey is said to be one of the inverse problem.

The inverse problems are classified into many categories (Kubo 1992). Before showing this classification, let us remind what the forward or direct problems are (see Fig.1).

The forward problem for the unknown variable ϕ is defined under the following informed conditions:

a) the domain Ω and its boundary $\partial\Omega$,

b) the governing equation such as

$$L(\boldsymbol{P})\phi = f \quad \text{in } \Omega,$$

c) the boundary conditions giving ϕ, its derivatives and/or the initial conditions if needed,

d) the internal excitation f, and

e) the distribution of the material parameter(s) $\boldsymbol{P}$.

If one of these conditions is lacked, the forward problem is not able to be solved. The inverse problem is defined by the procedure to guess one or combination of these conditions under some additional information. Thus, corresponding to the above conditions the inverse problem is classified into

A) the domain identification problem,

B) the governing equation identification problem,

C) the boundary/initial condition identification problem,

D) the internal excitation identification problem, and

E) the material parameter identification problem.

It is recognized that the inverse analysis plays an important role in the geotechnical engineering because the domain is essentially inhomogeneous and the material properties are unknown. The inverse problem appeared in the geotechnical engineering field is mainly concerned with the material parameter identification. These methods are classified into the inverse formulation (Cividini, Jurina and Gioda 1981, Sakurai and Takeuchi 1983), and the direct formulation (Gioda and Maier 1980).

These methods are, however, known to be insufficient. That is, for example, the Young's modulus and the Poisson's ratio are not able to be identified simultaneously.

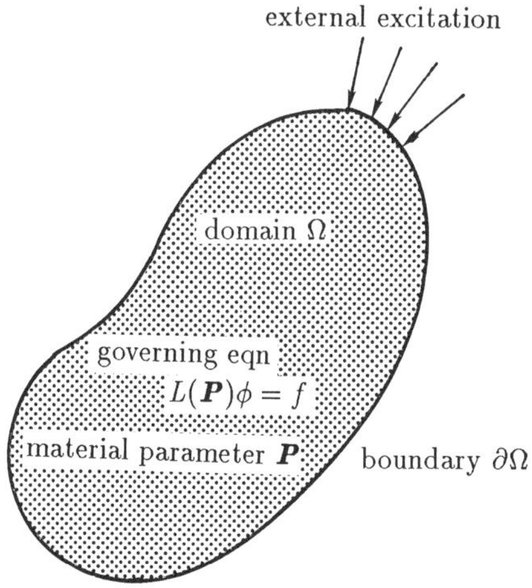

Figure 1. Forward problem.

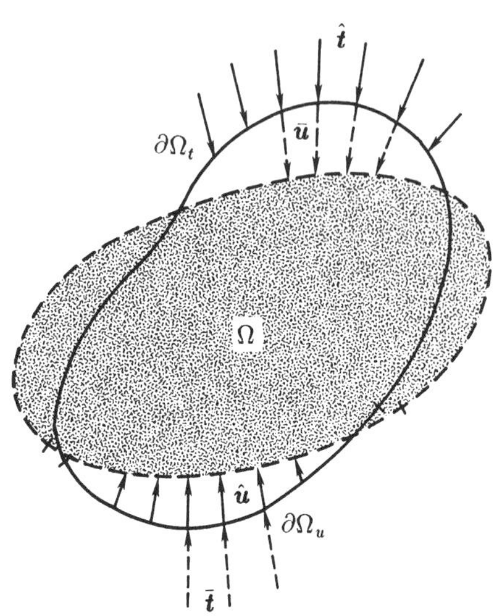

Figure 2. Dual boundary control problem.

We have presented a new methodology of the material parameter identification problem for the linear elastic material (Ichikawa and Ohkami 1992), for the visco-elastic material (Ichikawa and Ohkami 1991; Ohkami and Ichikawa 1993), for the damaged material with distributed discontinuities (Tuchiyama, Ohkami and Ichikawa 1993), and for the wave propagation problem (Ichikawa, Kyoya, Ohkami and Wu 1992). This method which is based on the dual boundary control concept is simple but it gives fairly correct results. We here present the key concept of this procedure.

2. PARAMETER IDENTIFICATION PROCEDURE FOR THE ELASTIC PROBLEM

Let the domain Ω and its boundary $\partial\Omega$ be fixed (Fig.2). Note that $\partial\Omega$ consists of the displacement boundary $\partial\Omega_u$ on which a forced displacement $\hat{\boldsymbol{u}}$ is given and the traction boundary $\partial\Omega_t$ on which a traction force $\hat{\boldsymbol{t}}$ is given. Then, the governing equation and its boundary conditions are written as

(Governing equation)

$$\nabla \cdot \boldsymbol{\sigma} = \boldsymbol{0} \qquad \text{in } \Omega \subset \boldsymbol{R}^n (n = 1, 2, 3) \tag{1}$$

(Given boundary conditions)

$$\boldsymbol{u} = \hat{\boldsymbol{u}} \qquad \text{on } \partial\Omega_u \tag{2}$$

$$\boldsymbol{\sigma n} = \hat{\boldsymbol{t}} \qquad \text{on } \partial\Omega_t \tag{3}$$

where $\boldsymbol{\sigma}$ is the stress tensor, $\boldsymbol{u}$ the displacement vector, and $\boldsymbol{n}$ the outward unit normal vector on the boundary $\partial\Omega$.

We assume that the material behavior can be described by the Hookean law (that is, a linear elastic material).

(Constitutive equation: Hooke's law)

$$\boldsymbol{\sigma} = \boldsymbol{D\varepsilon} \tag{4}$$

where $\boldsymbol{\varepsilon} = [(\nabla \boldsymbol{u}) + (\nabla \boldsymbol{u})^T]/2$ is the strain tensor, and $\boldsymbol{D}$ the Hookean tensor which involves Young's modulus E and Poisson's ratio ν for linear isotropic materials.

Under a dual boundary control concept the material parameter identification problem is formulated as followings: In order to identify the material parameter $\boldsymbol{P}$ which consists of Young's moduli E and Poisson's ratios ν for the elastic problem with multiple layers, we give an external excitation $\hat{\boldsymbol{u}}$ on $\partial\Omega_u$, and observe its (dual) response $\bar{\boldsymbol{t}}$ on $\partial\Omega_{u'}$ which is a subset of $\partial\Omega_u$. Or conversely we give an external excitation $\hat{\boldsymbol{t}}$ on $\partial\Omega_t$, and observe its (dual) response $\bar{\boldsymbol{u}}$ on $\partial\Omega_{t'}$ which is a subset of $\partial\Omega_t$. These conditions are written as

(Observed or control boundary conditions)

$$\boldsymbol{u} = \bar{\boldsymbol{u}} \qquad \text{on } \partial\Omega_{t'} \subset \partial\Omega_t \tag{5}$$

$$\boldsymbol{\sigma n} = \bar{\boldsymbol{t}} \qquad \text{on } \partial\Omega_{u'} \subset \partial\Omega_u \tag{6}$$

By using these additional information $\bar{\boldsymbol{t}}$ and/or $\bar{\boldsymbol{u}}$ we may determine the material parameter $\boldsymbol{P}$.

The virtual work equation equivalent to (1)-(3) is given by

$$\int_\Omega \boldsymbol{\sigma} : \delta\boldsymbol{\varepsilon} dv - \int_{\partial\Omega_t} \hat{\boldsymbol{t}} \cdot \delta\boldsymbol{u} dS = 0 \tag{7}$$

The finite element approximation

$$u \simeq u_h = NU \tag{8}$$

is applied to Eqn(7), and get a following nonlinear simultaneous equation;

$$K(P)U = F \tag{9}$$

$$K = \int_\Omega B^T D(P) B dv$$

$$F = \int_{\partial\Omega_t} N^T \hat{t} dS$$

where N is the matrix of shape functions, and B the strain-displacement matrix.

The observed boundary conditions (5) and (6) are discretized as

$$S_u U = \bar{U} \tag{10}$$

$$S_t F = \int_{\partial\Omega_u} N^T \bar{t} dS = \bar{F} \tag{11}$$

where S_u and S_t are diagonal matrices for choosing observational boundary nodes of displacement and traction, respectively. That is, if i-th node is the displacement observational boundary, and if j-th node is the traction one, we have

$$S_u = \begin{bmatrix} 0 & & & & \\ & \ddots & & \mathbf{o} & \\ & & 1 & & \\ & \mathbf{o} & & \ddots & \\ & & & & 0 \end{bmatrix} \cdots i\text{-th row}$$

i-th column

$$S_t = \begin{bmatrix} 0 & & & & \\ & \ddots & & \mathbf{o} & \\ & & 1 & & \\ & \mathbf{o} & & \ddots & \\ & & & & 0 \end{bmatrix} \cdots j\text{-th row}$$

j-th column

The Newton's iteration scheme for determining the material parameters P is applied for (9), (10) and (11). That is, at the k-th iteration step, we get

$$G dx = R \tag{12}$$

where

$$G = \begin{bmatrix} K(P^k) & -I & (\frac{\partial K}{\partial P} U)^k \\ S_u & \mathbf{o} & \mathbf{o} \\ \mathbf{o} & S_t & \mathbf{o} \end{bmatrix}$$

$$dx = \begin{bmatrix} dU^k \\ dF^k \\ dP^k \end{bmatrix} = \begin{bmatrix} U^{k+1} - U^k \\ F^{k+1} - F^k \\ P^{k+1} - P^k \end{bmatrix}$$

$$R = \begin{bmatrix} F^k - K(P^k)U^k \\ \bar{U} - S_u U^k \\ \bar{F} - S_t F^k \end{bmatrix}$$

Since (12) is an over-determined system, we solve the following normal equation:

$$G^T G dx = G^T R \tag{13}$$

We can now seek for the material properties P by the iteration scheme (13).

We here additionally introduce a linear searching method for P during the iteration step k. Let us define

$$\mathcal{W}(P) = \frac{1-\omega}{2}(U - \bar{U}) \cdot (U - \bar{U}) + \frac{\omega}{2}(F - \bar{F}) \cdot (F - \bar{F})$$

$$-\frac{1}{\varepsilon}E^- - \frac{1}{\varepsilon}\nu^- + \frac{1}{\varepsilon}(\nu - 0.5)^+ \tag{14}$$

where ω $(0 \leq \omega \leq 1)$ is a given constant, $0 < \varepsilon \ll 1$ a penalty constant, and

$$a^+ = \begin{cases} a & \text{if } a > 0 \\ 0 & \text{if } a < 0 \end{cases} \qquad a^- = \begin{cases} 0 & \text{if } a > 0 \\ a & \text{if } a < 0 \end{cases}$$

By using dU^k calculated by (13), we introduce the gradient α^k of P^k as

$$\alpha_i^k = \frac{dP_i^k}{P_i^k} \tag{15}$$

and we seek λ such that

$$\mathcal{W}(P^k + \lambda\alpha^k) \quad \rightarrow \quad min \tag{16}$$

(see Box, Davies and Swann 1969). At the iteration step $k+1$ in (34), we then set

$$P^{k+1} = P^k + \lambda\alpha^k \tag{17}$$

The convergence criterion for (18) is given if the following conditions are all satisfied:

$$\frac{\| U^{k+1} - U^k \|}{\| U^{k+1} \|} < \varepsilon_1$$
$$\frac{\| F^{k+1} - F^k \|}{\| F^{k+1} \|} < \varepsilon_2 \tag{18}$$
$$\frac{\| P^{k+1} - P^k \|}{\| P^{k+1} \|} < \varepsilon_3$$

where $0 < \varepsilon_1, \varepsilon_2, \varepsilon_3, \ll 1$ are constants, and we use the square norm $\| a \| = (a \cdot a)^{1/2}$.

3. NUMERICAL EXAMPLES

(1) *One layer problem*

Let us consider the problem consisted of an isotropic homogeneous layer (Fig.3). The top surface AB is a traction boundary $\partial\Omega_t$. At the center node, a concentrated force of $10^6(N)$ is given. Other node points on AB are traction-free, and at points 1 to 4 we observe displacements such as

$$u_1 = u_4 = 3.94, \quad u_2 = u_3 = 5.15 \quad (cm)$$
$$v_1 = v_4 = -10.51, \quad v_2 = v_3 = -18.76 \quad (cm)$$

Firstly, we identify the Young's modulus E only, and compare the results with ones of the inverse formulation method (Gioda and Maier 1980) and of the direct formulation method (Cividini, Jurina and Gioda 1981). We find the three methods give mostly similar results (Table 1).

Next by using the same model shown in Fig.3, we identify both the Young's modulus E and Poisson's ratio ν. The results are shown in Table 2. Here the direct formulation (1) implies that E and ν are simultaneously identified, while in the direct formulation (2), after E is roughly converged, the process for identifying ν is started. Note that the maximum number of iterations is set as 20, so that the inverse formulation and the direct formulation (2) are not convergent. It is said that the proposed method is accurate and effective. The accuracy of this method is also confirmed in Table 3 in which displacements of the control nodes 1 and 2 at the final iteration stage are shown. The converging processes of these four methods are found in Fig.4.

(2) *Multi-layer problem*

We next treat an excavation problem shown in Fig.5 (Arai, Ohta and Yasui 1983), which is formed by three layers. On the boundary 4-6, a distributed load due to excavation is given. The observed displacements are

$$v_1 = -0.014 \quad (\mathrm{cm})$$
$$u_4 = 2.16, \quad v_4 = 0.926 \quad (\mathrm{cm})$$
$$u_5 = 1.81, \quad v_5 = 0.521 \quad (\mathrm{cm})$$
$$u_6 = 0.646, \quad v_6 = 0.133 \quad (\mathrm{cm})$$
$$u_9 = 0.959, \quad v_9 = 0.344 \quad (\mathrm{cm})$$

We compare the results of the proposed method with ones of Arai, Ohta and Yasui (1983) in Fig.6. It is understood that the proposed method gives accurate results within 5 iterations, while the results of Arai et al. are not sufficiently accurate. Note that in Eqn(14) we choose $\omega = 0$, and the convergence criteria of Eqn(18) are set as $\varepsilon_1 = \varepsilon_3 = 0.01$.

4. CONCLUSIONS

By applying the concept of the inverse analysis methods we have a possibility to re-organize the procedures of surveying in geotechnical engineering. Among the inverse analysis methods the material parameter identification plays an important role in this field. However since this problem is ill-posed, it is difficult to establish a sufficiently accurate method. On the basis of a dual boundary control concept we have developed a new and generalized method to solve this difficult problem. Of course, the domain identification which is important in the geotechnical engineering is

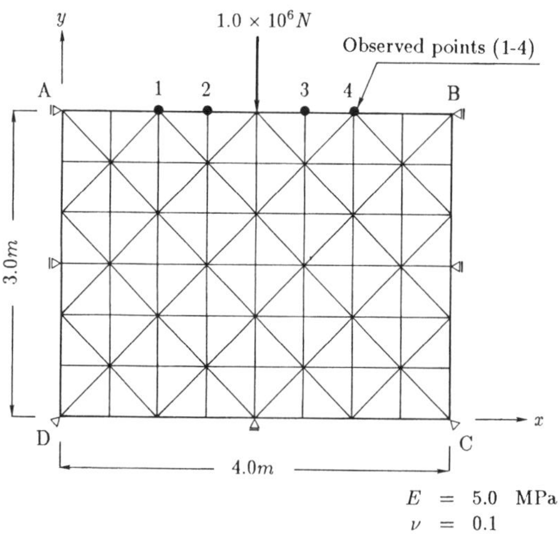

Figure 3. One layer model.

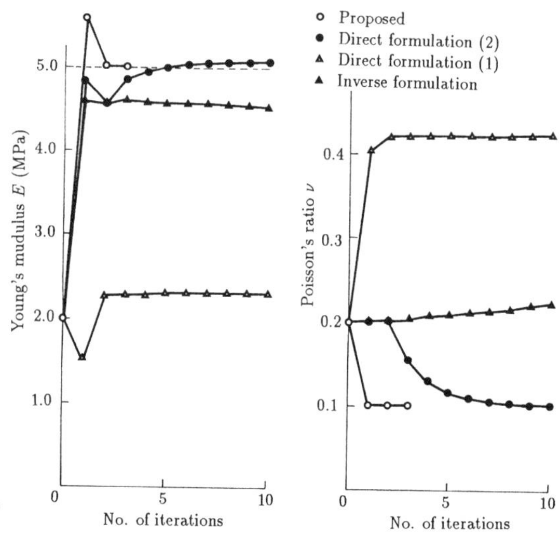

Figure 4. Iteration procedures for one layer problem identifying E and ν.

remained, and further research works are required in this material parameter identification problem such as an existence proof of the solution, a data sensitivity analysis, etc.

The proposed procedure is based on the deterministic way. On the other hand, since the measured data always involve an observed noise, or since the system equation (that is, the governing equation) may be beset with a system noise, we should establish a stochastic method on the basis of the proposed dual boundary control concept.

REFERENCES

1) Box, M.J., Davis, D., and Swann, W.H. (1969): *Nonlinear Optimization Techniques,* ICI Ltd., Monograph No.5, Oliver and Boyd, Ltd., Edinburgh.

2) Cividini, A., Jurina, L., and Gioda, G. (1981): "Some Aspects of Characterization Problem in Geomechanics," *Int. J. Rock Mech. Min. Sci. and Geomech. Abstr.*, Vol.18, pp.487-503.

3) Gioda, G., and Maier, G. (1980): "Direct Search Solution of an Inverse Problem in Elastoplasticity: Identification of Cohesion, Friction Angle and *in situ* Stress by Pressure Tunnel Tests," *Int. J. Numer. Methods Eng.*, Vol.15, pp.1823-1848.

4) Ichikawa, Y., and Ohkami, T.(1992): "A Parameter Identification Procedure as a Dual Boundary Control Problem for Linear Elastic Materials," *Soils and Foundations*, Vol. 32, No. 2, pp.35-44.

5) Ichikawa, Y., Kyoya, T., Ohkami, T., and Xu Wu (1992): "A Parameter Identification Procedure as a Dual Boundary Control Concept for Wave-Propagation Problem," *Proc. IUTAM Symp. on Inverse Problems*, Tokyo, 1922.

6) Kubo, S.(1992): *Inverse Problems*, Baifukan Pr. (in Japanese).

7) Ohkami, T., and Ichikawa, Y. (1993): "A Parameter Identification Procedure as a Dual Boundary Control Problem for Viscoelastic Materials," *Int. J. Num. Anal. Meth. Geomech.*, presented.

8) Tsuchiyama, S., Ohkami, T., and Ichikawa, Y. (1993): "Identification of Damage Tensor for Jointed Rock Mass and Deformation Prediction during Excavation of Underground Cavern," *Proc. JSCE*, accepted (in Japanese).

Table 1: Results for one layer problem identifying E

	Initial	True	Inverse form.	Direct form.	Proposed
E(MPa)	2.0	5.0	5.000	5.000	5.000
No. of iterations			2	3	3

Table 2: Results for one layer problem identifying E and ν

	Initial	True	Inverse form.	Direct form. (1)	Direct form. (2)	Proposed
E(MPa)	2.0	5.0	4.077	2.268	5.056	5.000
ν	0.2	0.1	0.280	0.418	0.102	0.100
No. of iterations			20	12	20	3

Table 3: Final values calculated for the control displacement boundary

Measured data	True (cm)	Inverse form. (Gioda)	Direct form. (1)	Direct form. (2)	Proposed
u_1	3.94	3.34	3.07	3.89	3.94
v_1	-10.51	-9.21	-3.51	-10.38	-10.51
u_2	5.15	4.43	3.66	5.08	5.15
v_2	-18.76	-18.87	-21.04	-18.54	-18.76

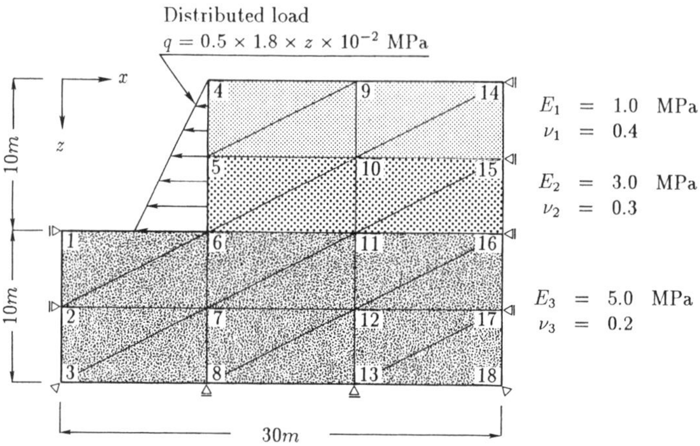

Figure 5. Three layer model.

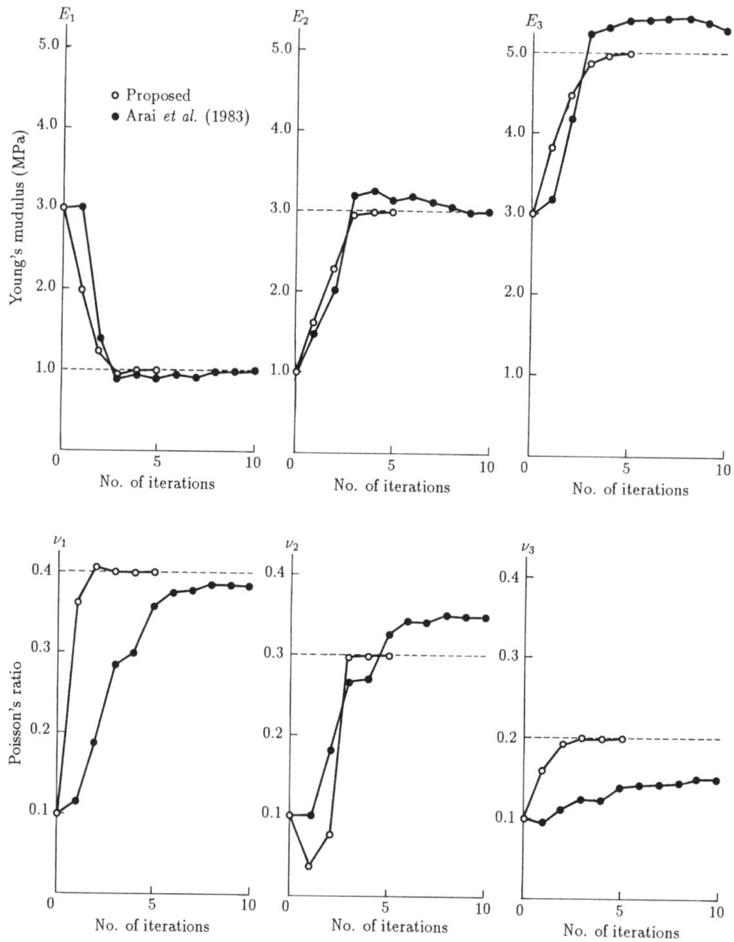

Figure 6. Iteration procedures for three layer model.

Impact of Computational Mechanics on Engineering Problems, Pulmano & Murti (eds)
 ISBN 90 5410 324 8

Fundamental study on in-situ fault triaxial compression test of rock mass

K. Ishibashi
Saga University, Japan

ABSTRACT: It is scarcely ever to be done in-situ triaxial compression test of rock mass, since this test needs a great deal of work and cost for its performence. As a way to clear the above mentioned disadvantages, the author fixed his eyes upon in-situ fault triaxial compression test (FTCT) of rock mass. The results obtained from FEM analysis about small-scaled FTCT of rock model are summarized in this note.

1 INTRODUCTION

In-situ triaxial compression test of rock mass is uncommon since it takes much time and costs a great deal of labor. As a way to dissolve these disadvantages, the author fixed his eyes upon the fault triaxial compression test (FTCT).

2 FAULT TRIAXIAL COMPRESSION TEST

The conception of FTCT is illustrated in Fig. 1. The testing method of FTCT is essentially similar to that of conventional triaxial compression test (CTCT) of intact rock. Cylinder specimen is continued to bedrock at bottom end and subjected to axial load at top end.

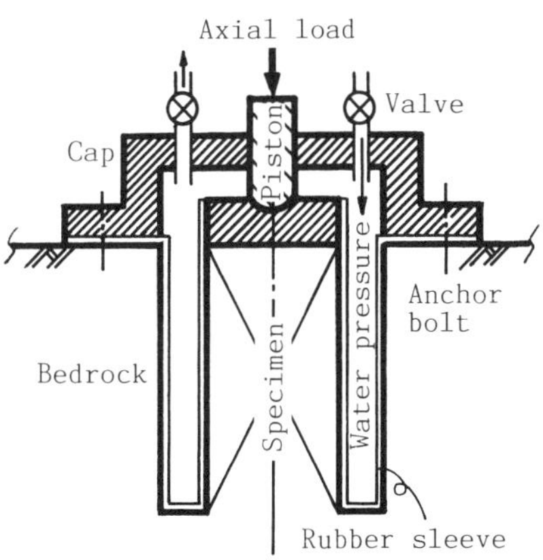

Fig. 1 The conception of FTCT.

3 FAILURE ANALYSIS OF MODEL SPECIMENS

Elasto-plastic analysis about rock model was conducted using Finite Element Nonlinear Structural Analysis System (FINAS), which was developed by Power Reactor and Nuclear Fuel Development Corporation in Japan.

3.1 ANALYTICAL MODEL AND CONDITIONS

Taking into account the results of preliminary model experiment about unconfined

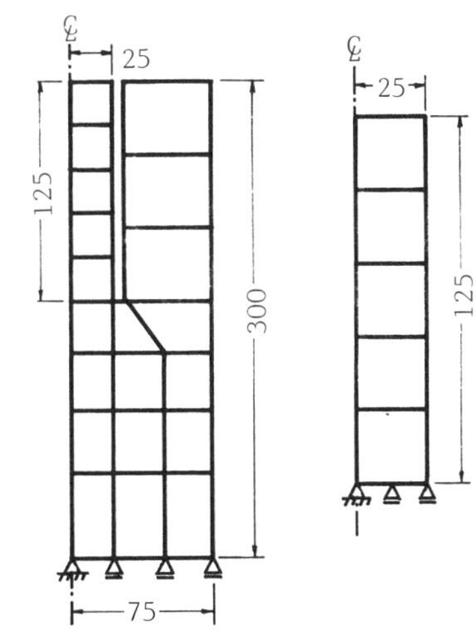

(a) For FTCT (b) For CTCT

Fig. 2 The analytical models.

condition, the specimen with 125 mm in cutting depth was assigned as objective case for analysis. Five values of confining pressures, 0, 0.49, 0.98, 1.96 and 3.92 MPa were considered in the analysis. Fig. 2 shows analytical models of FTCT and CTCT. Axisymmetric quadrilateral element with eight nodes was employed.

3.2 RESULTS AND DISCUSSIONS

Making reference to the results, Mohr's stress circles can be described as figs. 3 and 4. Failure envelope of FTCT specimens is drawn outside that of CTCT specimens. This means the influence of continuation at bottom end in FTCT specimen. Table 1 shows shearing strengths and internal friction angles graphically derived from both figures. As for as this result, FTCT yields about 15 % of larger shearing strength than that of CTCT and almost same internal friction angle with that of CTCT. It is regarded that FTCT estimats somewhat larger shearing strength than that of CTCT.

Table 1 Shearing strength and internal friction angle derived from figs. 3 and 4.

	Input Value	Specimen for	
		CTCT	FTCT
Shearing strength (MPa)	9.57	9.51	10.9
Internal friction angle (deg.)	46.0	46.0	46.5

Failure form is deduced by the locations of plasticized zone. Fig. 5 shows an example of final plasticized area. In the case that 3.92 MPa of confining pressure was applied to the FTCT specimen, macroscopical failure mechanism was developed in the lower part of fictitious specimen's boundary. This implies that existence of ceiling limits about confining pressure level in the FTCT.

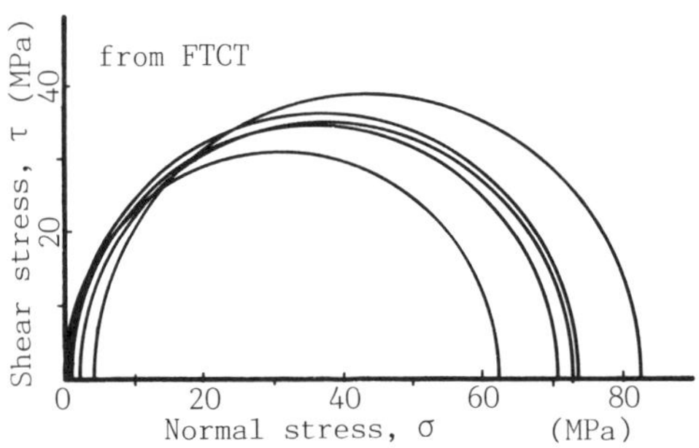

Fig. 3 Mohr's stress circles.

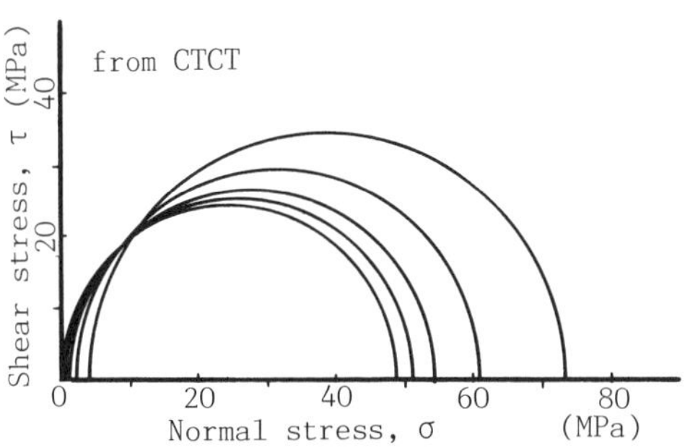

Fig. 4 Mohr's stress circles.

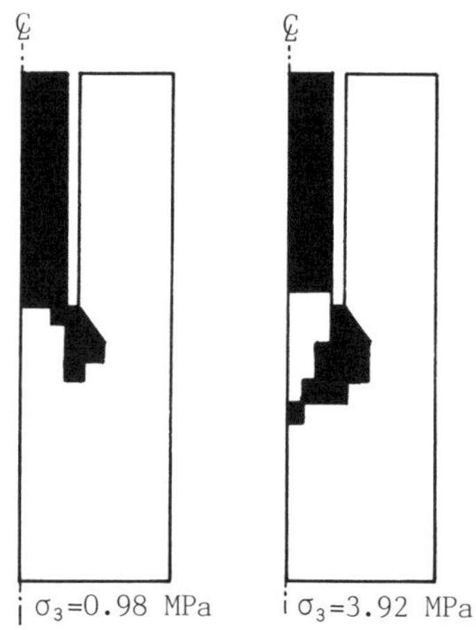

Fig. 5 Plasticized zone.

4 SUMMARY AND CONCLUSIONS

More and detailed investigations about FTCT are necessary to clarify the characteristics of this test in terms of both experimental and analytical approachs. At present state, FTCT looks upon as the test that shearing strength is evaluated appreciably larger than that by CTCT. The author is conducting farther research on this test.

Impact of Computational Mechanics on Engineering Problems, Pulmano & Murti (eds)
© 1993 Balkema, Rotterdam, ISBN 90 5410 324 8

Analogy between analysis of flow in pipe networks and slope-deflection in rigid frames

M. Kanoh
Kyushu Sangyo University, Japan

T. Kuroki
Fukuoka University, Japan

ABSTRACT: H. Cross method of flow in pipe networks and slope-deflection method of rigid frames have been used for long time in each related field. The slope-deflection method was developed to moment distribution method for calculation by hand. Since computer become widely used, the method is directed to matrix displacement method and connected successfully with finite element method. On the other hand it is difficult to analyze the flow in pipe networks by finite element method. The main reason of the difficulty may be the fact that the relation of water heads to discharges in pipe networks is nonlinear. Investigating the analogy between analysis of flow in a pipe and slope-deflection in a rigid frame, we try to linearyze the relation of water heads to discharges in pipe networks and succeed to solve the flow by finite element method.

1 GOVERNING EQUATIONS

1.1 Flow equations

The governing equation for the flow in pipe networks is given by

$$h = rQ^m \qquad , \qquad (1)$$

where h is the friction loss water head of the pipe, r is the resistance of the pipe, Q is the discharge in the pipe, m is the exponent. Here r and m are given by

$$r = 10.666C^{-1.85}D^{-4.87}l \ , \quad m=1.85 \ , \qquad (2)$$

in Hazen-Williams equation,

$$r = 10.296n^2D^{-5.33}l \ , \qquad m=2.0, \qquad (3)$$

in Manning equation,

with the velocity coefficient C, Manning's roughness coefficient n, the diameter D and the length of the pipe l.
From the law of the mass conservation, the discharges on each nodal point of pipe networks are governed by

$$\sum_{j=1}^{s} Q_{ij} = Q_{i\ out} \qquad , \qquad (4)$$

where Q_{ij} is the discharge of the j pipe which flows into i nodal point. Here s is the number of the pipes which connect with i nodal point and $Q_{i\ out}$ is the discharge which flows out from i nodal point.

1.2 End moment equations

Referring to Fig. 1(a),(b) the governing equation for the end moment and slope of the rigid frame is given by

$$M_A = \frac{2EI}{l}(2\theta_A + \theta_B) \quad ,$$

$$M_B = \frac{2EI}{l}(\theta_A + 2\theta_B) \quad , \qquad (5)$$

where M_A, M_B, θ_A and θ_B are the bending moments and slopes on the point A and B.

Here E is the Young's modulus, I is the moment of inertia of area and l is the length of the member.

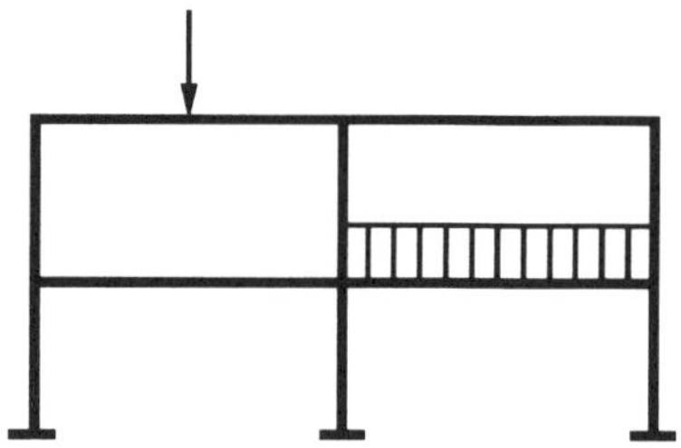

Fig.1(a). Rigid frame.

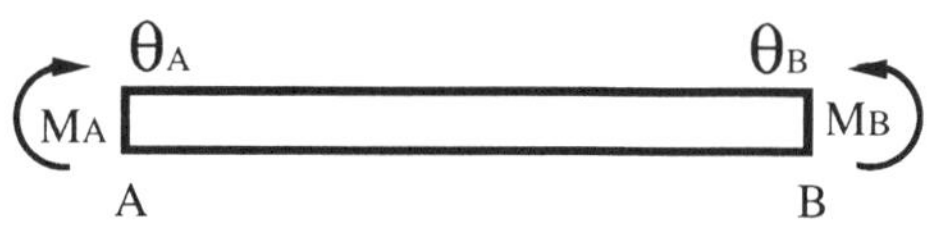

Fig.1(b). Moment and slope for a member.

The Eq.(5) yields

$$\theta_A = \frac{l}{6EI}(2M_A - M_B) ,$$
$$\theta_B = \frac{l}{6EI}(-M_A + 2M_B) . \quad (6)$$

It can be noted that the end slope is linearly related to the bending moment.

2 THE RELATION OF SLOPE–DEFLECTION METHOD TO FINITE ELEMENT METHOD

The slope-deflection method was developed to moment distribution method for calculation by hand 1930. The method is very convenient but is not inappropriate for a general-purpose program of computer.
Then the slope-deflection method is modified to matrix displacement method and connected successfully with finite element method.
In the slope-deflection method the bending moment and slope are balanced on each member or beam by Eq.(6). On the other hand, with the matrix displacement method the bending moments are considered to be balanced on each nodal point as follows

$$\sum_{j=1}^{s} M_{ij} = 0 , \quad (7)$$

where M_{ij} is the bending moment of the j member which connects with i nodal point. Here s is the number of the members which connect with i nodal point.
Namely considering the equilibrium or balance of the bending moments on the nodal point instead of the ones in the member, we become able to solve the stresses and strains in rigid frames by finite element method.

3 ANALOGY BETWEEN MOMENT DISTRIBUTION AND H. CROSS METHOD

3.1 Variables in pipes or rigid frames

Referring to the governing equations(1) and (6), the relations between the main variables in pipes or rigid frames are similarly simple. However the variables in pipes or rigid frames, for example the discharges in pipes or the bending moments in members, are strongly affected each other and the number of the variables becomes large in practical problems. So it is not easy to analyze many variables which are influenced each other.

3.2 Iteration technique

The iteration technique is employed in both H. Cross's analysis of flow in pipes and moment distribution methods in rigid frames, although the ways to assume the approximate values are different in the two methods. It is difficult to make a general-purpose program of computer for the both methods. The main reason of the difficulty may be the fact that the variables in the element are used for analyzing the problem without converting them to the related nodal points.

4 FINITE ELEMENT ANALYSIS OF FLOW IN PIPE NETWORKS

In order to analyze the flow in pipe networks by finite element method, it is necessary to accomplish two techniques. The former is to convert the friction loss water head and discharge in the pipe to the two variables on the nodal points which exist at the ends of the pipe. The latter is to linearyze the relation with the two variables on the nodal points.

4.1 Conversion of the friction loss water head and discharge in pipes to the variables on nodal points

Referring Fig.2 we separate a pipe network into the nodal points and the elements. Regarding to an element shown in Fig.3, we convert the friction loss water head(h_{e1}) and the discharge(Q_{e1}) in the element to the water heads(H_1 , H_2) and the discharges (Q_{12} , Q_{21}) on the nodal points (1 , 2).

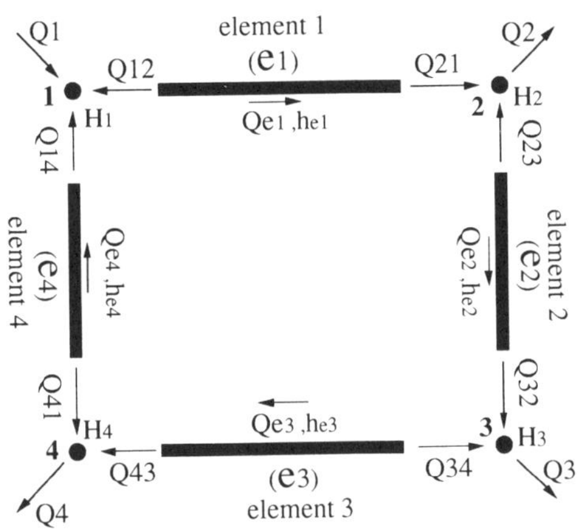

Fig.2. Seperation of a pipe network into nodal points and elements.

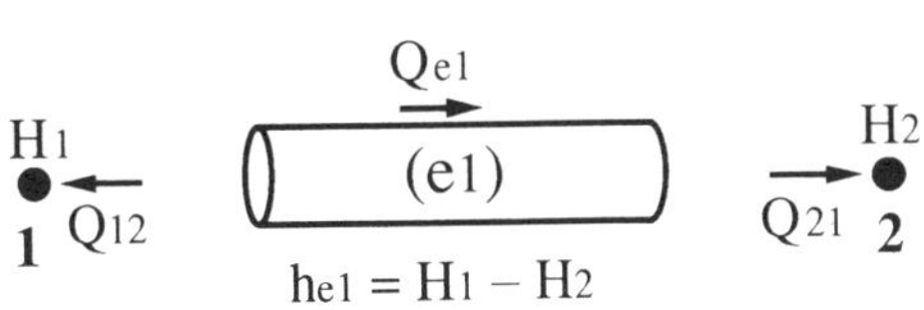

Fig.3. Conversion of h and Q in an element to H and Q on nodal points.

The relations are given by

$$h_{e1} = H_1 - H_2 \quad , \tag{8}$$

$$Q_{e1} = -Q_{12} = Q_{21} \quad . \tag{9}$$

Eqs.(1),(8) and (9) yield

$$h_{e1} = H_1 - H_2 = r\,Q_{e1}^m \quad . \tag{10}$$

4.2 Linearization

Referring Eq.(10), for an element we try to linearyze the relation as

$$h_{e1} = H_1 - H_2 = rQ_{e1}[Q_{e1}^{m-1}] \, , \tag{11}$$

where the term $[Q_{e1}^{m-1}]$ is considered to be known. The known term $[Q_{e1}^{m-1}]$ can be calculated by

$$[Q_{e1}^{m-1}] = [\hat{Q}_{e1}^{(m-1)\gamma} \cdot \hat{\hat{Q}}_{e1}^{(m-1)(1-\gamma)}] \, , \tag{12}$$

where the first and second terms in the right hand side are the approximate values obtained by the last iteration and the iteration before last [1]. Here γ means the ratio of the first to the second term.
Eqs.(9), (11) and (12) yield

$$Q_{12} = \frac{(H_2 - H_1)}{r_{e1}[\hat{Q}_{e1}^{(m-1)\gamma} \cdot \hat{\hat{Q}}_{e1}^{(m-1)(1-\gamma)}]} \, ,$$
$$Q_{21} = \frac{(H_1 - H_2)}{r_{e1}[\hat{Q}_{e1}^{(m-1)\gamma} \cdot \hat{\hat{Q}}_{e1}^{(m-1)(1-\gamma)}]} \, . \tag{13}$$

4.3 Element stiffness matrix

From Eq.(13) for an element we have

$$\begin{bmatrix} Q_{ij} \\ Q_{ji} \end{bmatrix} = \begin{bmatrix} -k_{ei} & k_{ei} \\ k_{ei} & -k_{ei} \end{bmatrix} \begin{bmatrix} H_i \\ H_j \end{bmatrix} \, , \tag{14}$$

where the first term of right hand side is the element stiffness matrix for the problem. Here k_{ei} , that means the known term of i element, is written by

$$k_{ei} = \frac{1}{r_{ei}[\hat{Q}_{ei}^{(m-1)\gamma} \cdot \hat{\hat{Q}}_{ei}^{(m-1)(1-\gamma)}]} \quad . \tag{15}$$

Once the element stiffness matrix is fixed, with Eq.(4) we can construct the finite element method for the analysis of flow in pipe networks.

4.4 Iteration

As the element stiffness matrix contains the approximate discharge of the pipe, it is necessary to use the iteration technique and assume the approximate values at the beginning of the analysis. Using the finite element method the first approximate discharge values can be set 1.0 for the all pipes in the networks.

5 NUMERICAL EXAMPLES AND RESULTS

First, the appropriate value of γ in the element stiffness matrix of the method is evaluated. Second, by the finite element method we calculate the appropriate discharge values in the pipe networks as shown in Fig.4 and 5. Then the accuracy and convergence of the method are investigated comparing the results with the solutions by H. Cross method.

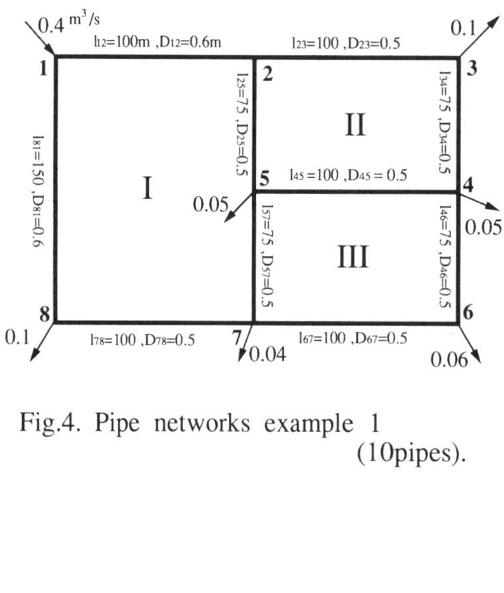

Fig.4. Pipe networks example 1 (10pipes).

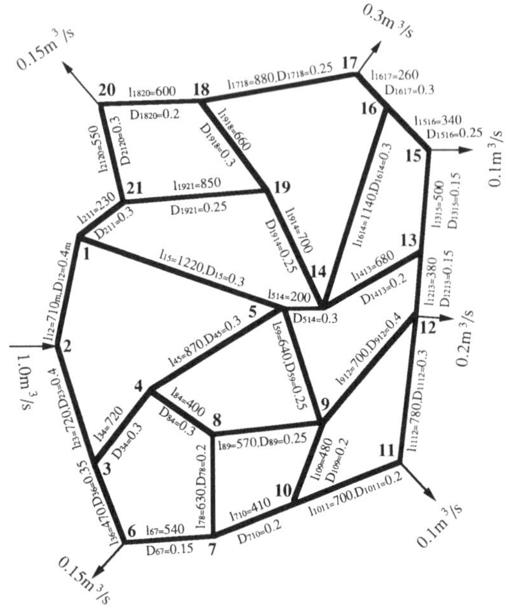

Fig.5. Pipe networks example 2 (30pipes).

5.1 Evaluation of γ value

We test twelve values of γ in the element stiffness matrix for analyzing the flow of the pipe networks. Figs.6 and 7 illustrate the relations between the values of γ and the number of the iteration times for obtaining the relative errors less than 0.1~5%. For instance if $\gamma = 1$ in Manning equation is used, the solution oscillates. In case $\gamma = 0.6 \sim 0.8$ the calculated results in the pipe networks may be stable with both Manning and Hazen-Williams equations.

5.2 Convergence and accuracy

Figs.6 and 7 show that in case $\gamma = 0.8$ the number of the iteration times for obtaining the relative errors less than 1% may be 4 or 5 with our finite element analysis. On the other hand with H.Cross method the number of the iteration times depends on the values of the first approximate discharges. For instance the first approximate discharge values are assumed

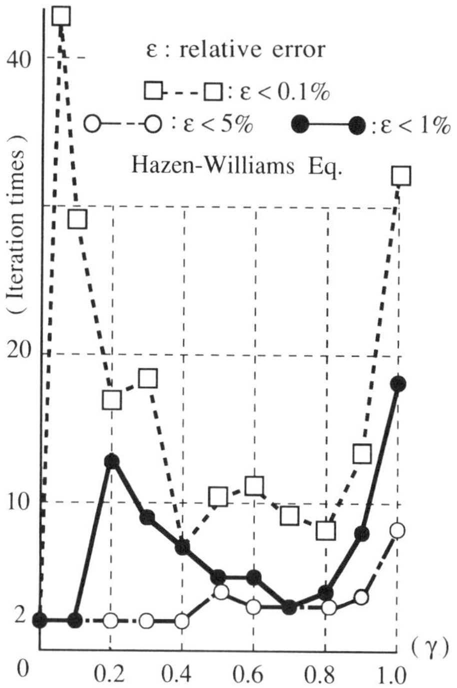

Fig.6. Relation between γ and iteretion times. (Hazen-Williams Eq.)

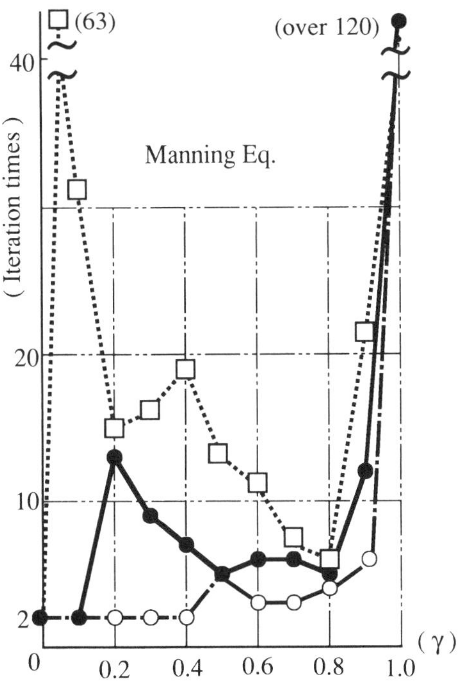

Fig.7. Relation between γ and iteretion times. (Manning Eq.)

1.0 for the all pipes in the example 2 by H. Cross method with Hazen-Williams equation, the number of the iteration times for obtaining the relative errors less than 1 % becomes 12. Namely this method can yield better convergence than H. Cross method. Referring Table 1 and 2 the computed results by the finite element method are as accurate as the results by H.Cross method.

Table 1. Computed results of example 1 (10pipes).

Pipe NO.	1 2	2 5	5 7	7 8	8 1
H.Cross Method	0.2183	0.1086	0.0065	0.0817	0.1817
FEM	0.2168	0.1087	0.0078	0.0832	0.1832
Pipe NO.	2 3	3 4	4 5	4 6	6 7
H.Cross Method	0.1097	0.0097	0.0521	0.0118	0.0482
FEM	0.1081	0.0081	0.0509	0.0089	0.0078

Table 2. Computed results of example 2 (30pipes).

Pipe NO.	1 2	2 3	3 4	4 5	5 1
H.Cross Method	−0.545	0.455	0.263	0.124	−0.184
FEM	−0.533	0.467	0.272	0.126	−0.191
Pipe NO.	3 6	6 7	7 8	8 4	8 9
H.Cross Method	0.192	0.042	−0.022	0.138	0.116
FEM	0.195	0.045	−0.024	0.146	0.122
Pipe NO.	9 5	7 10	10 9	10 11	11 12
H.Cross Method	−0.103	0.065	0.019	0.046	−0.054
FEM	−0.100	0.069	0.023	0.047	−0.053
Pipe NO.	12 9	12 13	13 14	14 5	13 15
H.Cross Method	−0.237	−0.017	−0.055	−0.206	0.038
FEM	−0.244	−0.009	−0.051	−0.216	0.042
Pipe NO.	15 16	16 14	16 17	17 18	18 19
H.Cross Method	−0.062	−0.203	0.141	−0.159	−0.087
FEM	−0.058	−0.207	0.149	−0.151	−0.089
Pipe NO.	19 14	18 20	20 21	21 19	21 1
H.Cross Method	0.052	−0.073	−0.223	0.139	−0.361
FEM	0.042	−0.061	−0.211	0.131	−0.342

6 CONCLUSIONS

First, we investigate the analogy between analysis of flow in pipe networks and slope-deflection in rigid frames. Second,we convert the friction loss water head and the discharge in a pipe to the two variables on the nodal points which exist at the both ends of the pipe. Third, we linearyze the relation with the two variables on the nodal points. Then we can set the finite element method for solving the flow in pipe networks.

Using the finite element method the first approximate discharge values can be set 1.0 for the all pipes in the networks. This may save the time for preparing the input data for the analysis as compared with H. Cross method.

We test twelve sorts of values of γ in the element stiffness matrix. In case $\gamma = 0.6 \sim 0.8$ the calculated results in the pipe networks may be stable with both Manning and Hazen-Williams equations.

The finite element method for analysis of flow in pipe networks can yield better convergence than H. Cross method. The computed results by the finite element method are as accurate as the solutions by H.Cross method.

REFERENCES

1. Kanoh , M. and Kuroki , T. , 'Seminar book in Analysis of Civil Engineering Problems', pp.7 to 17, Riko Tosho Inc., 1993(in Japanese).

Impact of Computational Mechanics on Engineering Problems, Pulmano & Murti (eds)
© 1993 Balkema, Rotterdam, ISBN 90 5410 324 8

An efficient scheme for land subsidence and negative skin friction

P. Karasudhi
Asian Institute of Technology, Bangkok, Thailand

ABSTRACT: The soil media are governed by Biot's theory. Land subsidence and negative skin friction in piles, due to water loss from artesian wells, are analyzed by discretizing the near field into finite elements, and the rest into infinite elements. The Laplace transform of governing conditions are solved numerically first, then the inverse Laplace transform of any solution function is obtained by a collocation method.

1. SATURATED POROUS ELASTIC MEDIA

Biot (1941) proposed the following theory for *quasi-statics* of porous isotropic elastic solids completely saturated with fluids. In the absence of body forces, the equilibrium conditions can be written, in the *indicial notation* of Cartesian tensors (Karasudhi, 1991a), as

$$\sigma_{ij,j} + \delta_{ij} p_{f,j} = 0 \tag{1}$$

where σ_{ij} is the *effective stress* tensor, and p_f the *excess pore pressure* (positive if in tension). The stress-strain relationship is in the usual form

$$\sigma_{ij} = 2\mu\varepsilon_{ij} + \delta_{ij}\lambda\varepsilon_{kk} \tag{2}$$

where λ and μ are Lamé's constants of the bulk material. The strain tensor ε_{ij} and the dilatation ε_{kk} also take the common forms, i.e.

$$\varepsilon_{ij} = \frac{1}{2}(u_{i,j} + u_{j,i}) \tag{3a}$$

$$\varepsilon_{kk} = u_{k,k} \tag{3b}$$

where u_i is a displacement component of the solid matrix. The fluid flow is governed by *Darcy's law* in the form

$$\frac{\partial w_i}{\partial t} = k p_{f,i} \tag{4}$$

where t is time, k the *coefficient of permeability*, and

$$w_i = f(U_i - u_i) \tag{5}$$

in which U_i is a displacement component of the fluid, and f the *porosity*, i.e. the volume of the voids per unit total volume. The condition of perfect saturation with an incompressible fluid implies that

$$\varepsilon_{kk} = -w_{i,i} \tag{6}$$

So that Darcy's law becomes

$$(\lambda + 2\mu)k\nabla^2(w_{i,i}) = \frac{\partial w_{i,i}}{\partial t} \tag{7}$$

where ∇^2 is the Laplace operator, i.e. $\nabla^2 f = f_{,ii}$.

Substituting Eq. 2 into Eq. 1 yields the *Navier equations* for the saturated porous solid as

$$\mu\nabla^2 u_i + (\lambda + \mu)\varepsilon_{kk,i} + p_{f,i} = 0 \tag{8}$$

while Darcy's law (Eq. 7), due to Eq. 6, becomes

$$c\nabla^2\varepsilon_{kk} = \frac{\partial \varepsilon_{kk}}{\partial t} \tag{9}$$

where c is the *coefficient of consolidation*, i.e.

$$c = (\lambda + 2\mu)k \tag{10}$$

Note that the Navier equations for solids and those for saturated solids (Eq. 8) become identical if the body force X_i in the former is replaced by the *excess pore pressure gradient* $p_{f,i}$. Substituting Eq. 6 into Eq. 9, in view of Eq. 4, yields

$$c\nabla^2\varepsilon_{kk} = -k\nabla^2 p_f \tag{11a}$$

or

$$c\varepsilon_{kk} = -kp_f + \Phi \tag{11b}$$

where Φ is a harmonic function, i.e. $\nabla^2\Phi = 0$.

The well-posed boundary conditions can be obtained by means of a *variational principle*. They state that, at a certain point on the boundary S, either

$$u_i = \bar{u}_i \quad \text{or} \quad n_j(\sigma_{ij} + \delta_{ij}p_f) = \bar{X}_i \tag{12a}$$

and, either

$$p_f = \bar{p}_f \quad \text{or} \quad kp_{f,i}n_i = \bar{Q} \tag{12b}$$

Here the bar ($\bar{}$) denotes a prescribed quantity, n_i the outward unit normal to S, $\bar{X}_i$ the stress vector, and $\bar{Q}$ the prescribed rate of the outflow. It can be seen that there are four basic unknown functions u_1, u_2, u_3 and p_f, and four governing field equations Eqs. 8 and 9. The problem becomes identical to that of pure elastic solids if p_f tends to zero, which happens if the system is subjected to constant forces for a sufficiently large time.

2. FORMULATION OF PROPER FUNCTIONAL

Consider the following *functional* Π,

$$\Pi \equiv \int_V \left[\sigma_{ij}*\varepsilon_{ij} + p_f*\varepsilon_{ii} - H(t)*\frac{\partial w_i}{\partial t}*p_{f,i} \right] \mathrm{d}V + \int_S [H(t)*\overline{Q}*p_f - \overline{X}_i*u_i]\,\mathrm{d}S \quad (13)$$

where $H(t)$ is the *Heaviside step function*, V the volume completely enclosed by S, and

$$v_1*v_2 \equiv \int_{0^-}^{t} v_1(t-s)v_2(s)\mathrm{d}s, \quad v_1*v_2 = v_2*v_1 \quad (14a,b)$$

$$v_1*v_2*v_3 = v_1*(v_2*v_3) = (v_1*v_2)*v_3 \quad (14c)$$

in which 0^- is infinitesimally less than zero. Thus

$$\tilde{\Pi} = \int_V [\tilde{\sigma}_{ij}\tilde{u}_{i,j} + \tilde{p}_f \tilde{u}_{i,i} - \tilde{w}_i \tilde{p}_{f,i}]\mathrm{d}V + \int_S \left(p^{-1}\tilde{\overline{Q}}\tilde{p}_f - \tilde{\overline{X}}_i \tilde{u}_i \right) \mathrm{d}S \quad (15)$$

$$\delta\tilde{\Pi} = \int_V (\tilde{\sigma}_{ij}\delta\tilde{u}_{i,j} + \tilde{u}_{i,i}\delta\tilde{p}_f + \tilde{p}_f\delta\tilde{u}_{i,i} - \tilde{w}_i\delta\tilde{p}_{f,i})\mathrm{d}V + \int_S \left(p^{-1}\tilde{\overline{Q}}\delta\tilde{p}_f - \tilde{\overline{X}}_i\delta\tilde{u}_i \right)\mathrm{d}S \quad (16)$$

where u_i and p_f are selected as the *primary field variables* in the *variation* above, p is the *Laplace transform* parameter versus the time t, δ the *variational symbol* signifying a *virtual quantity*, $\tilde{v}(p)$ the Laplace transform of $v(t)$, i.e.

$$\tilde{v}(p) \equiv \int_{0^-}^{\infty} \exp(-pt)v(t)\mathrm{d}t \quad (17)$$

and $v(t) \equiv 0$ for $t < 0$. Note that

$$\int_V \tilde{\sigma}_{ij}\delta\tilde{u}_{i,j}\mathrm{d}V = \int_V [(\tilde{\sigma}_{ij}\delta\tilde{u}_i)_{,j} - \tilde{\sigma}_{ij,j}\delta\tilde{u}_i]\mathrm{d}V \quad (18)$$

Due to Gauss's divergence theorem, the equation above becomes

$$\int_V \tilde{\sigma}_{ij}\delta\tilde{u}_{i,j}\mathrm{d}V = \int_S \tilde{\sigma}_{ij}\delta\tilde{u}_i n_j \mathrm{d}S - \int_V \tilde{\sigma}_{ij,j}\delta\tilde{u}_i\mathrm{d}V \quad (19)$$

Similarly,

$$\int_V \tilde{p}_f\delta\tilde{u}_{i,i}\mathrm{d}V = \int_S \tilde{p}_f\delta\tilde{u}_i n_i\mathrm{d}S - \int_V \tilde{p}_{f,i}\delta\tilde{u}_i\mathrm{d}V \quad (20)$$

$$\int_V \tilde{w}_i\delta\tilde{p}_{f,i}\mathrm{d}V = \int_S \tilde{w}_i\delta\tilde{p}_f n_i\mathrm{d}S - \int_V \tilde{w}_{i,i}\delta\tilde{p}_f\mathrm{d}V \quad (21)$$

Substituting Eqs. 19 to 21 into Eq. 16 yields

$$\delta\tilde{\Pi} = \int_V [-(\tilde{\sigma}_{ij,j} + \tilde{p}_{f,i})\delta\tilde{u}_i + (\tilde{u}_{i,i} + \tilde{w}_{i,i})\delta\tilde{p}_f]\mathrm{d}V + \int_S \left[\left(-\tilde{w}_i n_i + p^{-1}\tilde{\overline{Q}} \right)\delta\tilde{p}_f + \left(\tilde{\sigma}_{ij}n_j + \tilde{p}_f n_i - \tilde{\overline{X}}_i \right)\delta\tilde{u}_i \right]\mathrm{d}S \quad (22)$$

Setting $\delta\tilde{\Pi} = 0$ leads to Eqs. 1 and 6 in V, and (in view of Eq. 4) Eqs. 12 on S. In other words, the equilibrium equations, the perfect saturation condition and the well-posed boundary conditions can be obtained simultaneously by incorporating Darcy's law in a variational principle.

3. DERIVATION OF FAR-FIELD FUNCTIONS

Cartesian coordinates (x, y, z), cylindrical coordinates (r, θ, z), and spherical coordinates (R, ϕ, θ) are depicted together in Fig. 1. In cylindrical coordinates: the stress-strain relationship (Eq. 2) becomes

$$(\sigma_{rr} \quad \sigma_{\theta\theta} \quad \sigma_{zz}) = 2\mu(\varepsilon_{rr} \quad \varepsilon_{\theta\theta} \quad \varepsilon_{zz}) + \lambda\varepsilon_{kk}(1 \quad 1 \quad 1) \quad (23a)$$

$$(\sigma_{r\theta} \quad \sigma_{\theta z} \quad \sigma_{zr}) = 2\mu(\varepsilon_{r\theta} \quad \varepsilon_{\theta z} \quad \varepsilon_{zr}) \quad (23b)$$

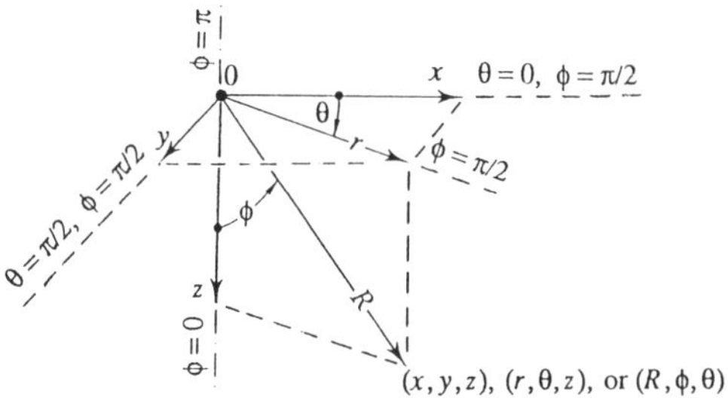

Fig. 1. Cartesian (x, y, z), cylindrical (r, θ, z) and spherical (R, ϕ, θ) coordinates.

and the strains and the dilatation in terms of the displacements (Eqs. 3) are

$$\varepsilon_{rr} = \frac{\partial u_r}{\partial r}, \quad \varepsilon_{\theta\theta} = \frac{1}{r}\left(u_r + \frac{\partial u_\theta}{\partial\theta} \right), \quad \varepsilon_{zz} = \frac{\partial u_z}{\partial z} \quad (24a,b,c)$$

$$\varepsilon_{r\theta} = \varepsilon_{\theta r} = \frac{1}{2}\left[\frac{\partial u_\theta}{\partial r} + \frac{1}{r}\left(\frac{\partial u_r}{\partial\theta} - u_\theta \right) \right] \quad (24d)$$

$$\varepsilon_{\theta z} = \varepsilon_{z\theta} = \frac{1}{2}\left(\frac{\partial u_z}{r\partial\theta} + \frac{\partial u_\theta}{\partial z} \right) \quad (24e)$$

$$\varepsilon_{zr} = \varepsilon_{rz} = \frac{1}{2}\left(\frac{\partial u_r}{\partial z} + \frac{\partial u_z}{\partial r} \right) \quad (24f)$$

$$\varepsilon_{kk} = \varepsilon_{rr} + \varepsilon_{\theta\theta} + \varepsilon_{zz} = \frac{1}{r}\left[\frac{\partial(ru_r)}{\partial r} + \frac{\partial u_\theta}{\partial\theta} \right] + \frac{\partial u_z}{\partial z} \quad (24g)$$

3.1 *Large space*, $0 \le r < \infty$

The general solution for the space where $0 \le r < \infty$ is in the form (Apirathvorakij and Karasudhi, 1980):

$$u_r(r,\theta,z,t) = \sum_{m=0}^{\infty} u_{rm}(r,z,t)\cos m\theta \quad (25a)$$

$$u_\theta(r,\theta,z,t) = \sum_{m=0}^{\infty} u_{\theta m}(r,z,t)\sin m\theta \quad (25b)$$

$$u_z(r,\theta,z,t)=\sum_{m=0}^{\infty} u_{zm}(r,z,t)\cos m\theta \quad (25c)$$

$$p_f(r,\theta,z,t)=\sum_{m=0}^{\infty} p_{fm}(r,z,t)\cos m\theta \quad (25d)$$

$$\sigma_{rr}(r,\theta,z,t)=\sum_{m=0}^{\infty} \sigma_{rrm}(r,z,t)\cos m\theta \quad (25e)$$

$$\sigma_{\theta\theta}(r,\theta,z,t)=\sum_{m=0}^{\infty} \sigma_{\theta\theta m}(r,z,t)\cos m\theta \quad (25f)$$

$$\sigma_{zz}(r,\theta,z,t)=\sum_{m=0}^{\infty} \sigma_{zzm}(r,z,t)\cos m\theta \quad (25g)$$

$$\sigma_{r\theta}(r,\theta,z,t)=\sum_{m=0}^{\infty} \sigma_{r\theta m}(r,z,t)\sin m\theta \quad (25h)$$

$$\sigma_{\theta z}(r,\theta,z,t)=\sum_{m=0}^{\infty} \sigma_{\theta zm}(r,z,t)\sin m\theta \quad (25i)$$

$$\sigma_{zr}(r,\theta,z,t)=\sum_{m=0}^{\infty} \sigma_{zrm}(r,z,t)\cos m\theta \quad (25j)$$

The Laplace transforms of the coefficients in the Fourier series above are

$$\begin{aligned}\bar{u}_{rm}(r,z,p)+\bar{u}_{\theta m}(r,z,p) &= \int_0^\infty \xi^2 J_{m+1}(\xi r)\,(A_m e^{-z\xi}+B_m e^{-z\gamma}-zE_m e^{-z\xi} \\ &\quad +2G_m e^{-z\xi}+C_m e^{z\xi}+D_m e^{z\gamma}-zF_m e^{z\xi} \\ &\quad +2H_m e^{z\xi})\,d\xi \end{aligned} \quad (26a)$$

$$\begin{aligned}\bar{u}_{rm}(r,z,p)-\bar{u}_{\theta m}(r,z,p) &= \int_0^\infty \xi^2 J_{m-1}(\xi r)\,(-A_m e^{-z\xi}-B_m e^{-z\gamma}+zE_m e^{-z\xi} \\ &\quad +2G_m e^{-z\xi}-C_m e^{z\xi}-D_m e^{z\gamma}+zF_m e^{z\xi} \\ &\quad +2H_m e^{z\xi})\,d\xi \end{aligned} \quad (26b)$$

$$\begin{aligned}\bar{u}_{zm}(r,z,p) &= \int_0^\infty \xi J_m(\xi r)\,[\xi A_m e^{-z\xi}+\gamma B_m e^{-z\gamma} \\ &\quad -(1+z\xi)E_m e^{-z\xi}-\xi C_m e^{z\xi}-\gamma D_m e^{z\gamma} \\ &\quad -(1-z\xi)F_m e^{z\xi}]\,d\xi \end{aligned} \quad (26c)$$

$$\begin{aligned}\frac{\bar{p}_{fm}(r,z,p)}{2\mu} &= \int_0^\infty \xi J_m(\xi r)\,(\eta c^{-1} p B_m e^{-z\gamma}+\xi E_m e^{-z\xi} \\ &\quad +\eta c^{-1} p D_m e^{z\gamma}-\xi F_m e^{z\xi})\,d\xi \end{aligned} \quad (26d)$$

$$\begin{aligned}\frac{\tilde{\sigma}_{rrm}(r,z,p)}{2\mu}+\frac{\bar{u}_{rm}(r,z,p)+m\,\bar{u}_{\theta m}(r,z,p)}{r} &= \int_0^\infty \xi J_m(\xi r)\,[\xi^2 A_m e^{-z\xi}+\gamma^2 B_m e^{-z\gamma}+\xi(1-z\xi)E_m e^{-z\xi} \\ &\quad +\xi^2 C_m e^{z\xi}+\gamma^2 D_m e^{z\gamma}-\xi(1+z\xi)F_m e^{z\xi}]\,d\xi \end{aligned} \quad (26e)$$

$$\begin{aligned}\frac{\tilde{\sigma}_{rrm}(r,z,p)+\tilde{\sigma}_{\theta\theta m}(r,z,p)}{2\mu} &= \int_0^\infty \xi J_m(\xi r)\,[\xi^2 A_m e^{-z\xi}+(\gamma^2+c^{-1}p)B_m e^{-z\gamma} \\ &\quad +\xi(2-z\xi)E_m e^{-z\xi}+\xi^2 C_m e^{z\xi}+(\gamma^2+c^{-1}p)D_m e^{z\gamma} \\ &\quad -\xi(2+z\xi)F_m e^{z\xi}]\,d\xi \end{aligned} \quad (26f)$$

$$\begin{aligned}\frac{\tilde{\sigma}_{zzm}(r,z,p)}{2\mu} &= \int_0^\infty \xi^2 J_m(\xi r)\,[-\xi A_m e^{-z\xi}-\xi B_m e^{-z\gamma} \\ &\quad +(1+z\xi)E_m e^{-z\xi}-\xi C_m e^{z\xi}-\xi D_m e^{z\gamma} \\ &\quad -(1-z\xi)F_m e^{z\xi}]\,d\xi \end{aligned} \quad (26g)$$

$$\begin{aligned}\frac{\tilde{\sigma}_{z\theta m}(r,z,p)+\tilde{\sigma}_{zrm}(r,z,p)}{2\mu} &= \int_0^\infty \xi^2 J_{m+1}(\xi r)\,(-\xi A_m e^{-z\xi}-\gamma B_m e^{-z\gamma}+z\xi E_m e^{-z\xi} \\ &\quad -\xi G_m e^{-z\xi}+\xi C_m e^{z\xi}+\gamma D_m e^{z\gamma} \\ &\quad -z\xi F_m e^{z\xi}+\xi H_m e^{z\xi})\,d\xi \end{aligned} \quad (26h)$$

$$\begin{aligned}\frac{\tilde{\sigma}_{z\theta m}(r,z,p)-\tilde{\sigma}_{zrm}(r,z,p)}{2\mu} &= \int_0^\infty \xi^2 J_{m-1}(\xi r)\,(-\xi A_m e^{-z\xi}-\gamma B_m e^{-z\gamma}+z\xi E_m e^{-z\xi} \\ &\quad +\xi G_m e^{-z\xi}+\xi C_m e^{z\xi}+\gamma D_m e^{z\gamma}-z\xi F_m e^{z\xi} \\ &\quad -\xi H_m e^{z\xi})\,d\xi \end{aligned} \quad (26i)$$

$$\frac{\tilde{\sigma}_{r\theta m}(r,z,p)}{2\mu}+\frac{m\,\bar{u}_{rm}(r,z,p)+\bar{u}_{\theta m}(r,z,p)}{r} = \int_0^\infty \xi^3 J_m(\xi r)\,(G_m e^{-z\xi}+H_m e^{z\xi})\,d\xi \quad (26j)$$

where

$$\gamma=(\xi^2+c^{-1}p)^{1/2}, \quad \eta=\frac{1-\nu}{1-2\nu} \quad (27a,b)$$

and ν is the Poisson's ratio. The arbitrary constants, A_m, B_m, C_m, D_m, E_m, F_m, G_m and H_m, are to be determined by boundary conditions at z = constants.

For a half space where $0 \le z < \infty$, $C_m = D_m = F_m = H_m = 0$. For axisymmetry, $u_\theta = \sigma_{r\theta} = \sigma_{\theta z} = m = G_m = H_m = 0$.

3.2 *Approximate far-field functions*

We shall involve the following *Hankel integrals,*

$$\int_0^\infty f(\xi)\,[\exp(-z\xi),\ \ \exp(-z\gamma)]\,J_m(\xi r)\,d\xi \quad (28)$$

Karasudhi (1991b) obtained the following approximation, which is not only very accurate but also exact at $\xi = 0$ and $\xi \to \infty$,

$$\exp(-z\gamma) = \exp[-h(\zeta^2+1)^{1/2}] \approx e^{-h(1+a\zeta)}$$

$$+e^{-h\zeta}(1-e^{-bh\zeta}) + c_1 h\zeta e^{-dh\zeta} + c_2(h\zeta)^n e^{-d_1 h\zeta} \quad (29)$$

where

$$\zeta = \xi(c/p)^{1/2}; \quad h = z(p/c)^{1/2} \quad (30a,b)$$

$$n, a, d, d_1 > 1; \quad b > 0 \quad (30c\text{-}g)$$

c_1 for the exact zero 'slope' at $\xi = 0$, and c_2 for the 'minimum' error. Karasudhi (1991b) also found that, for the far field (large r and/or large z),

$$\int_0^\infty f(\xi)\exp(-z\xi)J_m(\xi r)d\xi$$

$$= \int_0^\infty \left[\lim_{\xi\to 0} f(\xi)\right]\exp(-z\xi)J_m(\xi r)d\xi \quad (31)$$

provided that $f(\xi)$ has no branch nor pole in the range of integration, $0 \le \xi < \infty$.

3.3 *Point sink*

For a point sink on the surface of the half space at $r = z = 0$, the boundary conditions are

$$\sigma_{zz}(r,0,t) = \sigma_{zr}(r,0,t) = 0 \quad (32a)$$

$$-k\frac{\partial p_f(r,0,t)}{\partial z} = PH(t)\frac{\delta(r-0^+)}{2\pi r} \quad (32b)$$

where $\delta(\,)$ is the *Dirac-delta function*, and 0^+ is infinitesimally greater than zero. Equations 32 give, respectively,

$$\xi A_0 + \xi B_0 - E_0 = \xi A_0 + \gamma B_0 = 0 \quad (33a,b)$$

$$k(\eta c^{-1}p\gamma B_0 + \xi^2 E_0) = \frac{P}{4\mu\pi p} \quad (33c)$$

Due to Eqs. 28 to 31, the far-field solution corresponding to $\xi \to 0$ is associated with

$$B_0, E_0 \propto 1, \quad A_0 \propto \xi^{-1} \quad (34a,b,c)$$

Thus

$$u_z, p_f \propto zR^{-3}, \quad u_r \propto rR^{-3} \quad (35a,b,c)$$

The equations above can be shown to be the same as those for two different half spaces bonded together.

3.4 *Point forces*

For a point force in the z-direction on the surface of the half space at $r = z = 0$, the boundary conditions

$$\sigma_{zz}(r,0,t) = -PH(t)\frac{\delta(r-0^+)}{2\pi r},$$

$$\sigma_{zr}(r,0,t) = p_f(r,0,t) = 0,$$

give, respectively,

$$\xi A_0 + \xi B_0 - E_0 = \frac{P}{4\mu\pi p\xi} \quad (36a)$$

$$\xi A_0 + \gamma B_0 = 0, \quad \eta c^{-1}pB_0 + \xi E_0 = 0 \quad (36b,c)$$

The far-field solution is associated with

$$A_0, E_0 \propto \xi^{-1}, \quad B_0 \propto 1 \quad (37a,b,c)$$

Thus

$$u_r \propto rzR^{-3}, \quad u_z \propto R^{-1}, \quad p_f \propto zR^{-3} \quad (38a,b,c)$$

which is the same as obtained for two different half spaces bonded together. The final solution, $t \to \infty$, is equivalent to $p \to 0$ and $p_f \to 0$, and should be identical to that for elastostatics of pure solids, However, in the latter (Mindlin, 1936), Eq. 38b holds but Eq. 38a has to be changed into $u_r \propto rR^{-2}$.

For a point force in the x-direction at $r = z = 0$, the solution for the half space, where $0 \le r, z < \infty$, is given by Eqs. 25 and 26 with $m = 1$. The boundary conditions

$$\sigma_{zz}(r,\theta,0,t) = p_f(r,\theta,0,t) = 0,$$

give, respectively,

$$-\xi A_1 - \xi B_1 + E_1 = 0, \quad \eta c^{-1}pB_1 + \xi E_1 = 0 \quad (39a,b)$$

The conditions

$$\sigma_{z\theta}(r,\theta,z,t) = -\sin\theta\sigma_{zx}(r,\theta,z,t) + \cos\theta\sigma_{zy}(r,\theta,z,t),$$

$$\sigma_{zr}(r,\theta,z,t) = \cos\theta\sigma_{zx}(r,\theta,z,t) + \sin\theta\sigma_{zy}(r,\theta,z,t),$$

and the boundary conditions

$$\sigma_{zx}(r,\theta,0,t) = -PH(t)\frac{\delta(r-0^+)}{2\pi r}, \quad \sigma_{zy}(r,\theta,0,t) = 0,$$

give

$$\tilde{\sigma}_{z\theta 1}(r,0,p) + \tilde{\sigma}_{zr1}(r,0,p) = 0,$$

$$\tilde{\sigma}_{z\theta 1}(r,0,p) - \tilde{\sigma}_{zr1}(r,0,p) = \frac{P}{\pi};$$

or, respectively,

$$-\xi A_1 - \gamma B_1 - \xi G_1 = 0 \quad (40a)$$

$$-\xi A_1 - \gamma B_1 + \xi G_1 = \frac{P}{2\mu\pi\xi} \quad (40b)$$

Thus

$$A_1, G_1 \propto \xi^{-2}, \quad B_1 \propto 1, \quad E_1 \propto \xi^{-1} \quad (41a\text{-}d)$$

and the far-field functions can be shown to be

$$u_r, u_\theta \propto R^{-1}, \quad u_z \propto rR^{-2}, \quad p_f \propto rR^{-3} \quad (42a\text{-}d)$$

The solution for elastostatics of pure solids (Mindlin, 1936) is given by Eqs. 42a to c also.

4. FINITE AND INFINITE ELEMENTS

4.1 *Discretization of material domains*

Since an analytical method may be impractical if not impossible for problems of multilayered half spaces, a discretization of the material domains into finite elements in the near field and infinite elements in the rest is proposed as in Figs. 2. There are two types of infinite elements; horizontal elements (HE) in the horizontal layers in the upper part of such a

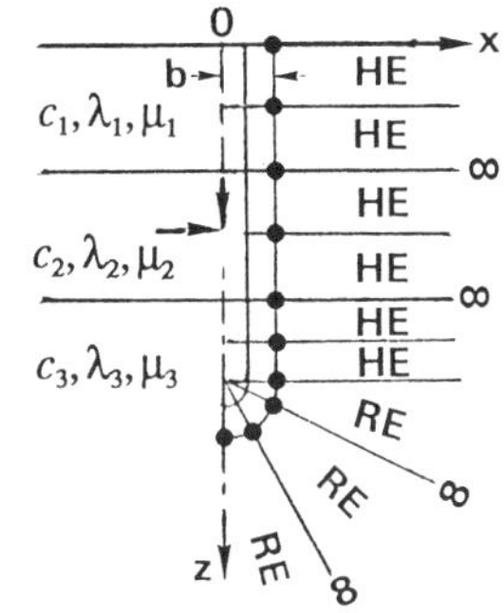

(a) Point forces.

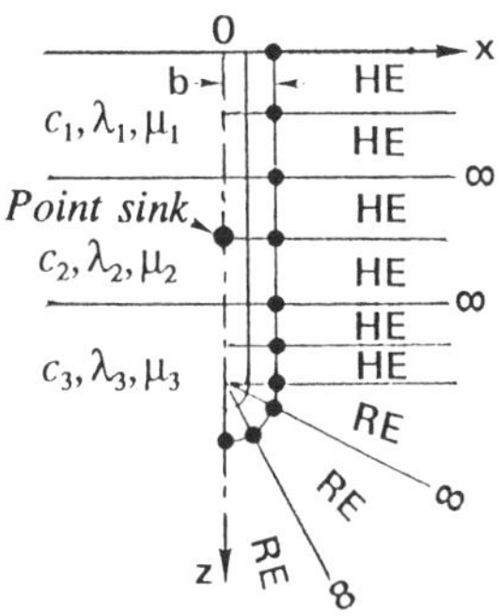

(b) Point sink.

Fig. 2. Discretization of multilayered half space.

multilayered half space, and radiating elements (RE) inside the underlying half space. Each infinite element has nodes only where it is connected to a finite element.

4.2 *Formulation of characteristic matrix*

The condition $\delta\tilde{\Pi} = 0$ can be utilized in formulating a *finite element* algorithm. In the form of matrices, Eq. 16 becomes

$$\delta\tilde{\Pi} = \int_V (\tilde{\mathbf{S}}\delta\tilde{\mathbf{E}} + \tilde{\varepsilon}_{kk}\delta\tilde{p}_f + \tilde{p}_f\delta\tilde{\varepsilon}_{kk} - kp^{-1}\tilde{\mathbf{P}}\delta\tilde{\mathbf{P}}^T)\mathrm{d}V + \int_S \left(p^{-1}\tilde{\overline{Q}}\delta\tilde{p}_f - \tilde{\overline{\mathbf{X}}}\delta\tilde{\mathbf{U}}\right)\mathrm{d}S \quad (43)$$

where $\mathbf{E}$ is the column matrix of strain components, $\mathbf{P}$ the row matrix of the gradients of the excess pore pressure, $\mathbf{S}$ the row matrix of stress components, $\mathbf{U}$ the column matrix of displacement components, and $\overline{\mathbf{X}}$ the row matrix of surface force components.

In cylindrical coordinates: those marices are

$$\mathbf{E} = (\varepsilon_{rr} \quad \varepsilon_{\theta\theta} \quad \varepsilon_{zz} \quad 2\varepsilon_{r\theta} \quad 2\varepsilon_{\theta z} \quad 2\varepsilon_{zr})^T \quad (44a)$$

$$\mathbf{P} = \left(\frac{\partial p_f}{\partial r} \quad \frac{\partial p_f}{r\partial\theta} \quad \frac{\partial p_f}{\partial z}\right) \quad (44b)$$

$$\mathbf{S} = (\sigma_{rr} \quad \sigma_{\theta\theta} \quad \sigma_{zz} \quad \sigma_{r\theta} \quad \sigma_{\theta z} \quad \sigma_{zr}) \quad (44c)$$

$$\mathbf{U} = (u_r \quad u_\theta \quad u_z)^T \quad (44d)$$

$$\overline{\mathbf{X}} = (\overline{X}_r \quad \overline{X}_\theta \quad \overline{X}_z) \quad (44e)$$

Expressing all functions involved in $\delta\tilde{\Pi}$ (Eq. 43) in terms of $\tilde{u}_r$, $\tilde{u}_\theta$, $\tilde{u}_z$and $\tilde{p}_f$ at nodes of each element; then setting $\delta\tilde{\Pi} = 0$; we get the *characteristic matrix*, which is equivalent to the stiffness matrix in pure solid problems, of that element.

4.3 *Mapping of infinite elements into finite elements*

Rajapakse and Karasudhi [3] presented a *contraction* to map an infinite element in the *rz*-plane into a finite element in the $\xi\eta$-plane by employing a simple *singular function* along $\xi = +1$ as shown in Fig. 3. The *coordinate transformation* for HE is

$$r = \sum_j 2(1-\xi)^{-1} L_j(\eta) r_j \quad (45a)$$

$$z = \sum_j L_j(\eta) z_j \quad (45b)$$

and for RE is

$$r = \sum_j 2(1-\xi)^{-1} L_j(\eta) r_j \quad (46a)$$

$$z = \sum_j 2(1-\xi)^{-1} L_j(\eta) z_j \quad (46b)$$

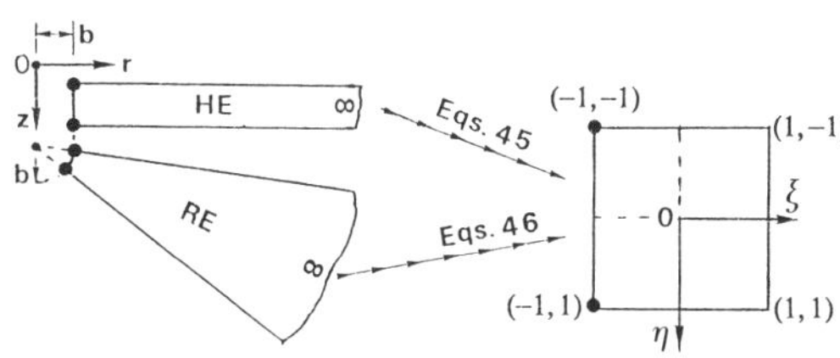

Fig. 3. Singular contraction of infinite elements into finite element.

where $L_j(\eta)$ is the *Lagrangian polynomial*. Accordingly, the nodes of the infinite elements are mapped on the $\xi = -1$ side of the finite element.

4.4 *Interpolation functions for infinite elements*

In view of the far-field functions in Eqs. 42, *interpolation functions* are selected for loading in the *x*-direction, for node *j* of any infinite element, as

$$N_j^{u_r}(\xi,\eta) = N_j^{u_\theta}(\xi,\eta) = N_j^{u_z}(\xi,\eta) = \frac{1-\xi}{2} L_j(\eta) \quad (47a,b,c)$$

$$N_j^{p_f}(\xi,\eta)=\frac{(1-\xi)^2}{4}L_j(\eta) \quad (47d)$$

For loading in the z-direction (Eqs. 38), for node j of an HE,

$$N_j^{u_r}(\xi,\eta)=\frac{(1-\xi)^2}{4}L_j(\eta) \quad (48a)$$

$$N_j^{u_z}(\xi,\eta)=\frac{1-\xi}{2}L_j(\eta) \quad (48b)$$

$$N_j^{p_f}(\xi,\eta)=\frac{(1-\xi)^3}{8}L_j(\eta) \quad (49c)$$

and for node j between two adjacent RE's, as

$$N_j^{u_r}(\xi,\eta)=N_j^{u_z}(\xi,\eta)=\frac{1-\xi}{2}L_j(\eta) \quad (50a,b)$$

$$N_j^{p_f}(\xi,\eta)=\frac{(1-\xi)^2}{4}L_j(\eta) \quad (50c)$$

For node j at the interface between an HE and an RE, the shape functions as Eqs. 49 hold.

For a point sink (Eqs. 35), for node j of an HE,

$$N_j^{u_r}(\xi,\eta)=\frac{(1-\xi)^2}{4}L_j(\eta) \quad (51a)$$

$$N_j^{u_z}(\xi,\eta)=N_j^{p_f}(\xi,\eta)=\frac{(1-\xi)^3}{8}L_j(\eta) \quad (51b,c)$$

and for node j between two adjacent RE's, as

$$N_j^{u_r}(\xi,\eta)=N_j^{u_z}(\xi,\eta)=N_j^{p_f}(\xi,\eta)=\frac{(1-\xi)^2}{4}L_j(\eta) \quad (52a,b,c)$$

For node j at the interface between an HE and an RE, the shape functions as Eqs. 51 hold.

4.5 *Proper integrals for infinite elements*

In the characteristic matrix of infinite elements, we shall involve integrals with respect to ξ,

$$\int_{-1}^{1}(1-\xi)^n d\xi = 2^{n+1}(n+1)^{-1}, \quad (n=1,2,3,\cdots) \quad (53)$$

which are *proper*, and have closed form values.

5. LAND SUBSIDENCE AND NEGATIVE SKIN FRICTION

Water withdrawal from deep artesian wells, symbolized by the point sink in Fig. 2b, can cause land subsidence, different consolidation at different depths in the soil mass, and *negative skin friction* in piles embedded in such soil media.

5.1 *Problem composition*

The problem of negative skin friction in piles can be treated as being composed of three systems, namely;

(1) free-field subsidence due to water withdrawal from artesian wells such as depicted in Fig. 2b,

(2) the half space subjected to the 'bond' forces q_i as shown in Fig. 4a, and

(3) the fictitious bar subjected to $-q_i$ as in Fig. 4b.

The fictitious bar has the same dimensions as the actual pile, but its stiffness is equal to the difference between those of the latter and the half space, i.e.

$$E^{(3)}(z)=E^{(p)}(z)-E^{(2)}(z) \quad (54)$$

where E denotes Young's moduli, a supernumeral within parentheses the system number, and the superscript p within parentheses the actual pile. The deformation at a certain point in system (3) is equal to the sum of those at the corresponding points in systems (1) and (2). The total forces or stresses are the appropriate sums of those in all three systems. Consider the fictitious pile as a uniaxial member and divide it into N elements with nodal points at z_i; i = 0 to N; z_0 = 0, z_N = pile length (Figs. 4). Accordingly, the equilibrium of axial forces acting on the fictitious bar gives

$$\sum_{i=0}^{N}\bar{q}_i=0 \quad (55a)$$

and its axial displacements at various values of z_i can be written as

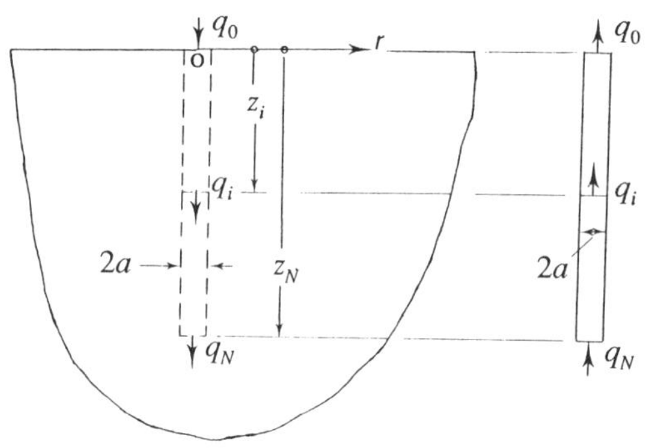

(a) Half space. (b) Fictitious bar.

Fig. 4. Interacting systems (2) and (3).

$$\bar{w}_0^{(3)}=\bar{w}_0^{(1)}+\sum_{i=0}^{N}b_{0i}\bar{q}_i \quad (55b)$$

$$\bar{w}_j^{(3)}=\bar{w}_{j-1}^{(3)}+\frac{z_j-z_{j-1}}{E_j^{(3)}}\sum_{k=0}^{j-1}\bar{q}_k=\bar{w}_j^{(1)}+\sum_{i=0}^{N}b_{ji}\bar{q}_i, \quad (j=1,2,3,\cdots,N) \quad (55c)$$

where b_{ji} = $\bar{w}_j^{(2)}$ due to unit $\bar{q}_i$. Equations 55 constitutes a set of $N+2$ equations for determining $N+1$ $\bar{q}_i$'s and one displacement, say $\bar{w}_0^{(3)}$.

5.2 *Actual negative skin friction*

The equilibrium of the axial forces acting upon the actual piles is depicted in Fig. 5, i.e.

$$\pi a^2 \frac{\partial \tilde{\sigma}_{zz}^{(p)}}{\partial z} + 2\pi a \tilde{\tau} = 0 \quad \text{or} \quad \tilde{\tau} = -\frac{a}{2} \frac{\partial \tilde{\sigma}_{zz}^{(p)}}{\partial z} \tag{56}$$

where τ is the negative skin friction. Rewriting the last equation in the form of finite differences, we obtain $\tilde{\tau}$ at various nodes as follows,

$$\tilde{\tau}_i = (\tilde{\sigma}_{zz,i-1}^{(p)} - \tilde{\sigma}_{zz,i+1}^{(p)}) \frac{a}{4h} + O(h^2) \tag{57}$$

where O denotes the order of the truncation error, and

$$h = z_i - z_{i-1} = z_{i+1} - z_i \tag{58}$$

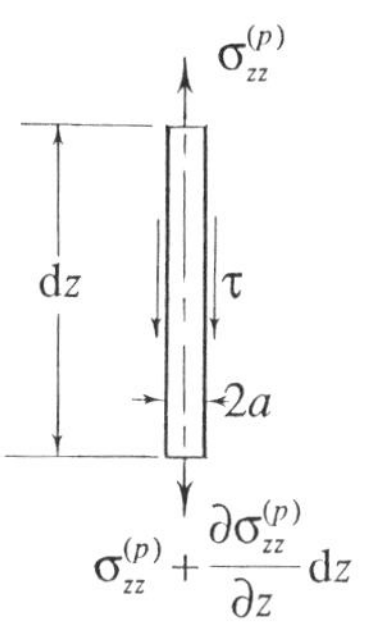

Fig. 5. Negative skin friction τ on actual pile.

6. APPROXIMATE INVERSE LAPLACE TRANSFORMS

It is proposed in this study that the Laplace transform of all governing conditions are satisfied first at each numerical value of the parameter p, then the inverse Laplace transform of any solution function is obtained as a function of time t by the following collocation method:

6.1 *Point sink problems*

For problems of point sinks with the outflow rate $\overline{Q} = PH(t)$, the solution is likely to be in the form

$$v(t) \cong v^*(t) = g_{-1}t + g_0 + \sum_{i=1}^{N} g_i \exp(-\alpha_i t) \tag{59}$$

where N is a positive integer and α_i are positive constants, g_i are unknown constants. The initial condition $v(0) = 0$ implies that

$$g_0 + \sum_{i=1}^{N} g_i = 0 \tag{60}$$

The Laplace transform of Eq. 59 is

$$\tilde{v}(p) = \frac{g_{-1}}{p^2} + \frac{g_0}{p} + \sum_{j=1}^{N} \frac{g_j}{p + \alpha_j} \tag{61}$$

The quantities g_i, $i = 1$ to N, are to be determined by the technique of the minimum square error, i.e. from the condition

$$\frac{\partial}{\partial g_i} \int_{0^-}^{\infty} [v(t) - v^*(t)]^2 dt = 0,$$

or

$$\int_{0^-}^{\infty} [v(t) - v^*(t)] \frac{\partial v^*(t)}{\partial g_i} dt = 0,$$

or, due to (59),

$$\int_{0^-}^{\infty} v(t) \exp(-\alpha_i t) dt = \int_{0^-}^{\infty} v^*(t) \exp(-\alpha_i t) dt,$$

$$(i = 1, 2, \cdots, N) \tag{62}$$

The equation above is just $\tilde{v}(\alpha_i) = \tilde{v}^*(\alpha_i)$ and identical to Eq. 61. Thus we can write

$$\tilde{v}(\alpha_i) = \frac{g_{-1}}{\alpha_i^2} + \frac{g_0}{\alpha_i} + \sum_{j=1}^{N} \frac{g_j}{\alpha_i + \alpha_j}, \quad (i = 1, 2, \cdots, N+1) \tag{63}$$

which together with Eq. 60 form a set of $N+2$ equations to determine g_{-1}, g_0, g_1 to g_N. The approximation is exact at all *pivotal points* $p = \alpha_i$, and $p = \infty$ (equivalent to $t = 0$).

6.2 *Force problems*

For problems of applied forces of the form $\overline{P} = PH(t)$, the solution is likely to be in the form

$$v(t) \cong v^*(t) = g_0 + \sum_{i=1}^{N} g_i \exp(-\alpha_i t) \tag{64}$$

As t tends to ∞, p_f vanishes, and $v(\infty)$ becomes the solution for pure solids, $v^{(ps)}$, i.e.

$$g_0 = v^{(ps)} \tag{65}$$

The Laplace transform of Eq. 65 as well as the technique of the minimum square error give

$$\tilde{v}(\alpha_i) = \frac{g_0}{\alpha_i} + \sum_{j=1}^{N} \frac{g_j}{\alpha_i + \alpha_j}, \quad (i = 1, 2, \cdots, N) \tag{66}$$

Thus we have $N+1$ equations, Eq. 65 and 66, to determine g_0, g_1 to g_N, and the approximation is exact at all pivotal points $p = \alpha_i$, and $p = 0$ ($t = \infty$).

7. CONCLUSIONS

The shape functions of infinite elements are derived analytically from simple problems. All completeness and compatibility requirements are satisfied by such elements. Standard finite element method programs are readily applicable, since all elements involved eventually are finite. The integrals with respect to ξ, which corresponds to the infinite dimension of the infinite elements, are proper and have closed form values. The Laplace transform of all governing conditions are satisfied first, almost in the same manner as in statical analyses (without Laplace transform) of pure elastic solids, at each numerical value of the Laplace transform parameter p. The approximation for the inverse Laplace transform is exact at all *pivotal points* $p = \alpha_i$, and $p = \infty$ (equivalent to $t = 0$) for problems of point sinks, and $p = 0$ ($t = \infty$) for problems of applied forces. The error in that approximation is 'minimum'.

REFERENCES

Apirathvorakij, V. & P. Karasudhi 1980. Quasi-static bending of a cylindrical elastic bar partially embedded in a saturated elastic half-space. *Int. J. Solids Structures*. 16:7:625-644.

Biot, M.A. 1941. General theory of three dimensional consolidation. *J. Appl. Phys.* 12:155-164.

Karasudhi, P. 1991a. *Foundations of solid mechanics*. p.5, 44, 108, 224, 378. Dordrecht: Kluwer.

Karasudhi, P. 1991b. A simple method for far-field functions in elastic and viscoelastic spaces. *Proc. U.S.-Korea-Japan Trilateral Seminar on Frontier Research and Development for Constructed Facilities:* p.326-340. Honolulu, Hawaii.

Mindlin, R.D. 1936. Force at a point in the interior of a semi-infinite solid. *Physics*. 7:195-202.

Rajapakse, R.K.N.D. & P. Karasudhi 1985. Elastostatic infinite elements for layered half spaces. *J. Eng. Mech., ASCE.* 111:9:1144-1158.

Impact of Computational Mechanics on Engineering Problems, Pulmano & Murti (eds)
© 1993 Balkema, Rotterdam, ISBN 90 5410 324 8

Some applications of homogenization method in rock mechanics

T. Kawamoto
Aichi Institute of Technology, Japan

T. Kyoya
Nagoya University, Japan

ABSTRACT: This paper presents an application of the Homogenization method as an effective mathematical tool to make quantitative evaluation of mechanical properties of rock and rock engineering structures which are characterized by their complex geometry of the structure, such as a porous rock and a rock mass reinforced by pattern bolting. The essential viewpoint required by the method is that such a rock and a rock mass are regarded as a composite material with geometric periodicity by repetitive appearance of a characteristic micro-structure. After the Homogenization method is briefly outlined and an illustrative example are given, two applications are presented: First, mechanical properties of Ryukyu limestone which is a porous soft rock distributed in Okinawa, Japan, are evaluated, and second, quantitative evaluation of the effect of pattern bolting installed into rock mass is made.

1. INTRODUCTION

Evaluation of characteristics of mechanical model of discontinuous rock mass are of great important for assessing the stability of rock engineering structures. The essential role of discontinuities in a rock mass has long been recognized by engineers engaged in rock mechanics. Because of the nature of geological discontinuities, it is necessary to distinguish clearly all material properties of an intact rock from the rock mass behavior.

For a rock mass involving a number of cracks which are relatively small in dimension compared with a rock structure, a distinct effect of individual cracks may be negligible. This implies that the rock mass is idealized as a continuum body whether isotropic or anisotropic. Several analytical and numerical tools are available for such idealizations.

The problem becomes important for the behavior of rock mass when intermediate sized discontinuities occur. Appropriate theory or workable numerical method such as damage mechanics[1,2,3] and method with fabric tensor[4,5] has been developed for this circumstance. That is the discontinuities can be regarded as damage of mass for constitutive modeling of highly joined rock mass. Then, the behavior of mass involving such cracks can conveniently be treated by describing the geometry of such cracks, and the properties of the intact rock, separately.

Recently, the constitutive modeling has been presented for the analysis of behavior of jointed rock mass based on Homogenization theory,. The Homogenization method is one of the mathematical methods which evaluated the average property of material with periodic micro-structure such as micro cracks or inhomogenious inclusions. Therefore, the Homogenization method may be applied not only to estimate the mechanical and hydraulic characteristics of highly jointed or porous rock masses, but to evaluate the equivalent characteristics of deformability of rock masses reinforcing with pattern-bolting, and ground with regularly piled system.

In this paper, a brief introduction to the mathematical formation of the homogenization method is explained briefly, and some illustrative applications are presented to the estimation of mechanical properties of Ryukyu limestone and the effect of pattern bolting in which predicted global properties are shown to be in good agreement with the results of the experimental investigations. Then, the change of elastic constant of the rock masses with distributed porosity or by installation of rock bolts is reasonably evaluated by the method.

2. OUTLINE OF THE HOMOGENIZATION METHOD

Let us consider a body characterized by the domain Ω, fixed on the boundary $\partial\Omega_d$, subjected to body force f in Ω, traction t on the boundary $\partial\Omega_t$ as shown in Fig.1. Further, we assume that the material is formed by the spatial repetition of a unit cell whose order is of order ε and is small enough compared to those of the body Ω.

The problem of the deformation of the body Ω in Fig.1 can be expressed by

$$\int_\Omega E^\varepsilon_{ijkl}(\boldsymbol{x})\frac{\partial u^\varepsilon_k(\boldsymbol{x})}{\partial x_l}\frac{\partial v^\varepsilon_i(\boldsymbol{x})}{\partial x_j}d\Omega = \int_{\partial\Omega} t_i v^\varepsilon_i(\boldsymbol{x})dS + \int_\Omega f^\varepsilon_i(\boldsymbol{x})v^\varepsilon_i(\boldsymbol{x})d\Omega, \quad \forall v^\varepsilon_i(\boldsymbol{x}) \tag{1}$$

where, the superscript ε indicate dependence of the micro-structure, and $E^\varepsilon_{ijkl}(\boldsymbol{x})$ is elasticity tensor, $f^\varepsilon_i(\boldsymbol{x})$ is body force and $u^\varepsilon_i(\boldsymbol{x})$ is displacement, which change their values with periodicity of order ε.

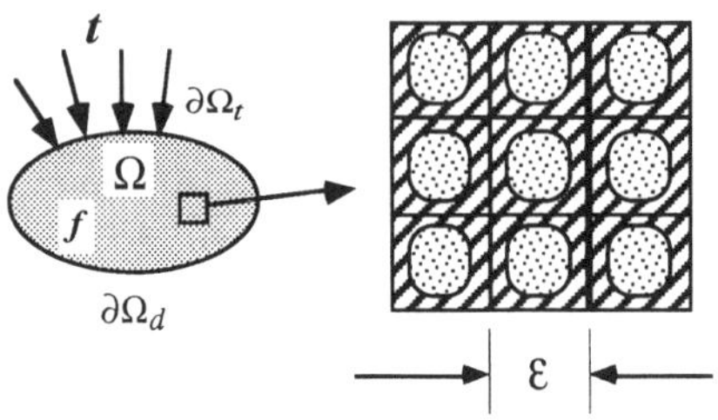

Fig. 1 Material body and its micro-structure

In order to handle such functions, we introduce a new variables $y_i = x_i/\varepsilon$ and rewrite the periodic functions as

$$\begin{aligned} E^\varepsilon_{ijkl}(\boldsymbol{x}) &= E_{ijkl}\left(\boldsymbol{x}, \frac{\boldsymbol{x}}{\varepsilon}\right) = E_{ijkl}(\boldsymbol{x},\boldsymbol{y}) \\ f^\varepsilon_i(\boldsymbol{x}) &= f_i\left(\boldsymbol{x}, \frac{\boldsymbol{x}}{\varepsilon}\right) = f_i(\boldsymbol{x},\boldsymbol{y}) \quad , \\ u^\varepsilon_i(\boldsymbol{x}) &= u_i\left(\boldsymbol{x}, \frac{\boldsymbol{x}}{\varepsilon}\right) = u_i(\boldsymbol{x},\boldsymbol{y}) \end{aligned} \tag{2}$$

Then, the above functions are said to be Y-periodic with respect to $y_i = x_i/\varepsilon$.

We search the displacement vector satisfying (1) in a perturbed form, called "two scale asymptotic expansion", as follow :

$$u^\varepsilon_i(\boldsymbol{x}) = u_i(\boldsymbol{x},\boldsymbol{y}) = u^0_i(\boldsymbol{x}) + \varepsilon u^1_i(\boldsymbol{x},\boldsymbol{y}) + \varepsilon^2 u^2_i(\boldsymbol{x},\boldsymbol{y}) + \cdots \tag{3}$$

Introducing the above expansion into (1) and taking the limit as $\varepsilon \to 0$, we obtain

$$\lim_{\varepsilon\to 0}\int_\Omega E_{ijkl}(\boldsymbol{x},\boldsymbol{y})\left(\frac{\partial u^0_k(\boldsymbol{x})}{\partial x_l} + \frac{\partial u^1_k(\boldsymbol{x},\boldsymbol{y})}{\partial y_l}\right)\frac{\partial v^1_i(\boldsymbol{x},\boldsymbol{y})}{\partial y_j}d\Omega = 0 \quad , \quad \text{for} \quad \forall v^1_i(\boldsymbol{x},\boldsymbol{y}) \tag{4}$$

$$\lim_{\varepsilon\to 0}\int_\Omega E_{ijkl}(\boldsymbol{x},\boldsymbol{y})\left(\frac{\partial u^0_k(\boldsymbol{x})}{\partial x_l} + \frac{\partial u^1_k(\boldsymbol{x},\boldsymbol{y})}{\partial y_l}\right)\frac{\partial v^0_i(\boldsymbol{x})}{\partial x_j}d\Omega = \int_{\partial\Omega} t_i v^0_i(\boldsymbol{x})dS + \lim_{\varepsilon\to 0}\int_\Omega f_i(\boldsymbol{x},\boldsymbol{y})v^0_i(\boldsymbol{x})d\Omega \quad \text{for} \quad \forall v^0_i(\boldsymbol{x}) \tag{5}$$

Noting that for a Y-periodic function $g(\boldsymbol{x},\boldsymbol{y})$,

$$\lim_{\varepsilon\to 0}\int_\Omega g(\boldsymbol{x},\boldsymbol{y})d\Omega = \int_\Omega \frac{1}{|Y|}\int_Y g(\boldsymbol{x},\boldsymbol{y})dY\,d\Omega \tag{6}$$

holds, where $|Y|$ is the volume of the unit cell, equations (4) and (5) are written as

$$\int_\Omega\left[\frac{1}{|Y|}\int_Y E_{ijkl}(\boldsymbol{x},\boldsymbol{y})\frac{\partial u^1_k(\boldsymbol{x},\boldsymbol{y})}{\partial y_l}\frac{\partial v^1_i(\boldsymbol{x},\boldsymbol{y})}{\partial y_j}dY\right]d\Omega = \int_\Omega\left[\frac{1}{|Y|}\int_Y \frac{\partial E_{ijkl}(\boldsymbol{x},\boldsymbol{y})}{\partial y_j}v^1_i(\boldsymbol{x},\boldsymbol{y})dY\right]\frac{\partial u^0_k(\boldsymbol{x})}{\partial x_l}d\Omega \quad \text{for} \quad \forall v^1_i(\boldsymbol{x},\boldsymbol{y}) \tag{7}$$

$$\int_\Omega\left[\frac{1}{|Y|}\int_Y E_{ijkl}(\boldsymbol{x},\boldsymbol{y})\left(\frac{\partial u^0_k(\boldsymbol{x})}{\partial x_l} + \frac{\partial u^1_k(\boldsymbol{x},\boldsymbol{y})}{\partial y_l}\right)dY\right]\frac{\partial v^0_i(\boldsymbol{x})}{\partial x_j}d\Omega = \int_\Omega\left[\frac{1}{|Y|}\int_Y f_i(\boldsymbol{x},\boldsymbol{y})dY\right]v^0_i(\boldsymbol{x})d\Omega + \int_{\partial\Omega} t_i v^0_i(\boldsymbol{x})dS \quad \text{for} \quad \forall v^0_i(\boldsymbol{x}) \tag{8}$$

The characteristic deformation $\chi^{kl}_p(\boldsymbol{x},\boldsymbol{y})$ which reflects the nature of the micro-structure is obtained by solving the following equation on the unit cell :

$$\int_Y E_{ijpq}(\boldsymbol{x},\boldsymbol{y})\frac{\partial \chi^{kl}_p(\boldsymbol{x},\boldsymbol{y})}{\partial y_q}\frac{\partial v_i(\boldsymbol{x},\boldsymbol{y})}{\partial y_j}dY = \int_Y E_{ijkl}(\boldsymbol{x},\boldsymbol{y})\frac{\partial v_i(\boldsymbol{x},\boldsymbol{y})}{\partial y_j}dY \quad , \quad \text{for} \ \forall v_i(\boldsymbol{x},\boldsymbol{y}) \tag{9}$$

Using the characteristic deformation, general form of the solution $u^1_i(\boldsymbol{x},\boldsymbol{y})$ for (7) can be written as

$$u^1_i(\boldsymbol{x},\boldsymbol{y}) = -\chi^{kl}_i(\boldsymbol{x},\boldsymbol{y})\frac{\partial u^0_k(\boldsymbol{x})}{\partial x_l} + \psi_i(\boldsymbol{x}) \tag{10}$$

in which $\psi_i(\boldsymbol{x})$ is an arbitrary vector depending only on $\boldsymbol{x}$. Then, we obtain the following equilibrium equation on the domain Ω by substituting (10) into (8) :

$$\int_\Omega E^H_{ijkl}\frac{\partial u^0_k(\boldsymbol{x})}{\partial x_l}\frac{\partial v^0_i(\boldsymbol{x})}{\partial x_j}d\Omega$$
$$=\int_\Omega f_i^H v_i^0(\boldsymbol{x})d\Omega+\int_{\partial\Omega} t_i v_i^0(\boldsymbol{x})dS \qquad (11)$$
$$\text{for}\quad \forall v_i^0(\boldsymbol{x})$$

where, the homogenized elasticity tensor is defined by

$$E^H_{ijkl}=\frac{1}{|Y|}\int_Y\left(E_{ijkl}(\boldsymbol{x},\boldsymbol{y})-E_{ijpq}(\boldsymbol{x},\boldsymbol{y})\frac{\partial\chi_p^{kl}(\boldsymbol{x},\boldsymbol{y})}{\partial y_q}\right)dY \qquad (12)$$

and the homogenized body force by

$$f_i^H=\frac{1}{|Y|}\int_Y f_i(\boldsymbol{x},\boldsymbol{y})dY \qquad (13)$$

The strain and the stress fields are evaluated as

$$e^\varepsilon_{ij}=\frac{1}{2}\left[\left(\delta_{ip}\delta_{jq}-\frac{\partial\chi_i^{pq}}{\partial y_j}\right)+\left(\delta_{jp}\delta_{iq}-\frac{\partial\chi_j^{pq}}{\partial y_i}\right)\right]\frac{\partial u^0_p}{\partial x_q}+O(\varepsilon) \qquad (14)$$

$$\sigma^\varepsilon_{ij}=E^\varepsilon_{ijkl}\left(\delta_{kp}\delta_{lq}-\frac{\partial\chi_k^{pq}}{\partial y_l}\right)\frac{\partial u^0_p}{\partial x_q}+O(\varepsilon) \qquad (15)$$

Using the characteristic deformations of the unit cell, the microscopic stress and strain field can be evaluated. We can obtain stress and strain fields both in the macroscopic level and the microscopic level.

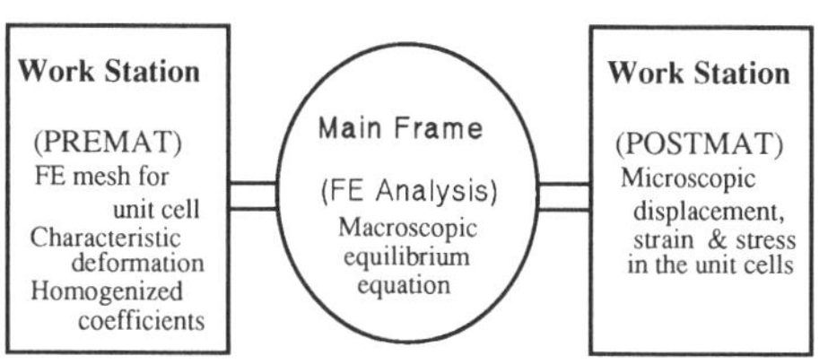

Fig. 2 The system for Homogenization analysis

The homogenization analysis mentioned the above is carried out in the system as shown in Fig.2. In each step, the following procedure is performed :

(1) PREMAT in the Work Station: A three dimensional finite element model for a unit cell is generated and the equation (12) is solved to obtain the characteristic deformations. Then, homogenized elasticity tensor and the body force are calculated. All of these computations are performed by the code PREMAT.

(2) FE analysis in a main computer : By using the homogenized values obtained in the above, the macroscopic equilibrium problem given by (11) is solved by an appropriate FEM code to obtain the averaged displacement, strain and stress .

(3) POSTMAT in the Work Station : By using the averaged macroscopic displacement, microscopic strain and stress in a unit cell at any spatial point are computed by (14) and (15), respectively. The results are presented in illustrative figures. All of these are carried out by the code POSTMAT.

3. UNIAXIAL COMPRESSION TESTS FOR SPECIMENS WITH CIRCULAR HOLES

In order to check the applicability of the Homogenization method, uniaxial compression tests for rectangular cement mortar specimens which have regularly arranged circular holes were carried out, and results obtained were compared with those by the Homogenization analysis and by the finite element analysis.

The cement mortar used is a 1:1:0.6 mixture of cement, water and sand. The rectangular specimens have a height of 40cm , a width of 20cm and a thickness of 10cm, and circular holes of which diameter is 5cm were arranged in three patterns as shown in Fig. 3. Through the uniaxial compression tests apparent stiffness of the specimens were measured, while the stiffness of the specimens were evaluated by the Homogenization analysis and by numerical simulations by 2-D finite element models, in which plane stress condition is assumed. In Fig. 4, an example of the unit cell for the Homogenization analysis is presented.

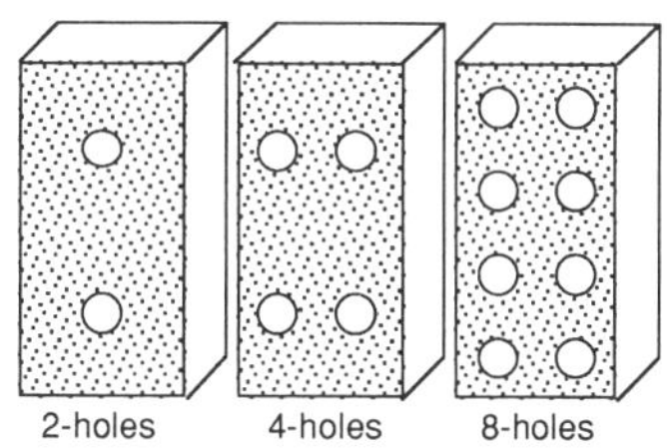

Fig.3 Specimens with circular holes

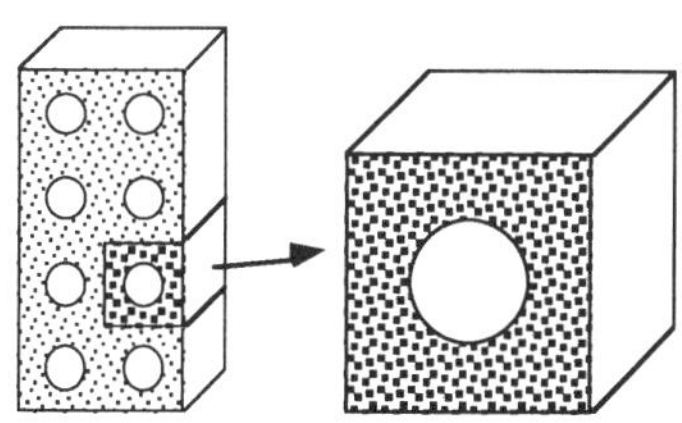

Fig.4 An example of the unit cell

The results obtained are compared in Fig. 5. The vertical axis represents the apparent stiffness in the loading direction of the specimen normalized by that of the intact(no hole) specimen , while the horizontal axis is the number of holes in a specimen. It is observed that the apparent stiffness obtained through the experiments are relatively lower than the predicted values. However, the tendency that the apparent stiffness decrease as number of holes increase are fairly well estimated by the homogenization analysis.

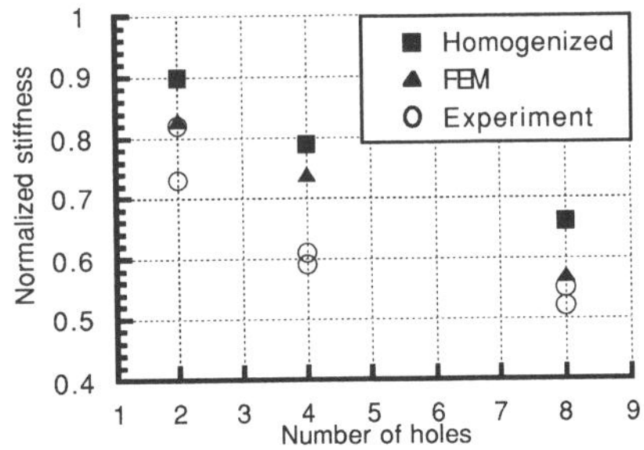

Fig.5 Stiffness of the specimens with circular holes

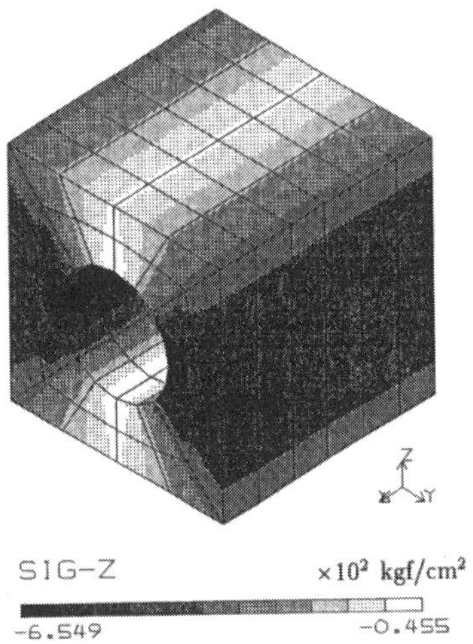

Fig. 6 Microscopic stress field around a hole in the eight holes specimen at the load level of 225kgf/cm^2

The difference between the values predicted by the Homogenization method and those by the FE analyses results from whether shape of the specimen's boundary is considered or not. In the FE analysis shape of a specimen was directly modeled, while in the homogenization analysis infinitely repetitive appearance of the unit cell was only assumed.

Taking into account that the experiment were carried out for the specimens which include at most eight holes and that if larger specimens which include many holes in them can be used in the test the experimental values are expected to show well agreement with results by the homogenization analysis, it can conclude that the homogenization method is quite useful in evaluation of averaged properties of rock-like materials including micro-structure.

Fig. 6 shows an example of the microscopic stress distribution around a hole. The stress field corresponds to the case of the eight holes specimen at the uniaxial load of 225 kgf/cm^2. In the figure, z-axis coincides with the loading axis. The figure shows that the homogenization method can also reasonably approximate the microscopic stress distribution in an unit cell of micro-structure.

4. EVALUATION OF MECHANICAL PROPERTIES OF POROUS ROCK

Porous rocks are consisted of complexity oriented pores and matrices, and they can be idealized to compose a periodic macro-structure. To such materials, the equivalent material properties can be evaluated by the homogenization method.

The first example of application is made to evaluate the effect of volume fraction of pores in Ryukyu limestone on its equivalent elastic modules and yield strength under uniaxial compression.

Ryukyu limestone is a porous soft rock which has a number of spherical or ellipsoidal pores. The diameter of the pores are distributed within a range from 1.8 mm to 2.5 mm, and the porosity (volumetric fraction of pores) is ranged from 5% to 15% for the cylindrical specimens with a diameter of 5cm and a height of 10cm. We assume that the Ryukyu limestone can be idealized as an assembly of a cubic unit cell which has a spherical pore in its center as shown in Fig. 7.

Fig.7 FE model for a unit cell of Ryukyu limestone

The relationships between equivalent elastic constants (Young's modulus E and Poison's ratio ν) and the volumetric fraction of spherical pores are presented in Fig.8 and Fig.9. In the calculation, the matrix of rock is assumed to be an isotropic elastic material of which Young's modulus, E_0, is fixed at 4.5×10^5kgf/cm^2, while the Poisson's ratio is changed from 0.1 to 0.4. In Fig.8, the equivalent Young's modulus is normalized by that of the matrix of rock.

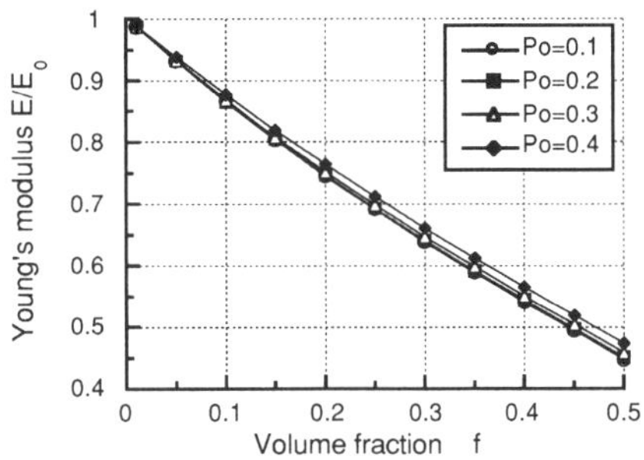

Fig.8 Relationship between normalized Young's modulus and volumetric fraction of spherical pores

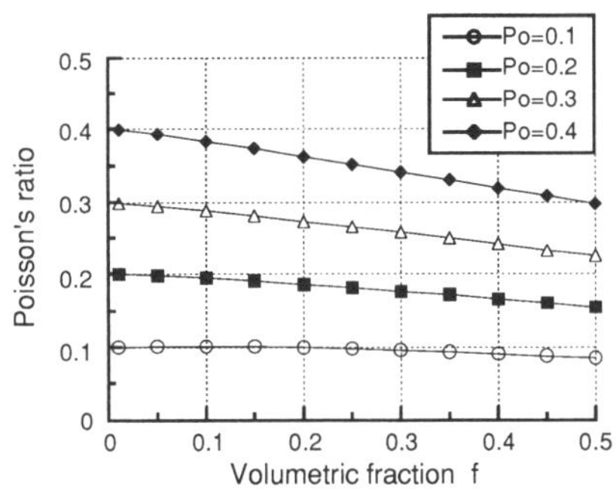

Fig.9 Relationship between Poisson's ratio and volumetric fraction of spherical pores

As can be seen from the figures, with increase in the volumetric fraction the equivalent Young's modulus and Poison's ratio decrease almost linearly. The change of the Poisson's ratio is more gentle than that of the Young's modulus. The rate of the change of the Poisson's ratio depends on Poisson's ratio of the matrix itself, while the rate of change of the Young's modulus is not so affected by value of the matrix's Poisson's ratio. As for Ryukyu limestone, it is found that the Poisson's ratio is almost equal to that of the matrix, while the Young's modulus is affected by the volumetric fraction of pores since its volumetric fraction distributes within 5% to 15%.

A series of uniaxial compression tests were carried out on cylindrical specimens of Ryukyu lime stone and various yield strengths of them were obtained.

Meanwhile, FE analyses for each specimen at the yielding uniaxial load were carried out, then assuming three kinds of porosity (volumetric fractions of pores), of 5%, 1% and 15%, the microscopic stress state in an unit cell is evaluated by POSTMAT. The unit cell is picked up from a point located near the lateral surface at middle height of the specimen. The obtained microscopic stress state are visualized as shown in Fig.10.

Table 1 summarizes the yield strength of each specimen obtained by uniaxial compression tests and calculated tensile stress around pores for various volumetric fraction, and Fig. 11 shows their plots. The results imply the scattering of the tensile strength of rock around pores.

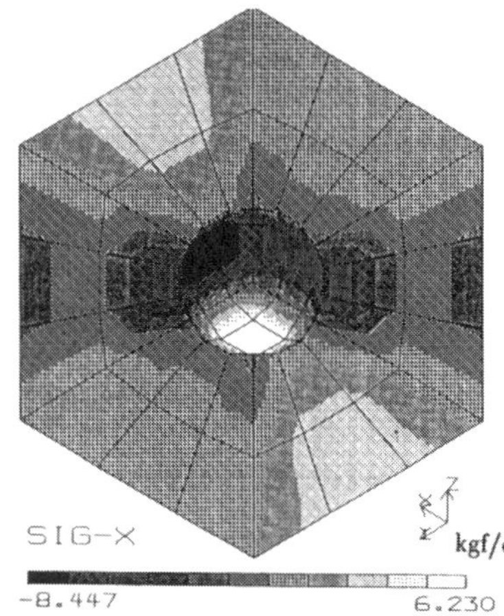

Fig.10 Microscopic stress distribution in a unit cell

Table 1 Yield strengths of Ryukyu limestone and maximum microscopic tensile stress around a spherical pore

Specimen No.	Yield strength (kgf/cm^2)	Tensile Stress (kgf/cm^2)		
		5%	10%	15%
B1	324	6.7	9.7	11.6
B2	409	8.5	12.2	14.7
B3	370	7.7	11.0	13.3
B4	300	6.2	9.0	10.8
B5	380	7.9	11.3	13.6
B6	280	5.8	8.4	10.0
B7	270	5.6	8.1	9.7
B8	460	9.6	13.7	16.5

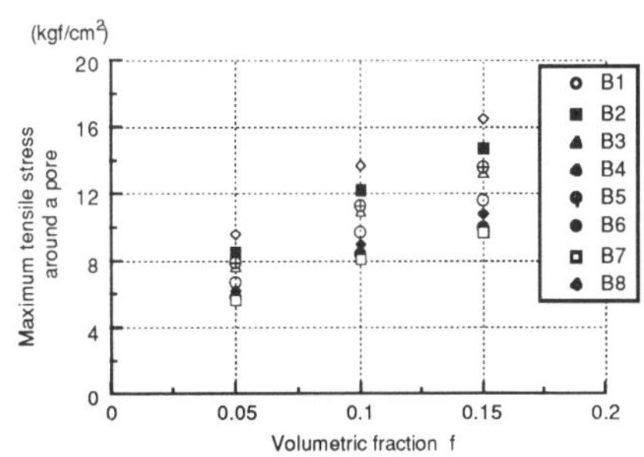

Fig.11 Maximum tensile stress around a pore

5. HOMOGENIZATION ANALYSIS OF EFFECT OF PATTERN BOLTING

Rock bolts have been in use for several decades as a practical, direct, and effective form of rock mass reinforcement for a wide range of geotechnical construction works. However, relatively limited efforts have been made, except for those approaches in which rock bolts are modeled explicitly in numerical simulation, to make quantitative evaluation of the effects of rock bolts as to how much additional stiffness they introduce to the reinforced rock mass.

In an effort to overcome this problem, we here present an application of the Homogenization method as an effective mathematical tool to make quantitative evaluation of effects of regularly installed rock bolts. The essential viewpoint is that a rock mass is regarded as a composite material with geometric periodicity introduced by repetitive appearance of rock bolts with certain interval and geometric regularity.

In this section increment of stiffness of rock mass by regularly installed rock bolts is precisely evaluated by the Homogenization method.

The rock mass is assumed to be isotropic elastic material. The Young's modulus of the rock mass is changed from 10,000 kgf/cm^2 to 100,000 kgf/cm^2 with appropriate intervals, while the Poisson's ratio is fixed at 0.25. As for the rock bolts, it is assumed that the Young's modulus is 2,000,000 kgf/cm^2 and the Poisson's ratio is 0.3, and that it have a diameter of 3 cm, and pitch of installation of the bolts is changed from 0.5 m to 2.33m.

Regarding the bolts as a micro-structure, the stiffness is calculated for each combination of the rock mass and the pitch of bolts. Local coordinate axes are set for the analyses as shown in Fig. 12.

A rock mass becomes an anisotropic medium by rock bolts. In Table 2, calculated results of the Young's moduli, the shearing stiffness and the Poisson's ratios of the rock masses changed by the rock bolts installed at the pitch of 0.5m, 1.0m, and 2.1m are summarized. They are the coefficients of the following compliance matrix:

$$[C]=\begin{bmatrix} \frac{1}{E_x} & -\frac{\nu_{yx}}{E_x} & -\frac{\nu_{zx}}{E_x} & & & \\ -\frac{\nu_{xy}}{E_y} & \frac{1}{E_y} & -\frac{\nu_{zy}}{E_x} & & & \\ -\frac{\nu_{xz}}{E_x} & -\frac{\nu_{yz}}{E_x} & \frac{1}{E_z} & & & \\ & & & \frac{1}{G_{yz}} & & \\ & & & & \frac{1}{G_{zx}} & \\ & & & & & \frac{1}{G_{xy}} \end{bmatrix}$$

In Fig. 13 to Fig. 18, change of the parameters are plotted in relation with the pitch of the rock bolts. In their legends E100 denotes that Young's modulus of the rock mass, E, is to be 100,000kgf/cm^2, and E50 to be E = 50,000 kgf/cm^2, E20 to be E = 20,000 kgf/cm^2, E10 to be E=10,000 kgf/cm^2, respectively.

It is observed, except for the Poisson's ratios ν_{zx} and ν_{xy}, that values of the parameters increase as the pitch of installation of the bolts decrease, and that, as you expected, the effects of rock bolts is dominant in the longitudinal direction of the bolts.

The amount of increment of the stiffness E_x seems to be slightly affected by the stiffness of the rock mass and takes almost constant value if the pitch of bolts is the same, while the other parameters do not so. However, one should note the difference between the amount of the increment of E_x and those of the other stiffness parameters.

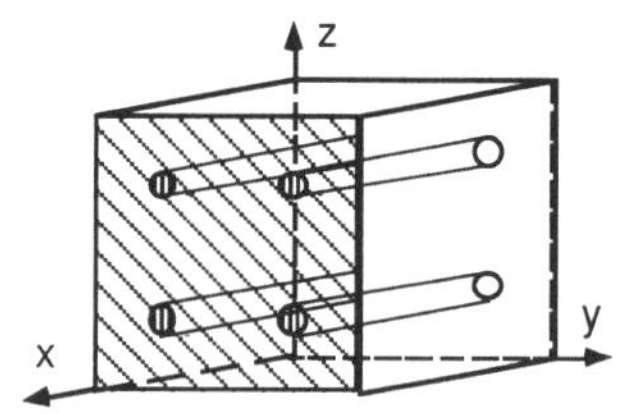

Fig.12 Coordinate axes for rock mass and rock bolts

Table 2 Change of mechanical parameters of rock mass by pattern rock bolting

Rock Mass E(kgf/cm^2) & ν=0.25	Bolt Pitch (m)	Ex (kgf/cm^2)	Ey, Ez (kgf/cm^2)	Gyz (kgf/cm^2)
E=100,000	0.5	103,640	100,490	40,120
G= 40,000	1.0	100,910	100,130	40,033
	2.1	100,210	100,030	40,008
E= 50,000	0.5	53,732	50,361	20,063
G= 20,000	1.0	50,933	50,096	20,017
	2.1	50,212	50,023	20,004
E= 20,000	0.5	23,789	20,261	8,026
G= 8,000	1.0	20,947	20,073	8,007
	2.1	20,215	20,017	8,002
E= 10,000	0.5	13,808	10,206	4,013
G= 4,000	1.0	10,952	10,063	4,004
	2.1	10,216	10,015	4,001

Rock Mass E(kgf/cm^2) & ν=0.25	Bolt Pitch (m)	Gzx, Gxy (kgf/cm^2)	νyz	νzx, νxy
E=100,000	0.5	40,147	0.252	0.243
G= 40,000	1.0	40,039	0.251	0.248
	2.1	40,010	0.250	0.250
E= 50,000	0.5	20,078	0.255	0.234
G= 20,000	1.0	20,021	0.251	0.246
	2.1	20,005	0.250	0.249
E= 20,000	0.5	8,032	0.262	0.213
G= 8,000	1.0	8,009	0.253	0.240
	2.1	8,002	0.251	0.248
E= 10,000	0.5	4,016	0.272	0.185
G= 4,000	1.0	4,006	0.257	0.230
	2.1	4,001	0.252	0.245

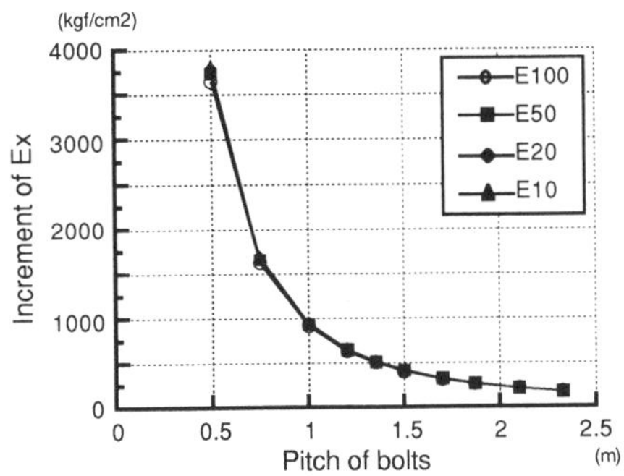

Fig. 13 Increment of the stiffness E_x

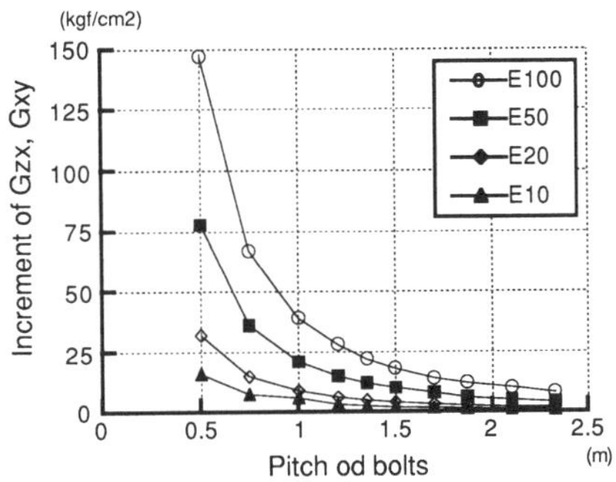

Fig. 16 Increment of the shearing stiffness G_{zx} and G_{xy}

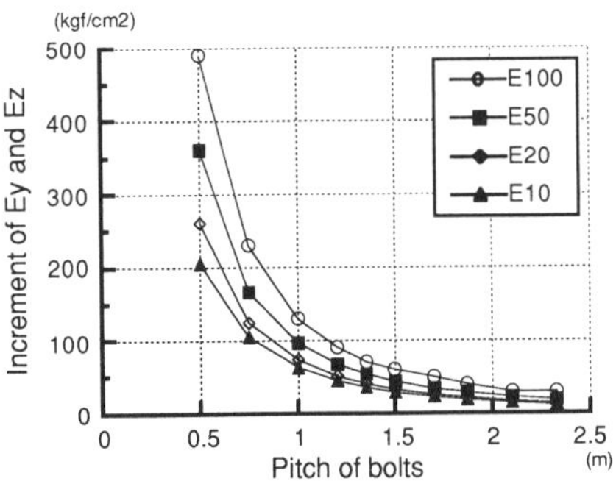

Fig. 14 Increment of the stiffness E_y and E_z

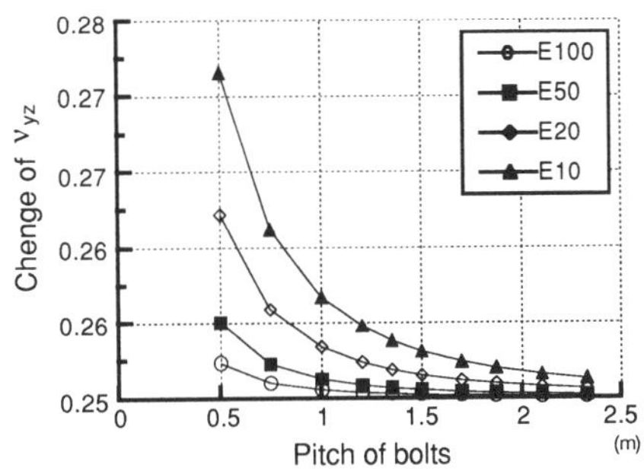

Fig. 17 Change of the Poisson's ratio ν_{yz}

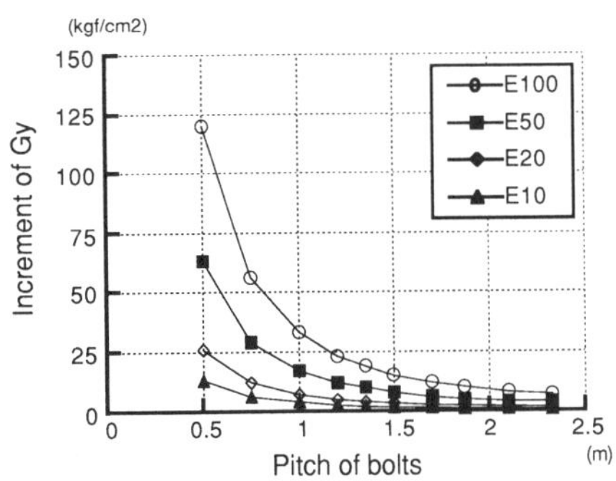

Fig. 15 Increment of the shearing stiffness G_{yz}

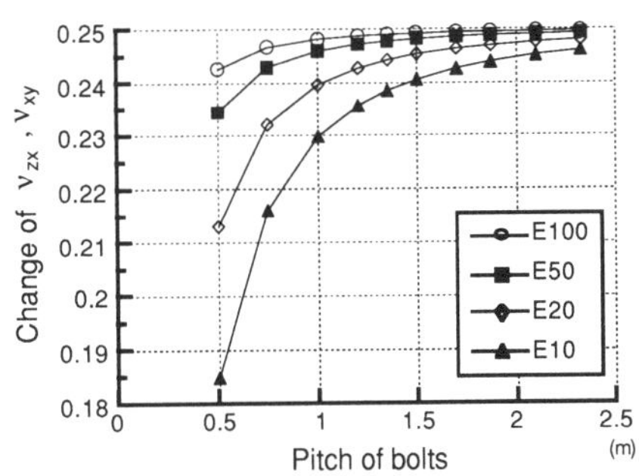

Fig. 18 Change of the Poisson's ratios ν_{zx} and ν_{xy}

It should be noted that value of the E_x can be well approximated by $E_x = (E_s A_s + E_r A_r)/(A_s + A_r)$, which is well known in the design of reinforced concrete, where subscripts s and r denote steel and rock, respectively, and A denotes the area occupied by the material in a cross section.

The change of Poisson's ratio is quite interesting. As the pitch of bolts decrease ν_{zx} and ν_{xy} decrease, while ν_{yz} increases, which is result of interaction between bolts($\nu = 0.3$) and rock mass($\nu = 0.25$). Different results are expected to be obtained for another combination.

6. CONCLUSION

In this paper, a mathematical tool, the Homogenization method, is applied to quantitative evaluation of mechanical properties of porous media and the effect of pattern bolting. After a brief outline of the Homogenization method is given, a illustrative example for verification of the method is presented. Then, the material constants of Ryukyu limestone are calculated by considering the shape and dimension of pores. Similar scheme is used to calculate the mechanical parameters of rock mass reinforced by rock bolts for various combinations of rock mass and pitch of installation of the bolts, and the results are illustratively summarized.

The authors consider that the homogenization technique is a powerful tool to calculate the macroscopic mechanical constants as well as microscopic scale stress concentrations which has important implications on the yielding phenomena of rock.

ACKNOWLEDGMENT

We would like to thank Prof. N. Kikuchi of the University of Michigan, U.S.A., who gave one of the authors a chance to study the Homogenization method. We would also like to thank Mr. K. Ishii and other members of Quint Ltd. in Tokyo, Japan for their kindly cooperation. The computer codes PREMAT and POSTMAT are developed by them.

REFERENCES

[1] T. Kyoya, Y. Ichikawa and T. Kawamoto, A damage mechanics theory for discontinuous rock mass, Proc. 5th Int. Conf. Num. Meth. Geomech, Nagoya, 1, pp.469-480,1985

[2] T. Kyoya, Y.Ichikawa, M. Kusabuka and T. Kawamoto, A damage mechanics analysis for underground excavetion in jointed rock mass, Proc. Int. Symp. Eng. Complex Rock Formations, Beijing, China, pp.506-513, 1986

[3] T. Kawamoto, Y.Ichikawa and T.Kyoya, Deformation and fracturing behaviour of discontinuous rock mass and damage mechanics theory, Int. J. Num. Anal. Meth. Geomech. 12(1), pp. 1-30, 1988

[4] M.Oda, K. Suzuki and T. Maeshibu, Elastic compliance for rock-like materials with cracks, Soils and Foundations, 24(3), pp. 27-40, 1984

[5] M.Oda, Permeability tensor for discontinuous rock mass, Geotechnique, 35(4), pp.483-495, 1985

[6] E. Sanchez-Palencia, Nonhomogeneous media and vibration theory, Lecture Note in Physics, No.127, Springer, Berlin (1980)

[7] J. L. Lions, Some methods in the mathematical analyses of systems and their control, Science Press, Beijing (1981)

[8] N. Bakhvalov and G. Panasenko, Homogenization: Averaging processes in periodic media, Kluwer Academic Pub. (1984)

[9] J. M. Guedes and N. Kikuchi, Preprocessing and postprocessing for materials based on the homogenization method with adaptive finite element methods, Computer Methods in Applied Mechanics and Engineering, 83, pp. 143-198 (1990)

[10] T. Kyoya, An application of the homogenization method to beam-like structures, Proc. 5th Symp. Computational Mechanics, Japan Association of Science and Technology, pp. 249-256 (1991). (in Japanese)

[11] T. Kyoya and N. Tokashiki, Evaluation of mechanical properties of porous Ryukyu limestone by the homogenization method, Proc. 2nd Workshop on Characterization of Subsurface Cracks, Mining and Mineral Processing Institute of Japan, pp. 92-97, (1992). (in Japanese)

[12] T. Kyoya and T. Kawamoto, Quantitative evaluation of effect of pattern bolting by the homogenization method, Int. Symp. Assessment and Prevention of Failure Phenomena in Rock Engineering, Istanbul, Turkey (in print).

Impact of Computational Mechanics on Engineering Problems, Pulmano & Murti (eds)

Application of numerical modelling to consequence-of-sliding analysis

N. Khalili-Naghadeh
University of New South Wales, N.S.W., Australia

W. Sheu & J. R. Boddy
Dames & Moore, Chicago, Ill., USA

ABSTRACT: Application of numerical modelling techniques to consequence-of-sliding analysis of earth structures is discussed. A case study involving a perimeter dike system of a tailings pond is presented. Limit equilibrium analysis of the perimeter dike system indicated potentially unstable dike conditions. However, a subsequent consequence-of-sliding analysis indicated that should a failure occur, it would be limited in extent and would not compromise the overall integrity of the tailings pond, rendering extensive remedial measures to enhance the stability of the dike unnecessary.

1 INTRODUCTION

Recent concerns on release of potentially undesirable materials to the environment have resulted in an increasing interest in the structural integrity of surface impoundments and tailings ponds. Although failure of a tailings pond constructed in a remote and sparsely populated area may not be significant in terms of loss of human life, it could be disastrous in terms of financial loss and damage to the environment.

Stability of earth structures and embankments is traditionally evaluated using the limit equilibrium methods of analysis; e.g. Fellinius method (1936), Janbu method (1954 and 1957), Bishop method (1955), Morgenstern and Price method (1965), and Spencer method (1967). These methods generally involve determination of a minimum factor of safety that reflects the most critical condition. Determination of the minimum factor of safety is typically performed through a two-step process. In the first step, a potential slip surface is assumed, and in the second step the limit equilibrium equations are solved for the factor of safety of the soil mass defined by the potential slip surface. This process is repeated until the minimum factor of safety against sliding, and the associated critical slip surface, are established. If the minimum factor of safety approaches unity, the section analyzed is considered unsafe, and often extensive remedial measures are adopted to enhance the stability of the slope; e.g. see Morgenstern (1982).

Limit equilibrium methods of analysis, despite their popularity in engineering practice, are associated with a number of shortcomings. These shortcomings are extensively discussed in many recent publications and will not be repeated here; e.g. see Wright et al (1973), Duncan and Wright (1980), Nash (1987), and Chang (1992). Further, it appears that the factor of safety against "sliding", as determined by the limit equilibrium methods of analysis, may not be an appropriate criterion for stability assessment of structures which may undergo large deformations without loss of integrity or functionality (e.g. tailings ponds). In these circumstances, a more appropriate design criterion may be the post-sliding (post-failure) deformation of the structure. It can be shown that in many earth structures, a factor of safety of less than one, as determined by the limit equilibrium methods of analysis, would only result in limited movements along the critical slip surface before the slope would attain a new stable geometry.

The concept of using deformation as a stability assessment criterion is not new, and was originally introduced by Newmark (1965) for analyzing stability of dams and embankments under earthquake loading [Newmark (1965)].

However, deformation approaches are considerably more involved than the frequently used limit equilibrium methods, requiring application of highly sophisticated numerical methods. As a result, these approaches have not been fully utilized in the past. To date, with the advances in computer technology, and improvements in numerical modelling techniques, the computational efforts associated with these methods are no longer a concern, and numerical models are increasingly applied to study slope stability problems.

This paper presents the results of the structural integrity assessment of a tailings pond located in the upper midwest of the United Sates. The integrity assessment included both the limit equilibrium and deformation (consequence-of-sliding) methods of analysis. The results of the limit equilibrium analysis

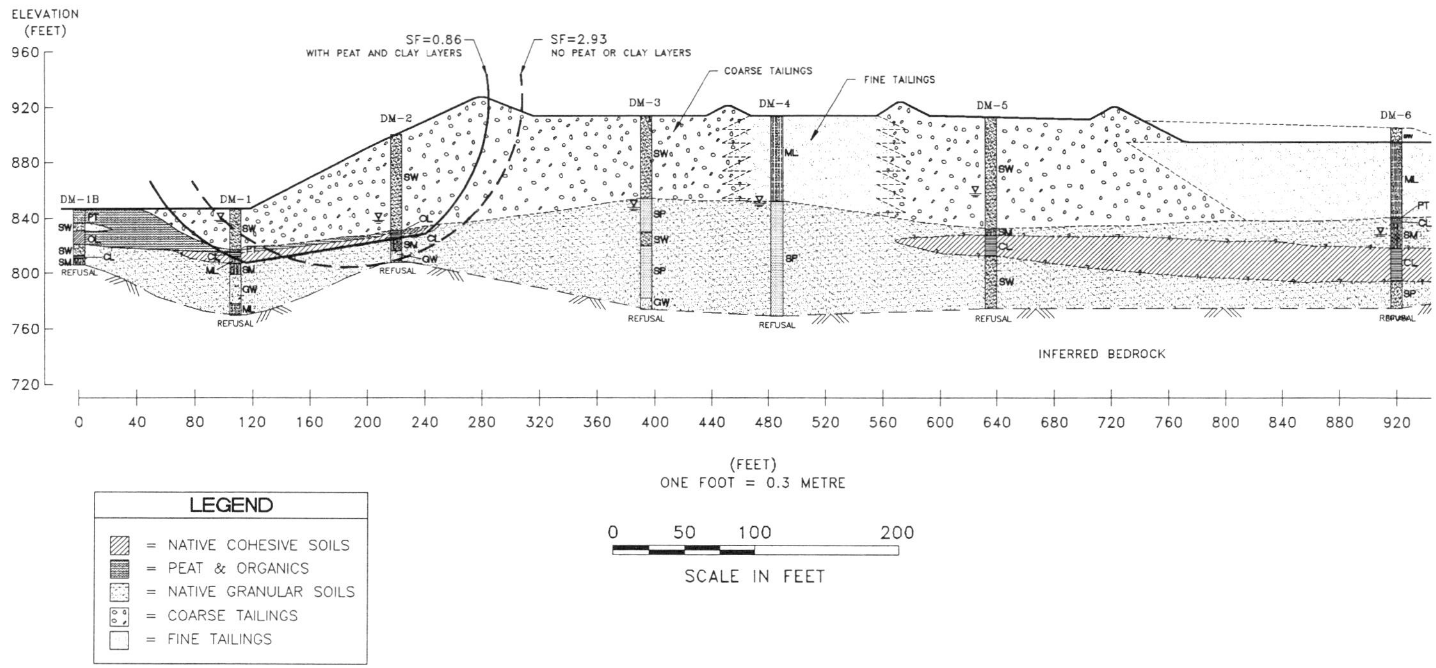

Fig.1 Critical dike section.

performed on several critical cross sections along the dike indicated marginally stable dike conditions. The results of the consequence-of-sliding, however, indicated that should a failure occur, it would be limited in extent and would not compromise the overall integrity of the tailings pond. As a result, it was concluded that remedial measures to enhance the stability of the dike system would not be necessary.

2 BACKGROUND INFORMATION

The subject tailings pond is part of an ongoing mining and milling activities in the area, and serves to store fine and coarse tailings waste from the ore processing operations. It is approximately 2500 acres in planar area, and is formed by a perimeter dike system approximately 41,000 feet long.

The perimeter dike consists of a relatively impermeable core of fine tailings (approximately 30 to 80 feet wide), and two coarse tailings outer shells (i.e. both upstream and downstream). The overall crest width of the perimeter dike is approximately 300 feet. The finished crest elevation of the dike is 910 feet, corresponding to a maximum crest height of 70 feet above the original ground surface.

The subsurface soils underlying the tailings pond consist of up to 10 feet thick post-glacial peat deposits (muskeg) overlying Pleistocene glacial deposits consisting of medium to fine sands and gravels, extending down to bedrock. Reportedly, muskeg was removed in most areas from the footprint of the perimeter dike system prior to the construction the pond. Underlying the glacial and post-glacial sediments is a massive watertight pink feldspar granite bedrock.

3 CRITICAL CROSS SECTION

To evaluate the overall stability of the perimeter dike system a total of three critical cross-sections were selected for analysis. Of the three cross sections selected, the section shown on Figure 1 (Critical Section) was considered the most critical based on the geometry, subsurface soil and rock conditions, and the consequence of failure. Thus, only the results of the stability analysis performed on this critical cross-section will be considered in this paper.

4 MATERIAL LAYERING

To evaluate the type and physical characteristics of the dike materials and the underlying soils, a total of seven soil borings (designated DM-1B to DM-6, see Figure 1) were drilled along the selected critical dike cross-section. Soil samples were collected at 5-foot intervals, or at the point of soil changes throughout each boring length. Disturbed samples were obtained from standard penetration tests split spoon sampler, and relatively undisturbed soil samples were obtained using a thin-walled split barrel sampler.

Based on the results of the field exploration program, and for the purpose of dike stability analysis, the subsurface soils encountered during the investigation were divided into the following five main categories:

1. Coarse Tailings: characterized in ASTM D-2487 as SW.
2. Fine Tailings: characterized in ASTM D-2487 as ML.
3. Granular Native Soils: native materials characterized in ASTM D-2487 as GP, GW, SP, SW, and SM.
4. Fine-Grained Native Soils: native materials characterized in ASTM D-2487 as CL.
5. Peat and Organic: native materials characterized in ASTM D-2487 as PT, and OL.

A graphical presentation of the subsurface soils encountered along the selected critical cross-sections is shown on Figure 1.

5 MATERIAL PROPERTIES

To evaluate the physical and engineering properties of the soil groups considered, an extensive laboratory testing program was performed including sieve analyses, Atterberg limits, moisture and density determination tests, consolidated undrained triaxial tests, unconsolidated undrained triaxial tests, unconfined compression tests, direct shear tests, and consolidation tests. As a supplement to the laboratory testing program, standard SPT blow count values ("N" values) and pocket penetrometer readings were also utilized to evaluate the engineering characteristics of the subsurface soils encountered. The physical and strength characteristics of the material categories used in the stability analysis are provided in Table 1.

6 STABILITY ANALYSIS

6.1 Limit Equilibrium Analysis

Using the strength parameters given in Table 1, the computer program PC-SLOPE was used to calculate the minimum factor of safety against sliding of the critical cross-section selected. PC-SLOPE is based on the theory of limit equilibrium, and is capable of calculating the factor of safety along both circular and non-circular (composite) slip surfaces, using 10 different methods of analysis. For the purpose of this investigation, the Bishop Simplified method was used, which has a degree of accuracy comparable with that of the input data.

A total of 216 slip surfaces were analyzed. Assuming a worst case scenario (i.e. the thin layers of peat/organic materials encountered in borings DM-1 and DM-2 are continuous and laterally extensive beneath the dike), the most critical slip surface resulted in a factor of safety of 0.86 (see Figure 1). Based on the results of the limit equilibrium analysis, the critical slip surface passes through the perimeter dike crest and along the interface between the peat and the underlying clayey layer, and then daylights at the toe of the dike. A factor of safety of 0.86 is not acceptable, and indicates a failure or a metastable slope condition. A subsequent walkover of the site, however, indicated no discernable sign of imminent dike instability, as may be identified by tension cracks and/or bulging at the toe of the dike. This suggests that the worst case scenario considered during the stability analysis (i.e. the thin layers of peat/organic materials encountered in borings DM-1 and DM-2 are continuous and laterally extensive beneath the dike) may not be realistic. It should also be noted that the stability analysis performed during the course of this investigation was a two-dimensional analysis, and did not take account of three dimensional end effects. By assuming that the peat and clay layers encountered are localized and limited in extent, a repeat analysis of the dike stability resulted in a factor of safety of 2.93 against a deep seated sliding. It is apparent the distribution of the peat and clay layers had a major impact on the stability of the perimeter dike, and additional site investigation was required to better define the extent of the peat and clay layers beneath the dike.

6.2 Consequence-Of-Sliding Analysis

To investigate potential post-sliding geometry of the critical cross-section (under a worst case scenario), and to justify the necessity for an additional and more comprehensive program of site investigation, a consequence-of-sliding analysis was performed using the computer program FLAC (Fast Langrangian Analysis of Continuum). FLAC is a general purpose, explicit finite difference program capable of simulating both material and geometry non-linearities. Due to the explicit (time marching) nature of the program, FLAC can be used to simulate and examine the development of yield or material collapse during any stage of the modelling.

The plane strain mesh used to model the critical dike cross-section is shown in Figure 2. A total of 960 continuum elements were used in this mesh to represent the tailing deposits and the underlying subsurface soils. The material layering used in the modelling is shown in Figure 3.

Using the material properties given in Table 1, the numerical modelling was performed in two stages:

(1) In the first stage, the current site conditions were simulated (to obtain in situ stress conditions) by assigning artificially high shear values to all materials.

(2) In the second stage, shear strengths were reduced to their realistic values (as given in Table 1) to simulate failure condition and to obtain displacement patterns due to the plastic deformation of the dike. The dike plastic deformations were simulated using the Mohr-Coulomb failure criterion, non-associated flow rule, and adjustable mesh concept (large deformations).

Figure 4 shows the plastic displacement pattern within the dike system at an intermediate stage immediately after the onset of dike failure [i.e. sliding along a critical slip surface(s)] ; the maximum displacement vector at this stage is approximately 11 feet. The failure mechanism is shown in Figure 5, indicating a slip pattern similar to that identified during the PC-SLOPE analysis. As shown, a continuous zone of yielded materials passes through the crest and along the peat/clayey material, causing the dike failure.

To further investigate the consequence of the dike failure, the numerical modelling was continued by allowing the unstable soil mass to move further down the slope along the slip surface. Figure 6 shows the post-failure (final) geometry of the dike. As it is evident, the failure has resulted in a limited loss of free board, and some bulging at the toe of the dike. According to Figure 7 the maximum displacement vector due to failure would be on the order of 24 feet. This indicated that the failure would not be catastrophic, and that the dike would attain a new stable geometry after a

Table 1. Material properties

Soil Categories	Unit weight (pcf)	Cohesion (psf)	Friction angle (degrees)	Elastic modulus (ksf)	Poisson's ratio
Coarse tailings	120	0	44	600	0.25
Fine tailings	125	60	8.5	220	0.30
Granular native soils	125	0	38	550	0.25
Fine-grained native soils	118	800	0	160	0.49
Peat and Organics	70	200	5	20	0.49

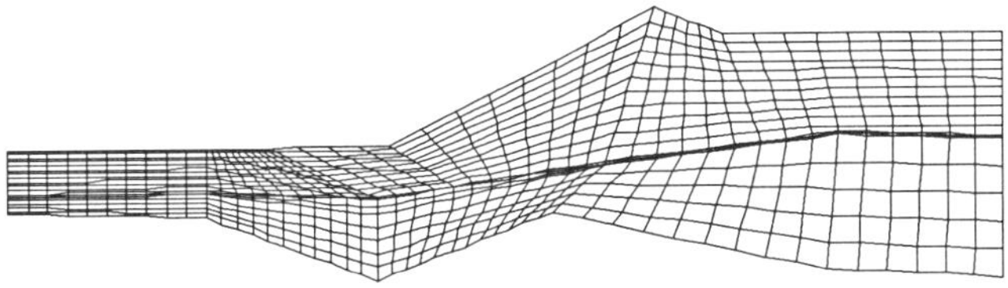

Fig.2 Finite difference mesh used to model critical cross-section.

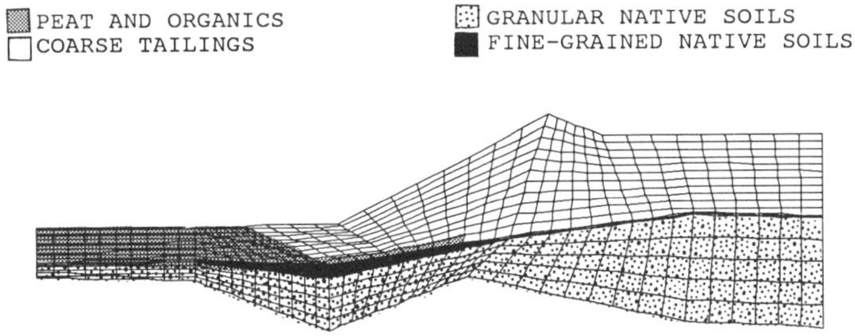

Fig.3 Material layering used in numerical modelling.

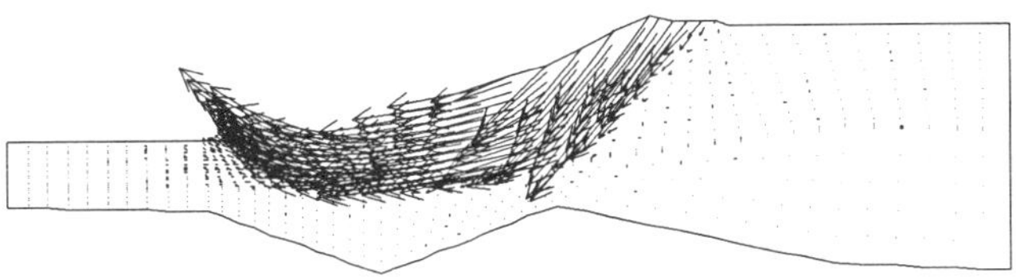

Fig.4 Displacement pattern during failure (maximum displacement vector = 11.3 feet).

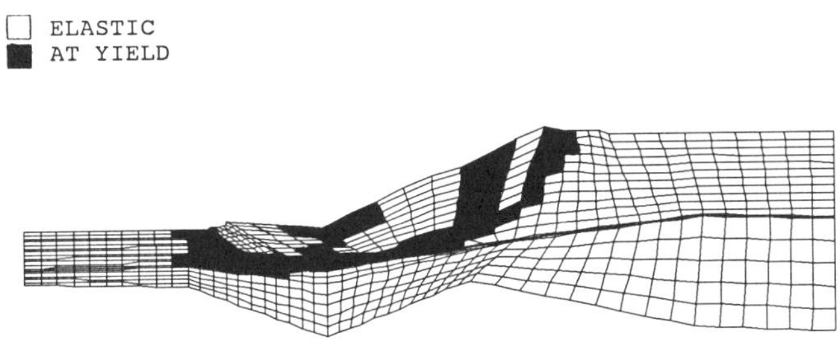

Fig.5 Zones at yield during failure.

limited plastic deformation, i.e. a failure in the downstream shell of the perimeter dike would not result in the progressive failure of the entire perimeter dike. The results also indicated that the stability of the perimeter dike would be independent of the geometry and/or material property of the relatively impervious core or the fine tailings behind the downstream dike shell.

In summary, based on the results of the consequence-of-sliding analysis performed, it was concluded that even under the worst case scenario a failure in the perimeter dike system would not be catastrophic, and would not compromise the overall

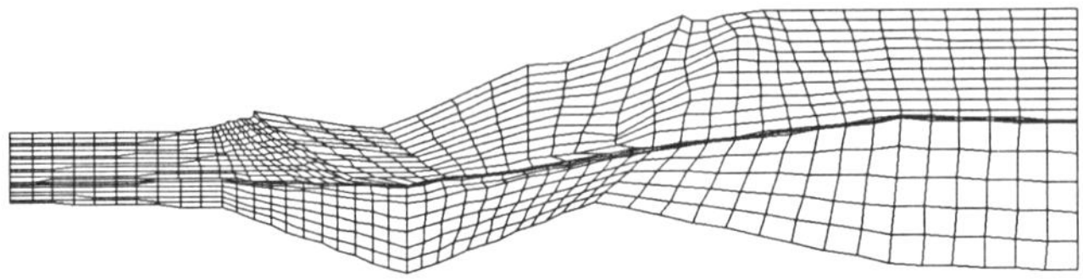

Fig.6 Post-failure (final) mesh geometry.

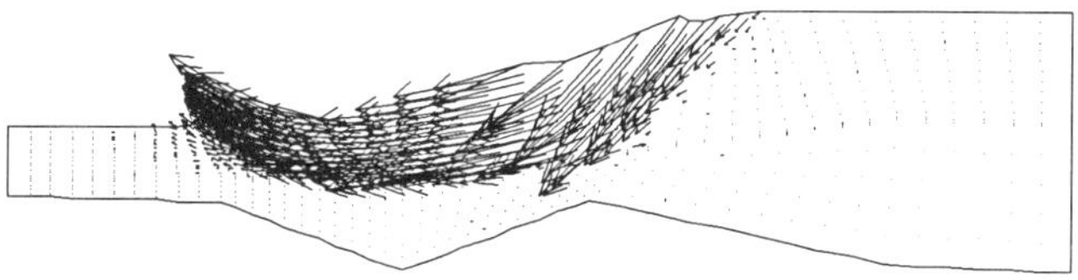

Fig.7 Displacement pattern after failure (maximum displacement vector = 23.8 feet).

structural integrity of the tailings pond, rendering a comprehensive program of field investigation and/or remedial measures unnecessary.

7 CONCLUSIONS

Both limit equilibrium and consequence-of-sliding methods of analysis were used to investigate perimeter dike stability of a tailings pond. The results of the limit equilibrium analyses indicated that depending on the continuity and the lateral extent of the peat and clay layers encountered beneath the perimeter dike, the downstream shell of the perimeter dike may be unstable or marginally stable. However, the results of consequence-of-sliding analysis (performed assuming the worst case scenario) indicated that:

1. A failure occurring along the most critical section would not be catastrophic and would attain a stable geometry after a limited down-slope movement.
2. A failure in the downstream shell of the perimeter dike would not result in the progressive failure of the entire dike cross-section.
3. The stability of the perimeter dike is independent of the geometry and material properties of the fine tailings core.

CONVERSION TO SI UNITS

ft = 304.8 mm
pcf = 160.2 N/m^3
psf = 48.8 N/m^2
ksf = 48.8 kN/m^2

REFERENCES

Bishop, A.W. 1955. The use of slip circle in the stability analysis of slopes. Geotechnique, 5(1), 7-17.

Chang, C.S. 1992. Discrete element method for slope stability analysis. J. Geotech. Engrg., ASCE, 118(12), 1889-1905.

Duncan, J.M., and Wright, S.G. 1980. The accuracy of equilibrium methods of slope stability analysis. Engrg. Geol., Amsterdam, the Netherlands, 16(1/2), 5-7.

Fellinius, W. 1936. Calculation of the stability of earth dams. Proc. Second Congress On large Dams, 4, 445-463.

Janbu, N. 1954. Application of composite slip surfaces for stability analysis. European Conf. on Stability of earth slopes, 3, 43-49.

Janbu, N. 1957. Earth pressure and bearing capacity by generalized procedure of slices. Proc. Fourth Int. Conf. Soil Mechanics, 2, 207-212.

Morgenstern, N.R. 1982. The analysis of wall supports to stabilize slopes. Application of walls to landslide control problems, R.B. Reeves, ed., ASCE national convention, Las Vegas, Nevada, April 29, 1982. 19-29.

Morgenstern, N.R., and Price, V.E. 1965. The analysis of the stability of general slip surfaces. Geotechnique, 15(1), 79-93.

Nash, D. T. F. 1987. A comprehensive review of limit equilibrium methods of slope stability analysis. Slope stability, M.G. Anderson and K.S. Richards, ed., John Wiley, New York, N.Y., 11-75

Spencer, E. 1967. A method of analysis of the stability of embankments assuming parallel interstice forces. Geotechnique, 17(1), 11-26

Wright, S.G., Kulhawy, F.H., and Duncan, J.M. 1973. Accuracy of equilibrium slope stability analysis. J. Soil Mech. Found. Div., ASCE, 99(10), 783-791.

Impact of Computational Mechanics on Engineering Problems, Pulmano & Murti (eds)
© 1993 Balkema, Rotterdam, ISBN 90 5410 324 8

Recent advances on numerical analysis of interaction problems: The Australian scene

Y.-C. Loo
University of Wollongong, N.S.W., Australia

ABSTRACT: This paper is an abridged version of a keynote paper prepared for the Chinese National Conference on Interaction Analysis to be held in Central China in June 1993. It reviews a total of over 75 technical papers and research reports on numerical analysis of interaction problems. Published by researchers who have worked and/or are working in Australia, the publications span the period mainly from 1985 to 1992. While this review is nowhere near exhaustive, it is believed that it gives a good indication of the many advances made by Australian researchers in the subject area.

INTRODUCTION

Interaction problems are encountered in many disciplines in the broad field of engineering and applied science. Thus, interaction analysis is an important area of research by academics and engineering practitioners alike worldwide.

During the past decade, researchers in Australia have made considerable advances in formulating and solving interaction problems using numerical methods. These advances cover topics in structural, mechanical, geotechnical and water engineering. This paper attempts to present a summary of these achievements. In all, over 75 papers and reports on interaction analysis published recently by researchers who have worked and/or are working in Australia have been reviewed. A brief description is made of each of these papers in the main text. This is followed by a bibliography: listed in alphabetical order of the first-named authors, these papers and research reports were published mainly between 1985 and 1992.

The main purpose of this report is to acquaint the reader with this considerable body of research work. These publications are classified and described herein under the following headings:

- Stability and large-deformation analyses of structures
- Impact
- Contact problems
- Fluid-structure interaction
- Flow of bulk solids through silos
- Soil-pile interaction
- Soil-plug analysis
- Raft foundations and shallow footings
- Soil-superstructure interaction and excavation problems
- Half-space simulation of soil mass
- Ground improvements

To cover a large collection of papers in a restricted number of pages, the description of each paper, by necessity, has to be brief. Further, it should be emphasised that this review is by no means exhaustive in coverage. Nevertheless, it should provide the reader with a good glimpse of the recent achievements made by Australian researchers in the area of numerical analysis of interaction problems.

STABILITY AND LARGE DEFORMATION ANALYSES OF STRUCTURES

Buckling, post-buckling and large-deformation analyses of structures and structural members involve the interaction of actions and deformations. This is still an important area for research. A finite element procedure was presented by Lu (1990) for tracing the inelastic pre- and post-buckling load-deformation path of tubular struts. The analysis can account for both geometric and material nonlinearities as well as the strain-unloading and strain-hardening effects. Murray (1991) studied the buckling of a mild steel pin-ended column of rectangular cross-section into the large-deflection range. Using the classical approach and taking into consideration the strain-hardening characteristics of mild steel, he analysed a range of cross-sectional stress conditions from elastic to strain hardening in both the compression and tension fibres. In the meantime, Wang,

Kitipornchai and Al-Bermani (1991) incorporated the effects of axial shortening on the elastic buckling and free vibration analyses of columns.

Al-Bermani and Kitipornchai (1990a, 1990b) presented a finite element procedure for the elasto-plastic, large deformation analysis of thin-walled structures. Their work was applied by Chan, Kitipornchai and Al-Bermani (1991) to cover box-beam-columns including local buckling of the constituent walls. Chin, Al-Bermani and Kitipornchai (1992) extended this procedure to analyse the stability of thin-walled beam-columns having a flexible web and arbitrary-shaped rigid flanges. Recently, this work was further developed by Al-Bermani and Kitipornchai for the analysis of space frames with flexible joints (1992a) as well as for predicting the ultimate behaviour of self-supporting transmission towers under static load conditions (1992b). Rasmussen and Hancock (1991), on the other hand, developed an elasto-plastic analysis of buckling and post-buckling behaviour of channel-section columns where local buckling may occur in the web. The analysis is also capable of predicting the ultimate behaviour of such structural components.

Hancock, Davids et al (1990) published a review paper in which they referred to over 40 papers, reports and books published mainly by Australian researchers from 1973 to 1989. The reader may refer to this paper for details of the earlier works on stability and large-deformation analyses.

When a steel beam-column which is bent about its major axis is inadequately braced, it may fail by deflecting laterally out of the plane of bending and twisting. This constitutes the so-called flexural-torsional buckling of the structural member. For beam-columns of intermediate slenderness, the in-plane actions cause yielding which reduces the resistance to lateral buckling. Cuk, Bradford and Trahair (1986) described a series of theoretical and experimental investigations of the lateral buckling of inelastic steel beam-columns. Bradford, Cuk et al (1987) employed the finite element method to analyse this flexural-torsional type of inelastic buckling of I-section beams and beam-columns. In the analysis, they modelled the three-dimensional I-member using a line element which is amenable to in-plane bending, lateral bending and torsional analyses. This line-modelling technique is considered comparatively efficient. Currently, an authoritative and comprehensive book is being finalised by Trahair (1993). Covering most aspects of the elastic and inelastic buckling problem, the book places special emphasis on the lateral or flexural-torsional buckling of steel structures.

IMPACT

Impact and energy absorption are important topics of research which have a wide range of applications including crash worthiness of steel structures and crack propagation of reinforced concrete beams.

Grzebieta and Murray (1985) investigated the energy-absorbing behaviour of a strut with initial deflection loaded statically. They (1986) extended their work to impact energy absorption of the strut with imperfection and later to include the design of crash barriers with road engineering applications (see Murray and Grzebieta, 1987). Under the direction of Murray (1992), similar research is continuing with special emphasis on automotive engineering.

In analysing the impact deflection of cracked reinforced and prestressed concrete beams, Hertz's contact law and vibration theory were adopted by Loo and Santos (1986). These led to an integral equation for the force history, the numerical solution of which produces eventually the deflection response of the beam. The laboratory verification of this numerical procedure has been conducted by Chau (1991) and Sari (1992)

CONTACT PROBLEMS

The theory of contact of elastic bodies was first developed by Hertz (1881). This has since served as the basis for the study of many contact problems in mechanical engineering. Thomas and Daniel (1990a, 1990b) presented a comprehensive review of dry unlubricated contact problems under static or rolling conditions. In their papers, they summarised the discussions on the subject related matters published in a total of 35 papers dating from 1881 to 1987.

In an attempt to study the fatigue life of rails, Hellier, Corderoy and Lakeland (1990) used a finite element procedure to model the wheel/rail contact conditions typical of those encountered in heavy-haul railway lines.

The interaction of a loaded solid body with a rigid surface was analysed by Kodikara (1991). A geometrically nonlinear finite element approach was employed in which a special procedure was incorporated to handle the rigid body movements of the solid at the solid and rigid boundary interface.

FLUID-STRUCTURE INTERACTION

An underwater explosion creates a huge air bubble. Amongst other researchers, Soh (1992a) studied the dynamics of such bubbles. In an earlier investigation, Blake and Gibson (1983) found that for a bubble near a rigid wall the momentum generated by the bubble is directed towards the wall. The bubble expands and collapses causing cavitation damages to the wall. The situation reverses for a free surface in which the momentum is directed away from such a surface. Under certain circumstances, a flexible (or compliant) surface behaves like a free surface (Gibson and Blake, 1980). In a recent paper, Soh (1992b) presented an energy approach for modelling the behaviour of cavitation bubbles near compliant surfaces. He confirmed the earlier findings of Gibson and Blake (1982) that the dynamics of a bubble near a compliant surface was dependent on the dynamic response of the surface. Some laboratory observations of underwater explosions using high speed photography have been made by Yu and Soh (1992).

A fluid element using displacements as nodal variables has been presented by Yeo (1991). A technique for applying this element was demonstrated by solving fluid-structure interaction problems. The application of this type of elements to the analysis of the transmission of pressure waves in elastic rods, tubes and fluids was earlier presented by Yeo and Schmid (1989).

FLOW OF BULK SOLIDS THROUGH SILOS

For the primary industry of a nation, silos for storing bulk solids like grains and ores form a vital part of the infrastructure. Thus, the safe and economical construction of this special class of structures is of great interest to researchers.

A finite element approach with velocity as the primary variable has been developed by Wu (1990) to model the flow of bulk solids through silos. In this study, the interaction of the flowing materials and the silo wall was coupled by invoking the contact condition of the two systems. This work in summary form has also been published (see Schmidt and Wu, 1991 and Wu and Schmidt, 1992).

Other recent work in this area includes that of Zhang (1991) who investigated the buckling of steel silos caused by the interaction forces from the stored bulk solids. On the other hand, Walters and McLean (1992) employed the classical Janssen equation to model the stresses during discharge in a silo fitted with an "anti-dynamic tube".

SOIL-PILE INTERACTION

Lee and Poulos (1990) presented a finite element procedure for the analysis of the axial response of piles embedded in multi-layered soils. The simplified form of boundary element analysis developed by Poulos and Davis (1980) was adopted in which the pile was modelled as an elastic cylinder and the surrounding soil mass as an elastic continuum. The analysis was extended to cover a group of two identical and equally loaded piles. In the determination of the group settlement interaction between piles embedded in multi-layered soils, an additional simple soil mass stiffness model was developed.

Chin and Poulos (1991) proposed a load-transfer approach for the quasi-static cyclic response analysis of axially loaded vertical single piles embedded in either homogeneous, two-layered or "Gibson" soil profiles. The soil medium along the embedded pile was represented by simple "t-z" curves which defined the shear stress-vertical displacement response of the soil at each particular depth. Under cyclic loading conditions, the well-known Masing's criteria governing the unloading and reloading responses were incorporated into the "t-z" curves.

A simplified method was presented by Poulos (1991) for analysing the behaviour of a piled strip foundation subjected to concentrated vertical and moment loading, distributed loading, and externally-imposed soil movements. A boundary element formulation was used in which the strip was modelled as a simple beam and the piles were simulated by spring elements. Interaction through the soil among the piles, and between the strip and the piles, were considered by treating the soil as an elastic continuum.

A finite element analysis of pile-raft foundation systems has been developed by Griffiths, Clancy and Randolph (1991). The piles were modelled by rod elements and the raft by 'thin' plate elements. The pile-soil-pile, pile-soil-raft and raft-soil-raft interactions were accounted for using the elastic theory of Mindlin. Incorporating field test data, Indraratna, Balasubramaniam et al (1992) developed a model for soil-pile interaction mechanisms using thin interface finite elements. The method was successfully used for the analysis of negative skin-friction developed at soil-pile interface in a soft compressible clay.

Piled bridge abutments on soft clay may be loaded laterally from horizontal soil movements generated by the approach embankment. Various methods of analysis hitherto published showed poor comparison with available experimental data. Stewart, Jewell and Randolph (1992a) described a finite element approach for analysing the response of pile groups influenced by lateral soil movements from a nearby surface load. The piles were represented by equivalent sheet-pile walls, and a plane strain analysis was performed. Results from this method were found to be in good agreement under some conditions with experimental data obtained by the same researchers (1992b).

Deeks and Randolph (1992) presented an analysis model of hammer impact which was based on lumped ram and anvil masses separated by a cushion with internal damping and connected to the pile. In this analysis, the pile was modelled as a dashpot. This method of analysis was used to explore the characteristics of hammer impact.

SOIL PLUG ANALYSIS

Open-ended steel pipes are commonly used for driven piles, particularly for offshore foundations. In rare cases, such as driving through dense sand layers, the pile may 'plug' during driving, with the soil plug becoming locked in the pile, and further penetration occurring as if the pile were closed ended. In such situations, the driving resistance increases sharply.

Leong and Randolph (1991) presented a finite element based method for assessing the end bearing capacity of the soil plug under loading conditions which ranged from undrained to fully drained. The soil plug was modelled as being either elastic, elastic-perfectly-plastic or elastoplastic and the soil-pile interface was simulated using elastic-perfectly plastic joint elements. Randolph, Leong and Houlsby (1990) also developed a simple one-dimensional method of analysis for the soil plug under partially drained condition. Murff, Raines and Randolph (1990), on the other hand, studied the soil plug behaviour of piles in sand.

RAFT FOUNDATIONS AND SHALLOW FOOTINGS

A finite layer method was proposed by Zhang and Small (1991) which could be used for the analysis of rafts constructed on soils having variable elastic properties. The proposed method is much more efficient than those which incorporate a full three dimensional finite element analysis of the soil. This is because it reduces the soil mass analysis to a two dimensional one. The finite layer approach was also applied by the same researchers (1992) to rectangular rafts subjected to concentrated loads. Similar to their earlier paper (1991), the raft was simulated by rectangular elements. In a paper to be published, Zhang and Small (1994) have further extended their method to the analysis of soil-raft-structure interaction.

The details of footing design, the soil stiffness and the soil resistance all affect the rotation characteristics of column bases of structural frames. A set of experimental results for shallow footing behaviour under load reversal was presented by Melchers (1992). Based on this work and on previous modelling attempts, he proposed a relatively simple theoretical model for predicting the rotation behaviour of simple pad footings under the "first loading" condition (ie. excluding hysteresis and unloading).

SOIL-SUPERSTRUCTURE INTERACTION AND EXCAVATION PROBLEMS

Progressively increased loading is inevitably experienced by the structural foundation system during construction. The effects of progressive loading on the interaction of framed structures and the foundation has been investigated by Brown and Yu (1986). They found that, for interaction analysis, the effective stiffness of a building loaded progressively during construction was about half the stiffness of the completed building.

Xu and Kwok (1991) described a theoretical investigation of the wind-induced motion of slender structures equipped with tuned mass dampers, taking into account the effect of soil compliancy under the footing. They used a transfer matrix formulation to analyse the soil-structure-mass damper interaction in the frequency domain.

For the dynamic analysis of a soil-structure system, Valliappan and Zhao (1991b) adopted the substructure and condensation techniques. In the analysis, the superstructure was modelled as a beam or a bar element, while the raft foundation was represented by the combination of thick plate and plane stress elements. The coupled model of finite and infinite elements developed by Zhao and Valliappan (1991), on the other hand, was used to simulate the behaviour of the soil mass. This method has the advantage that the natural foundation and the earthquake input mechanism can be modelled more realistically by the coupled element model and the related wave input procedure in the infinite medium. The wave input procedure has been detailed in a separate paper by Valliappan and Zhao (1991a).

An iterative process has been developed by Lai and Booker (1991) for solving elasto-plastic soil-structure interaction problems. Employing a hybrid 'residual force' method they considered the soil and the structure as separate bodies which were related only by compatibility of displacements and equilibrium of forces at the soil-structure interface. This scheme has enabled a significant improvement in numerical stability and rate of convergence over the conventional initial stress method.

In urban environments, temporary retaining wall systems are commonly used to maintain the stability of deep excavations. For such excavations in soft clay, bracing systems are normally used because of the low strengths of the soil and generally high water table. Conventional methods of predicting the basal stability of braced excavations were unable to take into consideration the stiffness of the retaining wall and the depth of penetration of the wall below the bottom of the excavation. A simple and improved procedure has been developed by Goh (1990) for predicting the stability of strutted excavations using the finite element method.

HALF-SPACE SIMULATION OF SOIL MASS

In an elastic analysis of soil-structure interaction, the soil mass may be assumed to be a full or half-space depending on the location of the structure concerned with respect to the ground level. Thambiratnam (1991) has developed a simplified method for the dynamic response analysis of rigid underground structures subjected to P-waves. This was done by discretising the soil-structure interface into a finite number of strips resulting in an analysis which involved only plane surfaces and plane waves.

Using a coupled model of three-dimensional finite and infinite elements, Zhao and Valliappan (1991) were able to simulate the near and far fields of a half-space which took into consideration the effects of radiation and material damping. They demonstrated the application of this coupled element model by analysing the dynamic response of a raft foundation on a viscoelastic half-space. A new wave input procedure for the finite element system was discussed by the same authors (Valliappan and Zhao, 1991a).

A finite element analysis of dynamic problems in full or half-space requires the use of transmitting boundaries to model the radiation of waves from the finite element mesh. The various traditional time-domain transmitting boundaries are unable to support a static load. Deeks and Randolph (1990) has presented an alternative time-domain transmitting boundary for analysing axisymmetric problems which is capable of supporting a static load. The boundary does not lead to unrealistic wave reflection; it is simple to implement and gives accurate results.

By modelling the soil bed using either Winkler springs or elastic half-space, Melerski analysed a number of soil-superstructure interaction problems. These include circular rafts (1990) and cylindrical storage tanks (1991) under axisymmetrical loading. The work was recently (1992a) extended to include the analysis of an elastic beam founded on a smooth, isotropic half-space. In each of these analyses, Melerski basically employed the displacement procedure of structural analysis in which the stiffness matrices of the superstructure and the foundation bed were combined to produce the stiffness matrix of the whole system. The solution of the overall stiffness equation led eventually to the evaluation of the internal actions in the superstructure as well as the contact pressure at the soil-structure interface.

Parallel to the above mentioned deterministic analyses, Melerski, by taking account of the randomness of soil subgrade modulus (1988), subgrade modulus and elastic properties of the superstructure (1992b, 1992c), carried out probabilistic studies of the various soil superstructure interaction problems.

GROUND IMPROVEMENTS

Based on the concepts of elasto-plasticity, a theory was developed by Indraratna and Kaiser (1990a) to consider the proper interaction mechanism between circular opening of a rock mass (underground) and the grouted, or friction, bolts. In this paper, they presented the derivation of the mathematical model and also illustrated the effect of bolts on the stress and displacement field near such an opening. The laboratory verification of their theory was presented in a separate paper by the researchers (1990b) in which the analysis was also carried out on a case history in Japan. A parallel study on reinforcement of planer discontinuities in (elastic) rock mass openings by grouted bolts was conducted by Indraratna and Nutalaya (1992).

A vast country like Australia which relies heavily on road transport, the proper design of road pavements has great economical implications. Wallace and Sapkota (1992) studied the interaction of swelling sub-grade and road pavement. They used a new mechanistic model of swell and swell suppression to study the behaviour of an expansive clay subgrade.

The analysis and design of reinforced earth, for a long while, have been based solely on semi-empirical formulas. In recent years, advances have been made to incorporate numerical methods to improve the accuracy and reliability of the calculations. In this regard, Fidler and Wallace (1992) have employed the well-established finite difference method in the design process and Arenicz and Chowdhury (1990) discussed the reliability analysis of amongst other things the location of the internal failure surfaces.

CONCLUDING REMARKS

In all, over 75 publications are herein surveyed and their contents, briefly discussed. Published by Australian researchers, mainly from 1985 to 1992, they deal with numerical analysis of the various interaction problems encountered in engineering and applied science. The quality and quantity of these Australian works can only be described as substantial. It is of little doubt that the 1990s will witness researchers in Australia making further advances in this important area of research.

ACKNOWLEDGMENTS

The author is grateful to his Australian colleagues for their enthusiastic response to the requests for the latest publications on the subject matter.

For obvious reasons, it is not easy to collect, in a short period of time, a large number of the latest publications from a variety of sources. The author regrets any omission of the works published by some Australian researchers. Further, he apologises for any misrepresentation of the contents of the publications he may have made. It should also be mentioned that because of the limitations stipulated by the title of the paper, some very worthwhile publications received by the author have regrettably been excluded.

Sanaul Chowdhury, a Research Assistant with the Structural Engineering and Construction Research Group at the University of Wollongong, is fondly remembered for his efforts in organising the bibliography.

BIBLIOGRAPHY

Al-Bermani, F.G.A. & S. Kitipornchai 1990a. Elasto-plastic large deformation analysis of thin-walled structures. *Engineering Structures*, Butterworth-Heinemann, Vol. 12, Jan, p. 28-36.

Al-Bermani, F.G.A. & S. Kitipornchai 1990b. Nonlinear analysis of thin-walled structures using least element/member. *J. Struct. Eng,* ASCE, Vol. 116, No. 1, Jan, p. 215-234.

Al-Bermani, F.G.A. & S. Kitipornchai 1992a. Elastoplastic nonlinear analysis of flexibly jointed space frames. *J. Struct. Eng.,* ASCE, Vol. 118, No. 1, Jan, p. 108-127.

Al-Bermani, F.G.A. & S. Kitipornchai 1992b. Nonlinear analysis of transmission towers. *Engineering Structures,* Vol. 14, No. 3, p. 139-151.

Arenicz, R.M. & R.N. Chowdhury 1990. Analysis of failure surfaces and their design implications in reinforced earth walls. *Procs, 10th Southeast Asian Geotech. Conf.,* Taipei, Taiwan, Apr. 16-20, p. 9-16.

Blake, J.R. & D.C. Gibson 1983. Cavitation bubbles near boundaries. *Annu. Rev. of Fluid Mech.,* Vol. 19, p. 99-123.

Bradford, M.A., Cuk, P.E., Glzejowski, M.A. & N.S. Trahair 1987. Inelastic lateral buckling of beam-columns. *J. Struct. Eng,* ASCE, Vol. 113, No. 11, Nov, p. 2259-2277.

Brown, P.T. & S.K.R. Yu 1986. Load sequence and structure-foundation interaction. *J. Struct. Eng,* ASCE,Vol. 112, No. 3, Mar, p. 481-488.

Chan, S.L., Kitipornchai, S. & F.G.A. Al-Bermani 1991. Elasto-plastic analysis of Box-Beam-Columns including local buckling effects. *J. Struct. Eng,* ASCE, Vol. 117, No. 7, July, p.1946-1962.

Chau, T.V. 1991. *Serviceability and reaction of partially prestressed concrete box beams under impact loading.* BE Thesis, Deptt. Civil & Mining Eng, Univ. of Wollongong, NSW, Australia.

Chin, C.K., Al-Bermani, F.G.A. & S. Kitipornchai 1992. Stability of thin-walled members having arbitrary flange shape and flexible web. *Engineering Structures,* Vol. 14, No. 2, p. 121-132.

Chin, J.T. & H.G. Poulos 1991. A "T-Z" approach for cyclic axial loading analysis of single piles. *Computers and Geotechnics,* Elsevier Science, Dec, p. 289-320.

Cuk, P.E., Bradford, M.A. & N.S. Trahair 1986. Inelastic lateral buckling of steel beam-columns. *Canadian J. Civil Eng,* Vol. 13, No. 6, Dec, p. 693-699.

Deeks, A.J. & M.F. Randolph 1990. An axisymmetric time-domain transmitting boundary. *Res. Report No. G1006,* Geomechanics Group, Univ. of Western Australia, Feb, p. 1-21.

Deeks, A.J. & M.F. Randolph 1992. Analytical modelling of hammer impact for pile driving. *Res. Report No. G1033,* Geomechanics Group, Univ. of Western Australia, Mar, p. 1-24.

Fidler, S.R. & K.B. Wallace 1992. Strain compatibility and design criteria for Reinforced Earth. *Procs, 6th Australia-New Zealand Geomech. Conf.,* Christchurch, New Zealand, Feb, p. 1-6.

Gibson, D.C. & J.R. Blake 1980. Growth and collapse of cavitation bubbles near flexible boundaries. *Procs, 7th Hydraulics and Fluid Mechs. Conf.,* Brisbane, Australia, August, Institution of Engineers Australia, p. 283-286.

Gibson, D.C. & J.R. Blake 1982. The growth and collapse of bubbles near deformable surfaces. *Applied Scientific Res.,* Vol. 38, p. 215-224.

Goh, A.T.C. 1990. Assessment of basal stability for braced excavation systems using the finite element method. *Computers and Geotechnics,* p. 325-338.

Griffiths, D.V., Clancy, P. & M.F. Randolph 1991. Piled raft foundation analysis by finite elements. *Procs, 7th Int'l. Conf. on Computer Methods and Advances in Geomechanics,* Cairns, Australia, May 6-10, Balkema, Rotterdam, p. 1153-1157.

Grzebieta, R.H. & N.W. Murray 1985. The static behaviour of struts with initial kinks at their centre point. *Int'l. J. Impact Eng.,* Pergamon Press, Vol. 3, No. 3, p. 155-165.

Grzebieta, R.H. & N.W. Murray 1986. Energy absorption of an initially imperfect strut subjected to an impact load. *Int'l. J. Impact Eng.,* Vol. 4, No. 3, p. 147-159.

Hancock, G.J., Davids, A.J., Key, P.W., Lau, S.C.W. & K.J.R. Rasmussen 1990. Recent developments in the buckling and nonlinear analysis of thin-walled structural members. *Thin-Walled Structures,* Elsevier Science, p. 309-338.

Hellier, A.K., Corderoy, D.J.H. & D. Lakeland 1990. A contact mechanics study of shelling under rolling contact. *Procs, Tribology Conference,* Brisbane, Dec. 3-5, Institution of Engineers Australia, p. 64-69.

Hertz, H. 1881. On the contact of elastic solids. *J. fur die reine und angewandte Mathematik,* Vol. 92, p. 156-171.

Indraratna, B., Balasubramaniam, A., Phamvan, P. & Y.K. Wong 1992. Development of negative skin friction on driven piles in soft clay.*Canadian Geotech. J.,* Vol. 29, Jun, p. 393-404.

Indraratna, B. & P.K. Kaiser 1990a. Analytical model for the design of grouted rock bolts. *Int'l. J. Numerical and Analytical Methods in Geomechs,* Vol. 14, p. 227-251.

Indraratna, B. & P.K. Kaiser 1990b. Design for grouted rock bolts based on the convergence control method. *Int'l. J. Rock Mechs, Mining Sciencs & Geomechs,* Vol. 27, No. 4, Pergamon Press, p. 269-281.

Indraratna, B. & P. Nutalaya 1992. Reinforcement of planar discontinuities in elastic rock. *Geotech. & Geolog. Eng,* Chapman & Hall, Vol. 10, p. 135-140.

Kodikara, J.K. 1991. Rigorous nonlinear interaction analysis of solids with rigid surfaces. *Procs, 6th Int'l. Conf. on Finite Element Methods,* Univ. of Sydney, Sydney, Australia, Jul. 8-10, p. 415-420.

Lai, J.Y. & J.R. Booker 1991. A residual force finite element approach to soil-structure interaction analysis. *Int'l. J. for Numerical and*

Analytical Methods in Geomech., Wiley & Sons, Vol. 15, p. 181-203.
Lee, C.Y. & H.G. Poulos 1990. Axial response analysis of piles in vertically and horizontally non-homogeneous soils. *Computers and Geotechnics,* Sept, p. 133-148.
Leong, E.C. & M.F. Randolph 1991. Finite element analyses of soil plug response. *Int'l. J. for Numerical and Analytical Methods in Geomech,* Vol. 15, p. 121-141.
Loo, Y.C. & Santos 1986. Impact deflection analysis of concrete beams. *J. Struct. Eng.,* ASCE, Vol. 112, No. 6, Jun, p. 1297-1312.
Lu, J.P. 1990. *Influence of strain hardening, srain aging and the Bauschinger effect on steel tubular strut load capacity.* Ph.D Thesis, Deptt. of Civil & Mining Eng, Univ. of Wollongong, NSW, Australia.
Melchers, R.E. 1992. Rotational stiffness of shallow footings. *Computers and Geotechnics,* p. 21-35.
Melerski, E.S. 1988. Thin shell foundation resting on stochastic soil. *J. Struct. Eng.,* ASCE, Vol. 114, No. 12, Dec., p. 2692-2709.
Melerski, E.S. 1990. Simple computer analysis of circular rafts under various axisymmetric loading and elastic foundation conditions. *Procs., Institution of Civil Engineers,* Vol. 89, Part 2, Sept., p. 407-431.
Melerski, E.S. 1991. Simple elastic analysis of axisymmetric cylindrical storage tanks. *J. Struct. Eng.,* ASCE, Vol. 117, No. 11, Nov., p. 3239-3260.
Melerski, E.S. 1992a. Elastostatic numerical analysis of beams resting on Half-space I : Deterministic studies. *Procs., Asia Pacific Conf. on Struct. Eng. & Construction,* Johor Bahru, Malaysia, Sept., p. 6-1 - 6-15.
Melerski, E.S. 1992b. Probabilistic computer analysis of circular rafts by the method of realizations. *Computers & Structures,* Pergamon Press, Vol. 43, No. 6, p. 1075-1083.
Melerski, E.S. 1992c. Elastostatic numerical analysis of beams resting on Half-Space II: Probabilistic studies. *Procs., Asia Pacific Conf. on Struct. Eng. & Construction,* Johor Bahru, Malaysia, Sept., p. 6-17 - 6-31.
Murff, J.D., Raines, R.D. & M.F. Randolph 1990. Soil plug behaviour of piles in sand. *Procs., 22nd Annual Offshore Technology Conf.,* Houston, USA, May 7-10, OTC 6421, p. 25-32.
Murray, N.W. 1991. The buckling of a pin-ended column of rectangular cross-section and strain hardening mild steel into the large-deflection range. *Thin-Walled Structures,* p. 85-114.
Murray, N.W. 1992. *Private Communication.*
Murray, N.W. & R.H. Grzebieta 1987. Some structural aspects in the design of crash barriers. *Procs., 1st National Struct. Eng. Conf.,* Melbourne, Australia, Aug. 26-28, p. 397-400.
Poulos, H.G. 1991. Analysis of piled strip foundations. *Proces., 7th Int'l. Conf. on Computer Methods and Advances in Geomech.,* Cairns, Australia, May 6-10, p. 183-191.
Poulos, H.G. & E.H. DAVIS 1980. *Pile foundation analysis and design.* John Wiley and Sons, New York.
Randolph, M.F., Leong, E.C. & G.T. Houlsby 1990. One dimensional analysis of soil plugs in pipe piles. *Res. Report No. G1005,* Geomechanics Group, Univ. of Western Australia, Jul, p. 1-18.
Rasmussen, K.J.R. & G.J. Hancock 1991. Nonlinear analyses of thin-walled channel section columns. *Thin-Walled Structures,* p. 145-176.
Sari, G. 1992. *Serviceability of reinforced concrete box beams under impact loading.* BE Thesis, Deptt. of Civil & Mining Eng., Univ. of Wollongong, NSW, Australia.
Schmidt, L.C. & Y.H. Wu 1991. Finite element modelling of the interaction of flowing granular materials and silo structures. *Procs., 6th Int'l. Conf. on Finite Element Methods,* Univ. of Sydney, Sydney, Australia, Jul. 8-10, p. 735-739.
Soh, W.K. 1992a. Dynamics of an underwater explosion bubble. *Procs., 5th Asian Congress of Fluid Mech.,* Taejon, Korea, Aug. 10-14, p. 1107-1110.
Soh, W.K. 1992b. An energy approach to cavitation bubbles near compliant surfaces.*Applied Mathematical Modelling,* Butterworth-Heinemann, Vol. 16, May, p. 263-268.
Stewart, D.P., Jewell, R.J. & M.F. Randolph 1992a. Piled bridge abutments on soft ground: II - Numerical Modelling. *Res. Report No. G1055,* Geomechanics Group, Univ. of Western Australia, p. 1-17.
Stewart, D.P., Jewell, R.J. & M.F. Randolph 1992b. Piled bridge abutments on soft ground : I - Physical Modelling. *Res. Report No. G1054,* Geomechanics Group, Univ. of Western Australia, p. 1-21.
Thambiratnam, D.P. 1991. Dynamic response of underground structures to pressure waves : the strip approximation. *J. Sound and Vibration,* Academic Press, Vol. 145, No. 3, p. 349-363.
Thomas, W.M. & W.J.T. Daniel 1990a. A review of unlubricated contact, Part I - Static contact. *Procs., Tribology Conf.,* Brisbane, Inst. of Engrs. Australia, Dec. 3-5, p. 156-160.
Thomas, W.M. & W.J.T. Daniel 1990b. A review of unlubricated contact, Part II - Rolling contact. *Procs., Tribology Conf.,* Brisbane, Inst. of Engrs. Australia, Dec. 3-5, p. 161-165.
Trahair, N.S. 1993. *Flexural-Torsional Buckling of Structures.* E & F N Spon, London.
Valliappan, S. & C. Zhao 1991a. An alternative wave input procedure in coupled finite and infinite model. *Computational Mechanics,* Vol. 1, Balkema, Rotterdam, p. 553-558.
Valliappan, S. & C. Zhao 1991b. Numerical simulation of dynamic soil-structure interaction problems. *Procs., 6th Int'l. Conf. on Finite*

Element Methods, Univ. of Sydney, Sydney, Australia, Jul. 8-10, p. 494-500.
Wallace, K.B. & B.K. Sapkota 1992. Distortion and cracking of pavements on swelling clay subgrades. *Procs.,16th Australian Road Res. Board Conf., Part 2,* Perth, Australia, Nov., p. 161-172.
Walters, J.K. & A.G. Mclean 1992. Stresses in a silo fitted with an anti-dynamic tube. *Procs., Int'l. Conf. on Bulk Materials Handling and Transportation; Symposium on Freight Pipelines,* Wollongong, Australia, Jul. 6-8, p. 443-447.
Wang, C.M., Kitipornchai, S. & F.G.A. Al-Bermani 1991. Buckling of Columns: Allowance for axial shortening. *Int'l. J. Mech. Sci.,* Pergamon Press, Vol. 33, No. 8, p. 613-622.
Wu, Y.H. 1990. *Static and dynamic analysis of the flow of bulk materials through silos.* Ph.D Thesis, Deptt. of Civil & Mining Eng, Univ. of Wollongong, NSW, Australia.
Wu, Y.H. & L.C. Schmidt 1992. Comparison of calculated and observed filling and flow pressures in silos and hoppers. *Procs., Int'l. Conf. on Bulk Materials Handling and Transportation; Symposium on Freight Pipelines,* Wollongong, Australia, Jul. 6-8, p. 449-452.
Xu, Y.L. & K.C.S. KWOK 1991. Soil-structure-damper interaction under wind loading. *Computational Mechanics,* Vol. 1, p. 529-534.
Yeo, M.F. 1991. Fluid element for static analysis using displacements as nodal variables. *Computational Mechanics,* Vol. 1, p. 547-552.
Yeo, M.F. & L.J. SCHMID 1989. Wave propagation in solid and fluid structures using finite element transfer matrices. *J. Sound & Vibration,* Vol. 130, No. 3, p. 439-452.
Yu, C.F. & W.K. Soh 1992. A flow visualisation study of underwater explosion. *Procs., 5th Asian Congress of Fluid Mech.,* Taejon, Korea, Aug. 10-14, p. 1177-1180.
Zhang, Q. 1991. *Stability behaviour of steel silos filled with bulk solids.* Ph.D Thesis, School of Civil and Mining Eng., Univ. of Sydney, Australia.
Zhang, B. Q. & J.C. Small 1991. Analysis of rafts on foundations of variable stiffness. *Procs., 7th Int'l. Conf. on Computer Methods and Advances in Geomech.,* Cairns, Australia, May 6-10, p. 1103-1108.
Zhang, B.Q. & J.C. Small 1992. The analysis of rectangular rafts of finite flexibility subjected to concentrated loads. *Procs., 6th Australia-New Zealand Conf. on Geomech.,* Christchurch, New Zealand, p. 205-210.
Zhang, B.Q. & J.C. Small 1994. Finite layer analysis of soil-raft-structure interaction.*Procs., 13th ICSMFE,* New Delhi, India (to be published).
Zhao, C. & S. Valliappan 1991. Vibration of three-dimensional rigid raft foundation on viscoelastic medium. *Earthquake Eng. & Struct. Dynamics,* John Willey & Sons, Vol. 20, p. 1159-1177.

Impact of Computational Mechanics on Engineering Problems, Pulmano & Murti (eds)
© 1993 Balkema, Rotterdam, ISBN 90 5410 324 8

Finite elements in biomechanics: Some new developments

J. Middleton, P. F. Hübsch & G. N. Pande
University College of Swansea, UK

Abstract

This study is concerned with the accuracy of numerical modelling in implant analysis. The discretisation error in a given finite element calculation is estimated and the mesh is refined accordingly. Furthermore the need for accurate modelling of the constitutive laws of the involved material is stressed. An isotropic model for bone is tested versus an orthotropic one. When modelling the setting process of a polymer based tooth filling, time dependent viscous flow occurring in the restorative material is taken into account and the results are compared to the linear case.

1 INTRODUCTION.

The finite element method has become a commonly used tool in biomechanics over the past decade or two. When compared to experimental techniques, numerical methods often prove to be the more economical choice, besides offering the possibility to investigate locations which are not accessible to experimental devices.

However, in order to produce meaningful results, it is important to estimate and to control the error in the calculation. Also, the material model used in the computation has to reproduce the (often rather complex) behaviour of the biological tissues and implant material very closely. Finally, realistic modelling of the boundary and load conditions as well as accurate geometric definition are of prime importance. This paper in mainly concerned with the first two of these points.

The method described by Zienkiewicz and Zhu [1] is used to estimate the error in the discretised domain. Once the error distribution is known, the mesh is refined accordingly and the analysis is restarted. The procedure is repeated until the error has been reduced to an acceptable level. This process is referred to as adaptivity. Apart from the error distribution, the procedure also yields a stress field of higher accuracy. Care is take to respect the natural discontinuity of the stresses at material interfaces.

The importance of accurate material modelling is demonstrated by means of two examples. For the analysis of a total hip replacement, an anisotropic material model is used for the representation of the bone and the results are compared to a calculation based on an isotropic model. To simulate the setting process of a dental restoration, a visco-elastic model which changes both its viscous and its elastic properties as time proceeds is adopted for the representation of the material behaviour of the cement.

2 THE USE OF ADAPTIVITY IN BIOMECHANICS.

Reference [1] describes a method to estimate the error distribution for a given finite element computation which is both efficient in terms of computational effort and straight forward to implement. It will only be briefly outlined here.

A global stress smoothing technique as introduced by Hinton and Campbell [2] is used to compute a stress field σ^* which is C^o continuous across element edges.

$$\sigma^* = N_\sigma \bar{\sigma} \qquad (1)$$

where N_σ is a set of shape functions used to interpolate the stresses and σ^* is the stress tensor at the nodal points. By forcing $\bar{\sigma}$ to approximate the stresses calculated using a conventional displacement based finite element method in the weak sense , the following condition can be introduced

$$\int_{\Omega} N_{\sigma}{}^{T}(\sigma^{*} - \hat{\sigma})d\Omega = 0 \tag{2}$$

This condition yields an equation which can be used to determine $\bar{\sigma}$

$$\int_{\Omega} N_{\sigma}{}^{T} N_{\sigma} d\Omega \cdot \bar{\sigma} = \int_{\Omega} N_{\sigma}{}^{T} \hat{\sigma} d\Omega \tag{3}$$

In the present implementation, bi-linear shape functions are used to interpolate the stresses while the displacement computation is based on a bi-quadratic approximation. Triangular elements with straight edges are used. This choice yields a computationally efficient procedure without unduly complicating the implementation.

In order to be able to use the above procedure on a domain consisting of several sub-domains featuring different material properties, the continuity condition for the stresses has to be relaxed across material boundaries while the same condition for the displacements remains enforced. The relaxation of the continuity of the stresses corresponds to the situation that the stress field is discontinuous across a material boundary whereas enforcement of the continuity of the displacements means that neither sliding or separation nor inter-penetration of the two materials can occur, which corresponds to perfect bonding.

These requirements are met by a simple procedure specifying double nodes along material interfaces [3]. That is, each node being located at the material boundary is allocated two node numbers. When assembling the matrices involved in the stress computation either one of these two node numbers is used depending on which material the considered element consists of. As opposed to this, when assembling the global stiffness matrix only one of the two node numbers is used regardless of which side of the boundary the element is located on.

As a virtually free by-product, the above procedure yields an error estimator which can be used as a basis for mesh adaptivity. Indeed the difference between the smoothed stress field σ^{*} and the un-smoothed stresses $\hat{\sigma}$ presents a fine approximation of the discretisation error, which can be written as

$$e = \sigma^{*} - \hat{\sigma} \tag{4}$$

For convenience, the error is measured in the L_2 norm rather then using a point-wise definition. The error integrated over a certain domain (say an element) then reads

$$\|e\|_{L_2} = \int_{\Omega} e^{T} e d\Omega \tag{5}$$

This quantity approximates the absolute value of the error over the domain Ω. This will usually be a maximum in the areas where the stresses are highest, which do not necessarily coincide with the regions where accurate results are most crucially needed. Considering the case of a domain consisting of sub-domains with different stiffness, the stresses will generally be highest in the sub-domain featuring the biggest stiffness. In the context of biomechanics this will most usually be associated with the metal and ceramic parts of an implant. However the more interesting phenomena like failure, creep or mechanically triggered biological reactions will occur in the softer components of the implant system (for example cements) and in the tissues surrounding the implant. In order to achieve accuracy of the results in these regions, it is necessary to use a measure of the relative error as a criterion for mesh refinement rather than its absolute value. Hence, the norm of the error estimator has to be normalised by the local stress level.

$$e_{Rel} = \frac{\|e\|_{L_2}}{\int_{\Omega} \sigma^{*T} \sigma^{*} d\Omega} \tag{6}$$

This error estimate is used to decide whether the mesh has to be refined at any particular location. If it exceeds the error value averaged over the whole mesh by a specified factor for a particular element, this element is then split along a line passing through the middle of its longest edge and the opposite corner node. This refinement strategy is chosen because it facilitates monitoring of the results at particular locations, namely material interfaces, which are defined in the original mesh and which by definition must be retained in all subsequent refinements.

2.1 Examples

A premolar was cut longitudinally in the buccolingual plane in order to obtain an accurate geometrical definition. In the crown, a cavity suitable for an inlay was defined. A glass ionomer liner was placed at the bottom of the cavity. A typical range of thickness for the cement lutes is 100 - 300 μm and in the model considered this was kept constant at 300 μm. The volumetric shrinkage of the cement lute was set to 3.6% which corresponds to the shrinkage of a commercially available restorative material between three and sixty minutes of curing time. Load is applied through polymerisation shrinkage only. This shrinkage was modelled as a case of initial strain. Fixity conditions were applied to the outer boundary nodes of the root of the tooth. This constraint is stiffer than the tooth's natural embedding in the periodontal ligament. The calculations however show that the stresses in the root of the tooth are so low that this choice of boundary condition is not likely to influence the stress field close to the restoration. The initial mesh was formed using a mesh generator based on Delaunay-triangulation [4] and the material parameters are displayed in table 1.

Figure 1 shows the original mesh together with the an automatically refined one. It can be seen that the refinement concentrates mainly around the restoration in the crown of the tooth. As load is applied through shrinkage of the cement lute only, this is the area where mechanical reactions are to be expected and a fine mesh is needed most crucially. Figure 2 reports the stress components along the interface between the cement and tooth, sampled from the side of the cement. A running coordinate is defined along that interface, being 0 respectively 1 at the points where the inter face meets the free surface. The figure clearly shows that the stress field computed on the refined mesh differs substantially from the one computed on the original mesh, especially around the sharp corners at the bottom of the cavity and close to both ends of the domain.

A second example, a model for a total hip replacement which will be described in more detail in section 3.1, is shown in figure 3. The initial mesh is characterised by a homogeneous mesh density. In the first adaptive cycle, mesh refinement is concentrated at the lower part of the stem and the adjoining part of the femur and at the neck of the stem. This is due to the load being transferred from the implant into the bone in these regions. This subsequently causes high local stress gradients which requires the mesh to be refined. Another zone affected by mesh refinement is located where the stem profile changes sharply and contact between the bone and the neck of the stem occurs. In the second adaptive cycle, the refinement is concentrated at both ends of the stem due to the singularities at these locations. The refinement in the inner part of the stem is due to the particular error definition chosen. As the major component of the load is bending the centre of the prosthesis stem bears the smaller part of the stress. Recalling that the error estimator is normalised by the local stress level as in equation 6, high error values can be expected in the unloaded zones. Figure 4 shows the stress components along the interface between the stem and the bone, sampled from the side of the bone. Similarly to figure 2,

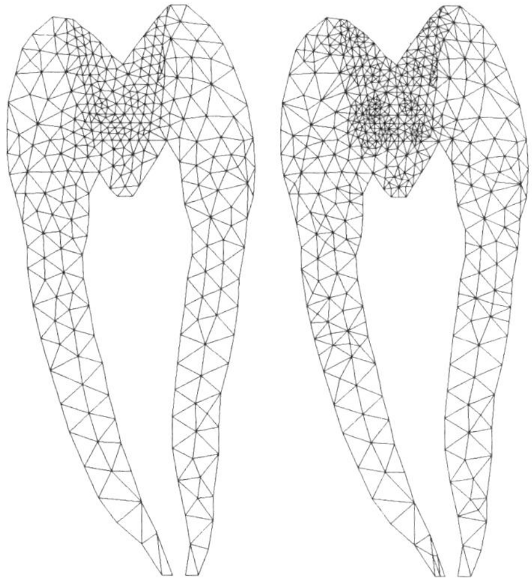

Figure 1: Automatically refined finite element mesh for a restored tooth.

a normalised running coordinate was defined along the interface, being 0 at the distal position and 1 at the proximal. The difference between the results from the initial and the final mesh can be found when comparing the normal stress component except towards the extremity of the domain. However, when comparing the tangential components, differences of over 20% can be found in the distal half of the domain. Similarly to the previous example, marked differences of the stresses close to the extremities of the domain were found. This can be explained by the fact that the refined mesh allows for a more exact representation of the singularities existing at these locations.

3 MATERIAL MODELLING.

3.1 Directional material properties

In order to demonstrate the influence of directional effects a computer model of a total hip replacement was developed. Computations were performed once ascribing isotropic properties to the bone and once using orthotropic properties using the same model to allow a direct comparision of results.

To avoid the huge computational effort associated with three dimensional finite element analysis without disregarding the important hoop effects, an axisymmetric finite element model with non-symmetric load conditions was adopted. This

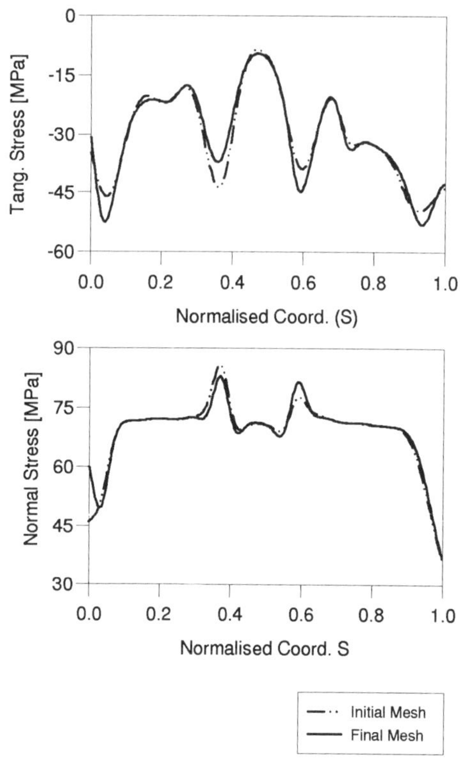

Figure 2: Normal and shear stress along the interface between the cement and the tooth, sampled from the side of the cement.

procedure is first described by Wilson [5] and is well documented in the finite element literature [6, 7]. In the field of implant analysis this technique has been advocated in the past by Huiskes [8, 9] but interest in the technique has faded over the last decade. However, the authors feel that the method is well suitable to demonstrate the qualitative difference between results obtained using an isotropic and an anisotropic material model.

The geometry of the femur considered was modelled from the data reported by Cowin in [10]. From the cross section and inertia data given in the reference, a model geometry (axisymmetric) which has the same stiffness in lateral bending and axial compression as the original (non symmetrical) bone was developed. The stem of the implant was chosen so as to fit this geometry perfectly. This together with the condition of perfect bonding can be thought of as a model for a coated implant, in which case a direct link between bone and prosthesis is presumed to be achieved. The initial mesh was generated using a Delaunay triangulation program [4].

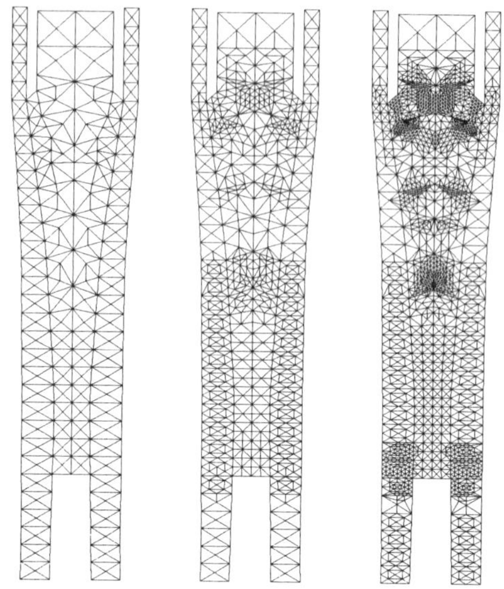

Figure 3: Automatically refined finite element mesh for a total hip replacement.

The derivation of the load conditions again follows that given in reference [10]. They simulate the force acting on the femur of a person of 1000 N standing on one leg. The muscular force acts on the greater trochanter at an angle of 20° while the femur is inclined at 11° with respect to the vertical. The resulting forces acting on the bone were transformed to the axisymmetric model geometry so as to recreate the same bending moment, longitudinal compression and axial force in the shaft of the femur. The model was constrained at the distal end of the bone.

The material behaviour of the bone was modelled as being orthotropic and linear elastic. The assumption of linear elasticity implies that the stress is related to the strain by Hook's law

$$\sigma_i = C_{ij}\epsilon_j \tag{7}$$

The orthotropic model requires nine independent constants to fully describe the material response. The values for the technical constants where taken from [11] and the resulting compliance- matrix is given by

$$\begin{bmatrix} 18000. & 10100. & 9980. & 0. & 0. & 0. \\ 10100. & 27600. & 10700. & 0. & 0. & 0. \\ 9980. & 10700. & 20200. & 0. & 0. & 0. \\ 0. & 0. & 0. & 5600. & 0. & 0. \\ 0. & 0. & 0. & 0. & 4520. & 0. \\ 0. & 0. & 0. & 0. & 0. & 6230. \end{bmatrix} \tag{8}$$

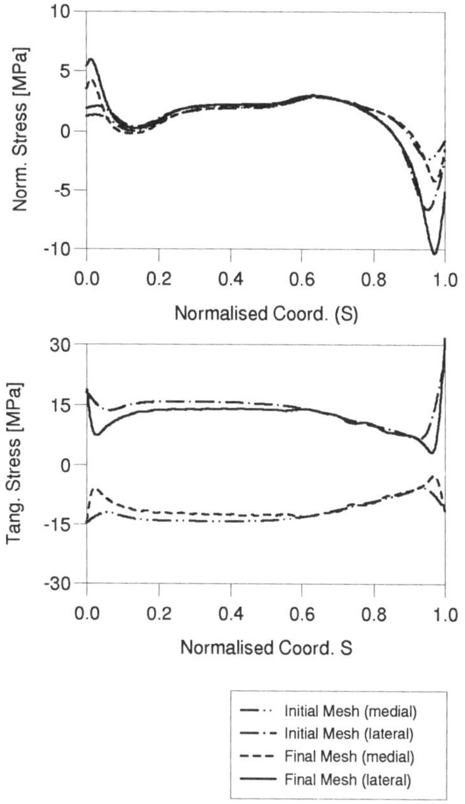

Figure 4: Normal and shear stress sampled along the interface between the prosthesis stem and the bone from the side of the bone.

In this compliance matrix, direction 1 refers to the radial, direction 2 to the longitudinal and direction 3 to the circumferential coordinates.

When using the isotropic model for bone, the elastic constants were chosen such that they reproduce the longitudinal stiffness and the interaction between longitudinal and radial direction of the orthotropic model. This condition leads to an equivalent Young's modulus of 15400 MPa and a Poisson's ratio of approximately 0.372. These constants yield the isotropic compliance matrix

$$\begin{bmatrix} 27601. & 16381. & 16381. & 0. & 0. & 0. \\ 16381. & 27601. & 16381. & 0. & 0. & 0. \\ 16381. & 16381. & 27601. & 0. & 0. & 0. \\ 0. & 0. & 0. & 5610. & 0. & 0. \\ 0. & 0. & 0. & 0. & 5610. & 0. \\ 0. & 0. & 0. & 0. & 0. & 5610. \end{bmatrix} \tag{9}$$

It was further assumed that the stem of the prosthesis consists of Titanium, which was modelled as an isotropic, linear elastic material with a Young's modulus of 102 GPa and a Poisson's constant of 0.3.

Figure 5 compares the tangental stress components along the interface between the bone and the stem sampled once from the stem and once from the bone computed using the isotropic material model for the bone as defined in equation 9 to those computed assuming an orthotropic compliance matrix as defined in equation 8. It is interesting to observe that, although producing a higher tangential stress in the stem, the isotropic model gives a lower stress in the bone. These (somewhat unexpected) results show that it is very important to use an accurate material model as no general statements about how the choice of the model will affect the results under specific load conditions can be made.

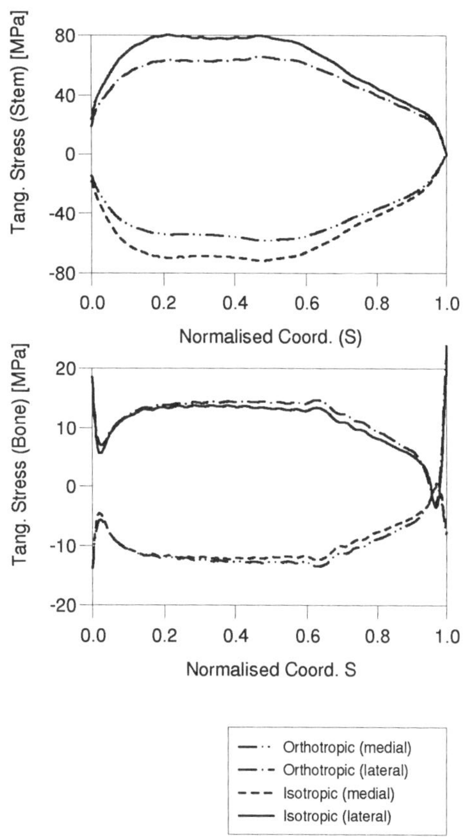

Figure 5: Interface stress components computed using an isotropic and an orthotropic material model for the representation of the bone.

3.2 Time dependent material properties

An other important feature associated with implant materials are time effects. Here, we will consider the case of polymer based dental restorations. The configuration analysed is the same as in section 2, but now time dependent viscous flow in the cement lute during the setting process is taken into consideration.

The composite cement modelled here consist of a polymeric resin matrix and finely ground glass particles acting as a filler. It can therefore be assumed that it exhibit a viscous behaviour of the same type as a polymer. In accordance with the literature [12] a visco-elastic model was chosen for the representation of the cement material. Such a model (Figure 6) can be fully described with a minimum of three parameters.

These parameters were extracted from data given by Feilzer et al. [13] who reported the time dependent elastic behaviour together with the polymerisation shrinkage and the stress field building up in a well defined experimental setup. They were approximated by a set of discrete values at specific points in time (given in figure 1 for "Brilliant", the material being modelled as an example) with linear variation between these points. The procedure relies on two assumptions regarding the strain rates at which the experiments were performed. It is also assumed that Poisson effects do not change with time. Though the accuracy of the extracted set of time dependent material parameters is not quite satisfactory, it seems to be the only set available from published data at present.

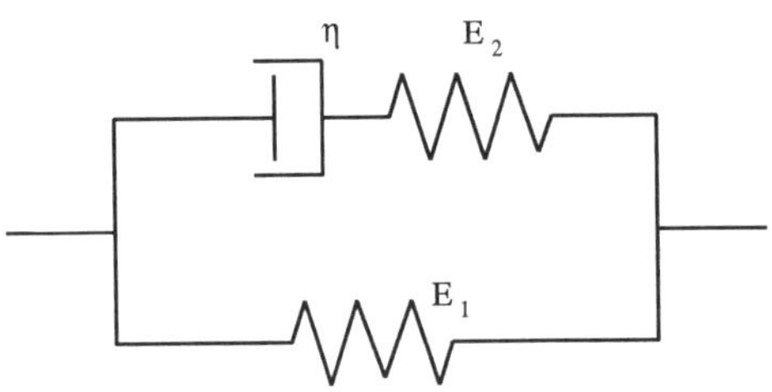

Figure 6: Rheological model used for the representation of the cement.

Table 1: Time dependent material parameters for the cement used in the computation.

t min	ϵ_{plo} %	E_1 MPa	E_2 MPa	η MPa*min
3	0.0	6.	600.0	0.1584
4	0.8	98.2	1627.7	0.0589
5	1.4	319.9	1990.0	0.0482
6	1.9	339.9	3470.0	0.0276
7	2.2	467.9	3551.0	0.0270
8	2.4	497.0	4381.9	0.0219
10	2.7	320.7	5865.2	0.0163
15	3.0	559.7	6997.2	0.0137
20	3.1	840.9	7523.0	0.0127
30	3.3	78.4	9512.5	0.0101
60	3.6	690.4	10331.5	0.0093

In the numerical procedure, the overlay-model developed by Pande et al [14] in combination with the initial strain method was adopted to model the visco-elastic behaviour of the cement. An explicit time integration scheme was used. Assuming a von Mises yield function and an associated flow rule together with a yield stress of zero, the maximal time step length for which numerical stability of the algorithm is guaranteed is given by [15]

$$\Delta t = \frac{2(1+\nu)}{3\eta E} \qquad (10)$$

Assuming a Poisson's ratio ν of 0.3 and considering the material properties displayed in table 1 it is clear that stability is only achieved for rather small time steps.

The material parameters are incremented at every time step to their value at the middle of the current time step and the global stiffness matrix is updated.

The loading through shrinkage of the cement lute is applied by the means of an initial strain condition. The load vector resulting from the shrinkage occurring during the considered time step is added to the global load vector.

Figure 7 shows how the stress components build up along the interface between the tooth and the cement as polymerisation proceeds. The lowest curve corresponds to 4 minutes of curing time, the topmost to one hour. The plots show that the stresses are increasing monotonously during the first hour of curing, a finding which is consistent with Feilzer's experimental findings [13]. The importance of taking the time dependent, non linear behaviour of the restorative material into account becomes clear when comparing these results to the results of linear elastic calculations reported in figure 2. Stresses along the interface are reduced by a factor of 5 to 10 through viscous flow. The stresses obtained from the non linear computation are of

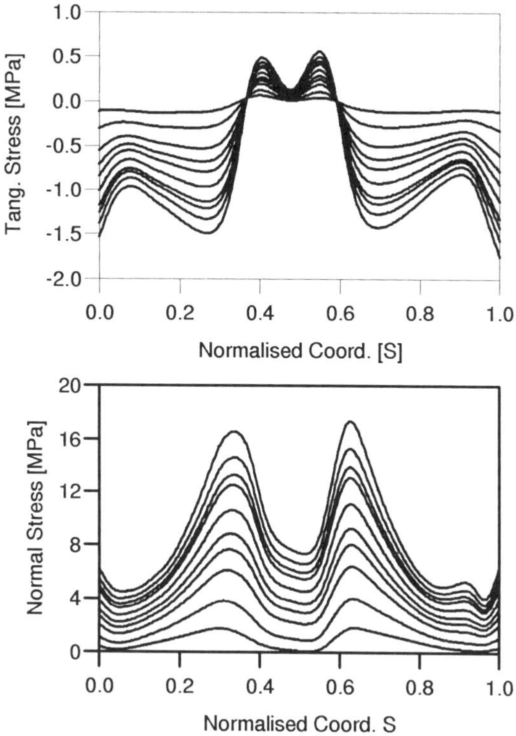

Figure 7: Stresses along the interface between tooth and cement for 4, 5, 6, 7, 8 ,10, 15, 20, 30 and 60 minutes of curing time.

the order as the bond strength between cement and tooth [16] where as they were consistently higher in the linear calculation, a fact which contradicts experimental evidence of the filling not dropping out of the cavity.

4 CONCLUSIONS.

A numerical model capable of accurately predicting stresses in biomedical implants was implemented. Error analysis of the finite element results is carried out and the mesh is refined accordingly. Significant improvement of the results are achieved through this method. The importance of accurately reproducing the material behaviour of the different components of an implant system was demonstrated by means of two examples. By neglecting the directionality of the constitutive law for bone, the predicted stress field suffers a severe loss in accuracy. By not taking into account viscous flow of dental cements during the setting process, over prediction of up to one order of magnitude of the stresses was found. The numerical approach presented can be used on other problems of implant analysis and is readily extendible into three dimensions.

References

[1] O. C. Zienkiewicz and J. Z. Zhu. A simple error estimator and adaptive procedure for practical engineering analysis. *Int. J. for Numerical Methods in Engineering*, 24:337–357, 1987.

[2] E. Hinton and J. S. Campbell. Local and global smoothing of discontinuous finite element functions using a least square method. *Int. J. for Numerical Methods in Engineering*, 8:461–480, 1974.

[3] P. F. Hübsch and J. Middleton. Numerical and analytical assessment of dental restorations. In *Recent Advances in Computer Methods in Biomechanics and Biomedical Engineering*. B.J.I., May 1992.

[4] N. P. Weatherill. A method for generating irregular computational grids in multiply connected planar domains. *Int. J. for Numerical Methods in Fluids*, 8:181:197, 1988.

[5] E. L. Wilson. Structural analysis of axisymmetric solids. *AIAA Journal*, 3:2269–2274, 1965.

[6] B. Irons and S. Ahmad. *Techniques of Finite Elements*. Ellis Horwood Ltd, 1980.

[7] K.-J. Bathe. *Finite Element Procedures in Engineering Analysis*. Prentice-Hall, 1982.

[8] R. Huiskes, P. T. Elangovan, J. P. A. Banens, and T. J. J. H. Slooff. Finite element computer methods for design and fixation problems of orthopaedic implants. In *Proc. of the* 6th *Int. Congress of Biomechanics, Kopenhagen, Denmark*, July 1977.

[9] R. Huiskes. Stress analysis of implanted orthopaedic joint prostheses for optimal design and fixation. In *Proc. of the European Society of Biomechanics, Brussels, Belgium*, May 1978.

[10] S. C. Cowin. *Bone Mechanics*. Boca Raion, Florida, 1989.

[11] A. N. Natali and E. A. Meroi. A review of the biomechanical properties of bone as a material. *J. Biomed. Eng.*, 11:266–276, 1989.

[12] R. G. C. Arridge. *Mechanics of Polymers*. Clarendon Press . Oxford, 1975.

[13] A. J. Feilzer, A. J. De Gee, and C. L. Davidson. Quantitative determination of stress reduction by flow in composite restorations. *Dental Materials*, 6:167–171, 1990.

[14] G. N. Pande, D. R. J. Owen, and O. C. Zienkiewicz. Overlay models in time-dependent non-linear material analysis. *Computer and Structures*, 7:435–443, 1977.

[15] I Cormeau. Numerical stability in quasi-static elasto/visco-plasticity. *Int. J. for Numerical Methods in Engineering*, 9:109–127, 1975.

[16] C.L. Davidson, A. J. De Gee, and A.J. Feilzer. The competition between the composite-dentin bond strength and the polymerisation contraction stress. *The Journal of Dental Research*, 63:1396–1399, 1984.

Impact of Computational Mechanics on Engineering Problems, Pulmano & Murti (eds)
© 1993 Balkema, Rotterdam, ISBN 90 5410 324 8

Development of a microscopic view traffic flow simulation model for expressway

H. Nakayama
Nihon University, Japan

ABSTRACT: Numerical modeling for traffic flow simulator based on microscopic view simulation model have been developed. Traffic flow is effected by many factors, for example driver's reactions, influenced characteristics each other and so on. In this paper, author investigated the influence factors of driving behaviors for traffic flow simulator and introduced spiral curve to numerical model.

1 INTRODUCTION

There are approximately over 60 million registered automotive vehicles in Japan. There are large road networks which total length is more than 1 million km without expressway. The total length of expressway is become more than 5000 km in 1992. There are many traffic jams occur on ordinary roads and expressways especially on Metropolitan expressways in Tokyo and it's become serious problems for economic activities. Traffic accidents also become serious problems in recent years. The number of deaths caused by traffic accidents exceeded 11,000 from 1989 to 1992 in Japan.

It is indispensable to investigate the traffic flow on expressway with microscopic view, because of traffic jam and accidents relate to human behaviors. The microscopic view simulation model is one of the tool for understanding it. As investigates traffic flow with microscopic view, drivers usually have individual driving characteristics that cause various traffic behaviors. The driver characteristics are the most influence factors for traffic flow behavior.

The purpose of this study is to make clear driver's characteristics and to get some suggestions for the numerical modeling of microscopic view flow model and development of simulation systems for traffic flow on expressway.

2 INVESTIGATION

The purpose of this investigation is obtain the fundamental factors for simulation model. The observation was done at three kinds of places. Those locations and the conditions are as follows:

(a) Takeishi interchange of Keiyo expressway
This location has two driving lanes and an entrance lane. It is assumed that the entrance vehicles affect on the other vehicles that running on slow lane.

(b) Hakozaki interchange on route 6 of Metropolitan expressway in Tokyo
The road alignment of this location is slight curve and there is a diverging point ahead of this location. It is assumed that the road alignment includes diverging point affect on the driving behavior.

(c) Higashiohjima on route 7 of Metropolitan expressway in Tokyo
The road alignment of this location is straight for 5 km. It is assumed that the vehicles are independent of road alignment. That is to say the driver's characteristics are expressed in driving behavior.

The traffic flow on the expressway was investigated using a video camera. The coordinates' data of the each vehicle, driving speed and space headway data are corrected using digitizer and computer. Especially about location (3), the situation data of surroundings of the vehicles was also corrected.

2.1 Personal space

The individual driver's characteristics that have direct effect up on driving behavior it could be as follows.

(a) An individual personal space.
(b) An individual desire speed.

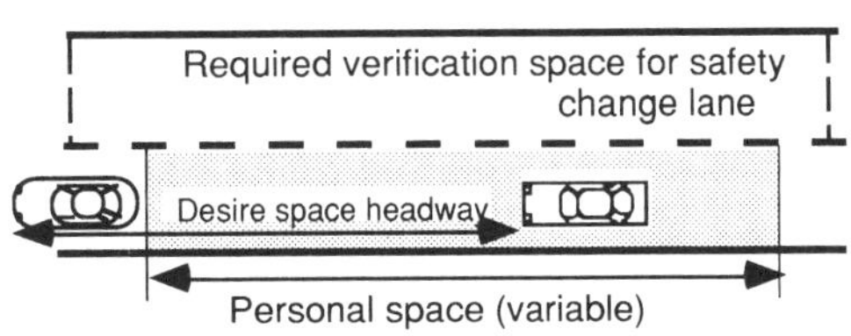

Figure 1 The concept figure for personal space

Figure 1 shows the concept figure of personal space. The personal space means specific area that is the driver wants to occupy the area around the vehicle. If other vehicles break into the area, the driver feels unpleasant and acts acceleration, deceleration or lane changes to keep own area. It is the variable area with velocity and traffic density.

2.2 Analysis of observed data

The driving behavior was analyzed from the recorded video tape. The coordinates of the vehicles were measured using digitizer. Some factors that the vehicle types, traveling lane and changed lane coordinates were collected for information of individual driving behavior. The situation of surroundings was also collected. It could be divided into the following class.

(a) Free traveling vehicles
There is no vehicle within 80m in front and rear of the object vehicle. It can be considered the free traveling vehicles driving one's desire speed.

(b) Driving lead the group of the vehicles
There is no vehicle within 80m in front of the object vehicle. Some vehicles were following object vehicle. It seems that the object vehicle driving one's desire speed.

(c) Following vehicles
There are some other vehicles within 80m in front of the object vehicle. Those drivers were driving to get a chance for change lane or who has not a mind to change lane.

Figure 2 shows the relationship between space headway and relative velocity at three observed locations. In this figure, when the relative velocity is extent in plus, it means that the leading vehicle is going away or the object vehicle decreases the speed. The location (a) shows the widely distributed data for relative velocity direction. It seems that the running vehicles were influenced by entrance vehicles. It caused very dangerous situation for traffic accidents. Some vehicles change lane or slowdown quickly to avoid those situations.

In location (b), the traffic volume is always heavy, and there is the diverging point ahead of this location. It seems that the degree of freedom of the vehicles is low. Almost the driver who driving this route knows diverging point ahead. They changed lanes before this location. Each vehicle is following the leading vehicle. Figure 2(b) expressed very low degree of freedom because of this situation. The data distributed very narrow range than the other locations. It means that the vehicles are following each other tightly and stability.

In figure 2(c), it is assumed that the vehicle is independent of road alignment because of the straight road. The driver's characteristics are expressed in driving behavior oneself. The data distribute more widely for space head way than the other locations. The unique relationships that desire speed distribution, space headway distribution and so on were collected from these data.

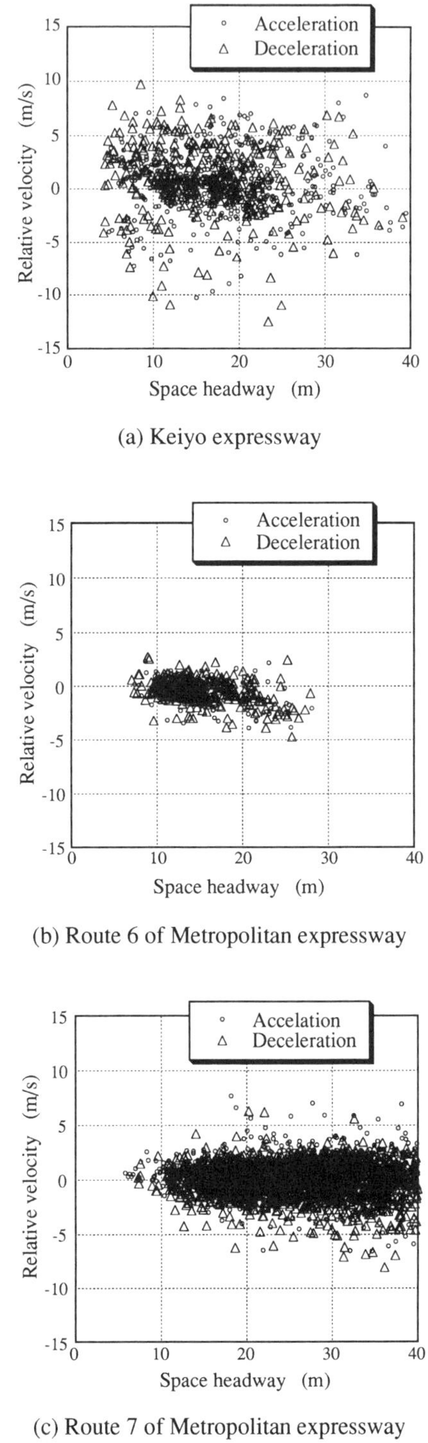

(a) Keiyo expressway

(b) Route 6 of Metropolitan expressway

(c) Route 7 of Metropolitan expressway

Figure 2 The relationship between space headway and relative velocity

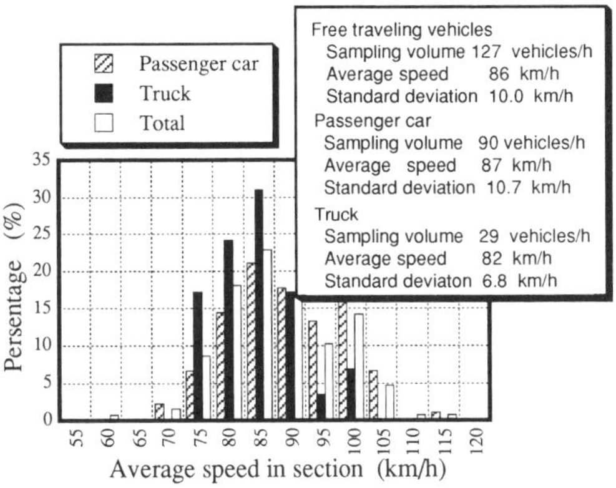

Figure 3 Average speed distribution each type of vehicle

2.3 Desire speed

Figure 3 shows the distribution of average speeds of free traveling vehicles as expressed in 2.2(a). It is assumed that the speed in the case of free traveling can be considered desire speed for each driver in the case of the route 7. The idea of desire speed is considered to the speed that is the drivers satisfied with this speed. In a strict sense, it is very difficult to find out the desire speed for each driver. In this research, author assumed the desire speed of the driver could define to investigate speed distribution of free traveling vehicles.

Figure 3 also shows the difference of average speed between passenger cars and trucks. The trucks were running lower speed than passenger car. It gives a good indication of the vehicle's condition. It means the performance of the vehicles as limit speed, acceleration and deceleration. It could be consider the one of main influence factor for decision making model of the simulator. The factor is very important to make clear for individual characteristics.

2.4 Border and critical line for lane change

Three characteristics driving behaviors are observed by recorded video tapes as shown in figure 4 (a), (b), (c). Those points were selected as these typical cases and show behavior of the following vehicles and leading vehicle that tandem drives each other on same lane every 0.5 second. There were many points near axis of relative velocity (0m/s), but those points were neglected in this figure for stability following cases. On the human engineering, drivers drive in one's response delay time. It is the one of the important factor for understanding the driving behavior. In this study, it was observed by this stability following cases. Those three cases show typically driving behaviors of each driver on the expressway as follows.

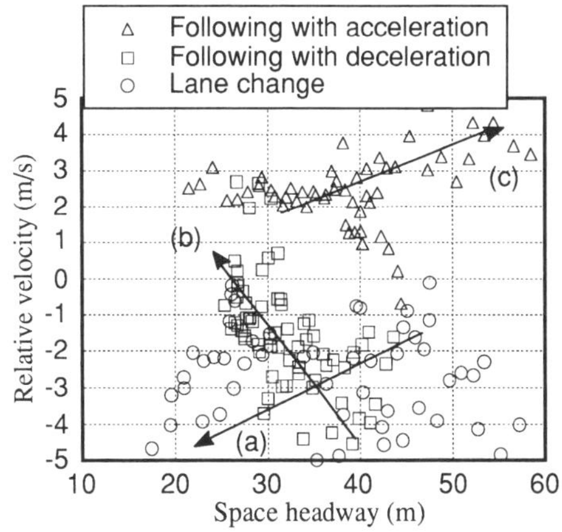

Figure 4 The tendency for following and lane change

Allow (a),(b),(c) show peculiarity tendency. Allow (a) indicates tendency for free driving. The driver wants to drive ones own driving speed. It means the driver shows the desire speed. Allow (b) indicates tendency to approach the leading vehicles and following. Those vehicles did not change the lane and keep ones own space headway with in front of the object vehicle. It means one of the case to keep ones in front of personal space. Allow (c) indicates the change lane vehicles.

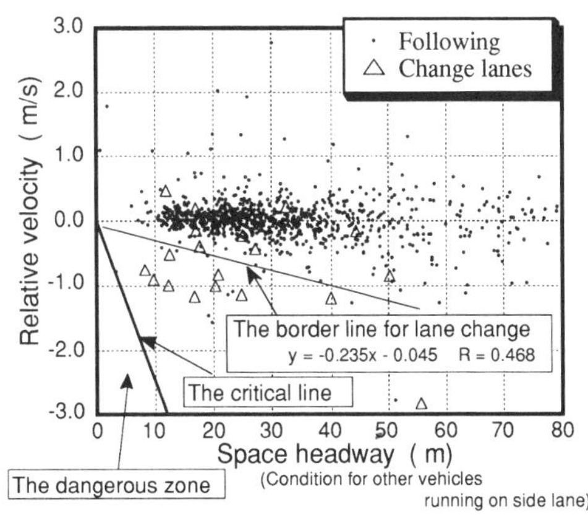

Figure 5 The mutual relation chart of space headway and relative velocity

Figure 5 shows the same relation as figure 4 especially within the vehicles following in front of the object vehicles and the vehicles that changed lane. There is a precise difference between them and it proved that there is the border line. This border line can be considered as the line for safety lane change (see thin line in figure 5). If the driving behavior close to this line, the driver begins to choice the action either changing lane or following.

There is another critical line in figure 5. It indicates thick line and the area of left side the line shows the dangerous zone. The meaning of this line is critical line for a rear-end collision. The driver must to change lane before this line to avoid the accident. It is supposed that the individual personal space and internal condition of vehicle affect on the critical line The personal space will change with speed, traffic density and environmental condition of an expressway.

2.5 Individual driving behavior

Let me discus the individual behavior about the relationship between space headway and relative velocity. The behavior can be classified into some groups. Figure 6 and 7 show this typical example.

The vehicles shown in figure 6 were changing lane on the point of the arrow. At this point, it is observed that the behavior of the vehicle transfer to the other phase. It is the new phase with a new vehicle that running with tandem driving each other. That situation made new relationship and following vehicle start to control one's desire speed and space headway.

The vehicles shown in figure 7 were follow the leading vehicle. As seen from these examples, it seems that the locus of vehicle running behavior on this graph is moving looks like a circle rotating clockwise and the driver's desire space headway as the center of the circle. It is also observed that the vehicle becomes stable around the driver's desire space headway. It seems that the following vehicle tried to keep distance to the leading vehicle. It caused the relative velocity change like a vibration.

The reaction time of each driver was also observed at same time. The mean reaction delay time of each driver is approximately 1 second. It means the driver's reaction for control the vehicle delays 1 second after understood the actions what happen in front of one's vehicle. It caused many behaviors that we could meet on the expressway such as traffic jam and waving phenomenon of traffic flow. In this study, author assumed delay time to 1 second, but actual delay time distributes for normal distribution.

Those loci look like spiral curve shown in figure 8. In this study, for develop the microscopic view simulation model, it is studied that the acceleration is decided from the spiral curve that introduced reaction delay time.

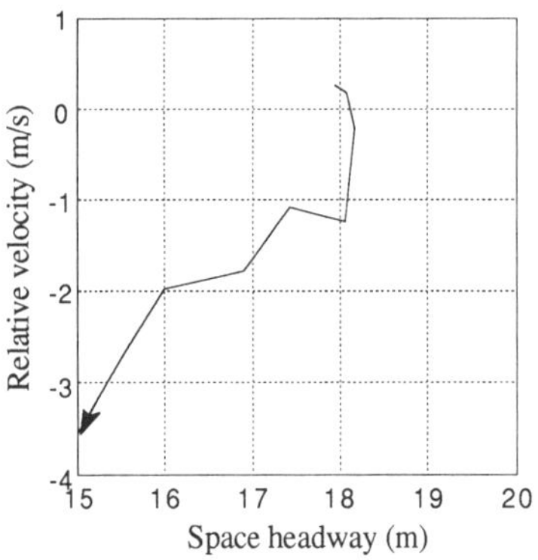

(a)

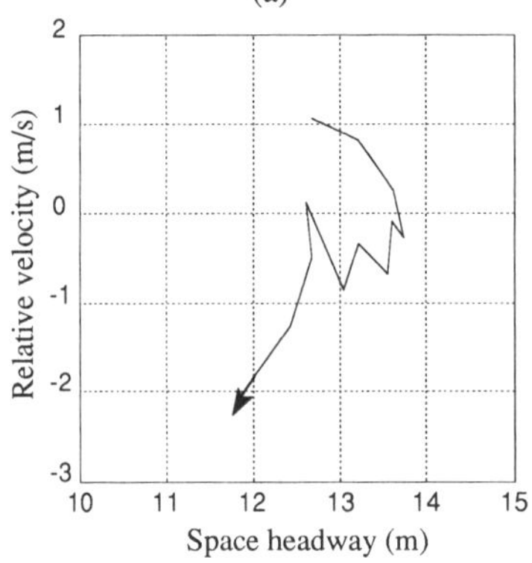

(b)

Figure 6 A trace chart of individual behavior (Change lane behavior)

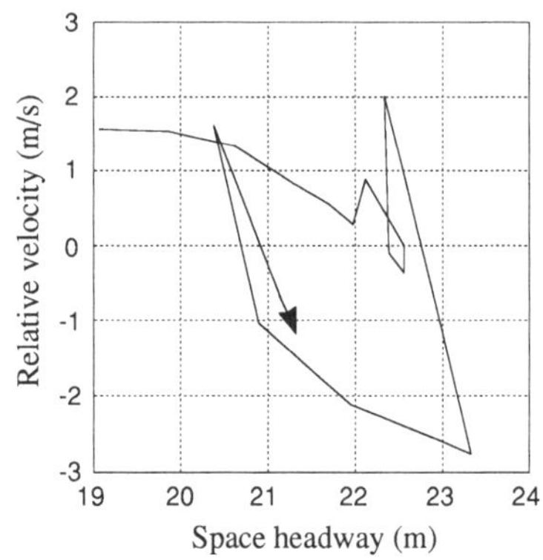

(a)

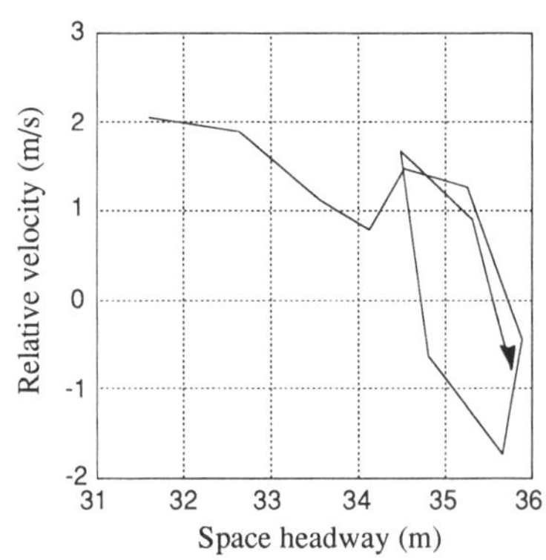

(b)

Figure 7 A trace chart of individual behavior (Following behavior)

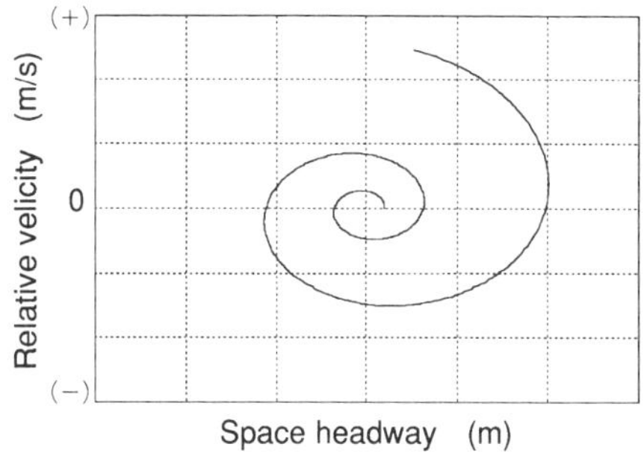

Figure 8 spiral curve

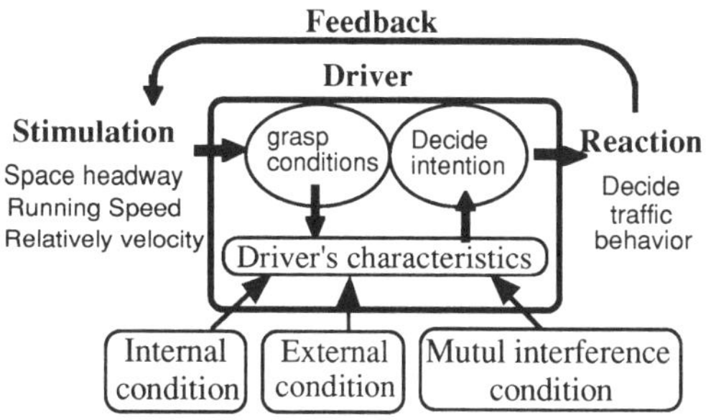

Figure 9 Decision making flow for driving behavior

3 DEVELOPMENT OF THE MODEL

3.1 Discuss about influence factors

In this study, author considered the driver's characteristics influenced to traffic flow phenomenon. For development of the traffic flow simulation model, driver characteristics are selected for one of the main influence factor. It is important thing that the influence factor would be selected as limitation for the simple modeling. Most ideal simulator is to offer to accurate outcome with several essence factors as traffic phenomenon, because many factors take a long period for research it and as well as calculation time. In this paper, determinations of influence factors for driving behavior are considered by following condition as shown in figure 9.

The driving behavior is stimulated by traffic condition, such as space headway, relative velocity and running speed, and is influenced by physical condition at the same time, and finally decide own behavior as reaction. Internal, external and mutual interference conditions are selected for influence factor of physical condition. Those factors cause variation of characteristic as follows.

(1) Internal condition: Performance of the vehicles such as speed and acceleration.

(2) Outside condition: Three conditions are picked up. First is physiognomy condition of road alignment and surface. Second is artificial condition of traffic regulation. Final is natural condition such as weather.

(3) Mutual interference: When some vehicles are running in the same road, the vehicles influenced each other.

The driver's characteristics are fundamental parameter that causes various driving behaviors. Three factors were defined for the driver characteristics.

(1) Desire speed: Desire speed is satisfactory driving speed of a driver. Decision of driving speed is influenced by gap between desire speed and driving speed. We estimate that desire speed is free traveling speed.

(2) Personal Space: Personal Space is an area that the driver wants to keep surrounding the vehicle. If the other vehicles invade his personal space, the driver starts reaction to keep one's space. It will suppose that each driver has different personal space. Personal space will be changed by driving speed and traffic density. It is difficult to measure this space, but personal space is defined by limit space at the moment of lane change shown in figure 10. This is the decision making model for driving behavior of running on expressways. It shows the border line for change lane and the critical line. The two lines show that the personal space is variable area with relative velocity as linear relation.

(3) Response characteristics of the driver: It is the characteristics for determination of ones driving behavior. For example, there are many types of drivers who want to drive very first, keep his own speed, want to change lane every time and so on. It is one of the parameter to express those responses.

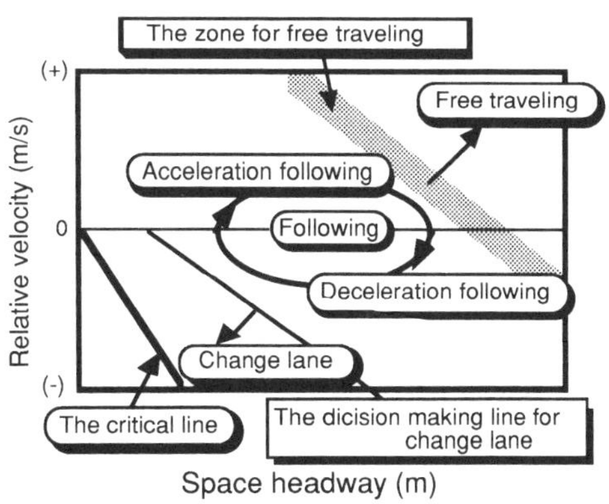

Figure 10 The decision making model for driving behavior

3.2 Microscopic view modeling

In this study, it was assumed that the acceleration is decided from the spiral curve. Figure 11 shows the relation of the spiral curve and acceleration. Spiral curve is represented by four parameters and desire space headway L. As the leading vehicle is running with constant speed, acceleration of the following vehicle is equal to the rate of change of the relative speed in unit time. It could be calculated the slope of the spiral curve.

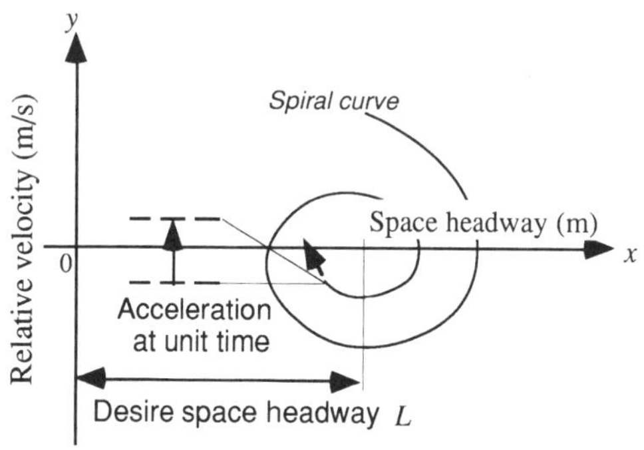

Figure 11 Spiral curve and acceleration

Acceleration of the following vehicle is given by tangent of spiral curve at the time. The equation of acceleration is given as follows:

$$\frac{dy}{dx} = \frac{\gamma(x-L)+\delta y}{\alpha(x-L)+\beta y}$$

$$= \frac{-a\,dt}{y_0 dt - (1/2)a\,dt^2} \quad \text{......................(eq.1)}$$

$$a = \frac{2y_0 \dfrac{dy}{dx}}{\dfrac{dy}{dx} - 2} \quad \text{....................(eq.2)}$$

Where:
x:Space headway
y:Relative velocity
L:Desire space headway
a:Acceleration
y_0 :Initial relative velocity
t :Time
$\alpha, \beta, \gamma, \delta$:Coefficient for spiral curve

The traffic flow behavior of the vehicles running on the expressways is modeled as concept figure shows in figure 9 and 10.

Figure 10 shows the decision making figure of traffic phenomenon for traffic flow simulation model. The feature of the concept for traffic behavior is divided into three phases by two border lines. The driving behavior is defined by three phases. First phase is the driver goes to stability conditions in nearly 0 m/s on relative velocity with reiterating acceleration and deceleration on clockwise rotating a circle. Second phase is the driver change lane and transforms the situation to new phase. Final phase transforms the situation to free driving. These actions are decided by border line based on relations between space headway and relative velocity. Personal space is the most important factor that defines movement, because border line is decided by limited distance in front of personal space and it is defined by investigation of the vehicles at the moment of change lane.

4 RESULT OF SIMULATION

It has been developed that the microscopic view traffic flow simulation model on expressway using C language and engineering work station SONY NEWS 3460 and Silicon Graphics IRIS Indigo XS24. This simulation model is based upon figure 9,10 and investigation data. Each driver has his desire speed, personal space (approximately equal to ones desire space headway for in front of it) and response characteristics. The driver keeps driving with desire speed as same as possible and is influenced by, three type conditions, internal, external and mutual interference. Traffic flow behavior is decided by stimulation and movement model that defined at figure 10, especially acceleration is decided by spiral curve that is defined in figure 11 and equation 1 and 2. The response delay time is installed by using mean data 1 second.

The conditions for simulation were selected as following. The condition of model expressway was set up straight alignment and it has two lanes and 2,400 m length. The simulation step time is applied 0.5 second. The appearance ratio between vehicle types set up "passenger vehicle : compact truck : truck : bus" as "4 : 3 : 2 : 1". Response characteristics of the driver are set into three types' drivers. One is the driver who runs following the leading vehicle (Type A). Second is the driver who tries to change lane if the vehicle appeared running slower than his desire speed or drives same or higher speed than his desire speed (Type B). Final is the driver who is driving with his desire speed (Type C). Table 1 shows own characteristics and probability of appearance each type. Desire speed of each type is distribute with normal distribution that standard deviation is 10 km/h.

Table 1. Characteristics of each type

Type	Meen desire speed (km/h)	Max speed (km/h)	Probability of apperance (%)
A	75	120	50
B	85	120	30
C	70	120	20

Figure 12 shows the one of the result of simulation. Each driver kept their personal space and desire speed. They were driving group formation with stability condition but some driver changed lane because of their response characteristics. Driving speed of the vehicles that running on the second lane is faster than the first lane. Almost the vehicles that running on the second lane are type B. The leading vehicles on the first lane are mostly type C. Those vehicles formed some groups and type A follow those leading vehicles. Following vehicles were doing delicate control their speed to keep their desire space headway. It is the merit of this following theory that introduced spiral curve. On the other hand, there were some irregular results in the case of using previous following theory. It shows that the vehicles running very short space headway situation.

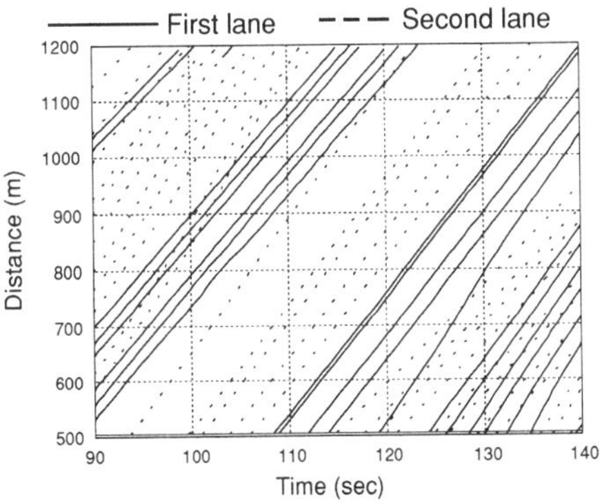

Figure 12 Result of time - distance chart calculated by simulation model

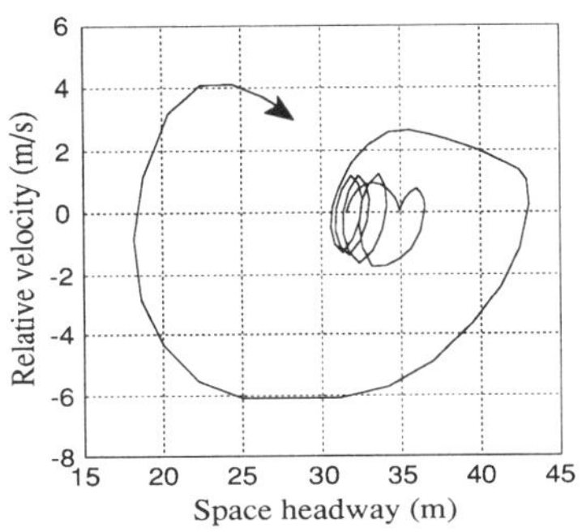

(a)

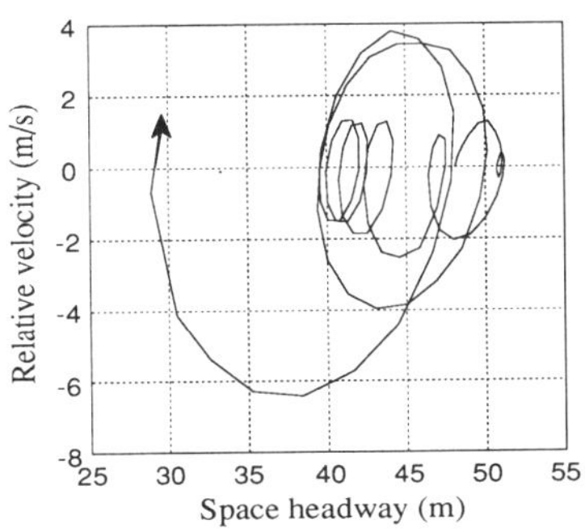

(b)

Figure 13 The locus of the following vehicle (From simulation)

Figure 13 shows the loci of the following vehicles that was calculated by this model. When the following vehicle has short space headway, the following vehicle decelerates ones speed to keep the safety headway. Those loci show the same as that of (a) and (b) in Figure 7. In figure 13 (a), it shows the stables driving in first stage and the vehicle that driving in front of object start accelerate, then object vehicle tried to follow within response delay time. It causes large diameter locus and going to his own space headway.

5 CONCLUSION

From those results of investigations and simulations, the conclusion could be expressed as follows.

(1) The desire speed of each driver can be observed on free traveling speed at straight road alignment condition.
(2) The border line for change lane was obtained by the data that the moment of lane change in consideration of response delay time of driver.
(3) The simulation model for vehicle running on expressway was developed using careful selected influence factors as internal, external and mutual interface condition.
(4) Personal space can be obtained by border line for change lane.
(5) Driving behavior on the space headway and relative velocity chart moved look like spiral curve rotating clockwise.
(6) Simulation model that introduced spiral curve has been development. It could be shows delicate control of vehicles.
(7) The individual driving behaviors could simulate many kinds of driving environments of road condition using this model and we could check it for understanding where is the dangerous point on the expressway and so on. It could use for the tool of safety driving.

REFERENCES

Azegami, Y. 1991. Traffic engineering. Riko Tosho : 73-75

Kondo,J. 1976. Numerical modeling. Maruzen : 233-242

Nakayama,H. 1992. Development of numerical modeling for traffic flow simulation. Proc. Electoronics Society meeting RTA-92 1992: 27-36

Impact of Computational Mechanics on Engineering Problems, Pulmano & Murti (eds)
© 1993 Balkema, Rotterdam, ISBN 90 5410 324 8

An adaptive *hp*-finite element method in frictional contact problems

J.T.Oden & C.Y.Lee
Texas Institute for Computational Mechanics, The University of Texas at Austin, Tex., USA

ABSTRACT: *Static* or *quasistatic* contact problems are described using the normal compliance law on the contact boundary and an *a posteriori* error estimator is developed for a nonlinear elliptic equation. Also an adaptive strategy is introduced and *hp*-finite element meshes are obtained through a procedure based on *a priori* and *a posteriori* error estimations. Numerical examples are given to support the theoretical results.

1 INTRODUCTION

Recently a new *a posteriori* error estimation theory was developed by Ainsworth and Oden [1, 2] which provided for the construction of element-wise *a posteriori* error estimates by a special dual formulation in which continuity conditions at interfaces are treated as constraints. An upper bound of the local energy norm of the error is obtained using a duality approach. This approach leads to error indicators which provide rigorous upper bounds of the global element errors and was extended to contact problems in [8]. Here, we introduce the adaptive strategy proposed by Oden, et al. [10] and show how good *hp*-finite element meshes can be obtained through a procedure based on *a priori* and *a posteriori* error estimations.

2 STATIC CONTACT PROBLEMS

We begin by considering the deformation of an elastic body unilaterally supported by a rigid foundation. The body occupies a domain Ω in $\mathbb{R}^N (N = 2, 3)$ which has a Lipschitz boundary Γ, and is subjected to body forces $\boldsymbol{f}$ throughout Ω and surface traction $\boldsymbol{t}$ applied to a portion Γ_F of Γ. The body is fixed along a portion Γ_D of Γ, and Γ_C denotes a candidate contact surface. The actual surface on which the body comes in contact with the foundation is not known in advance but is contained in Γ_C. The initial gap between the body and the foundation is defined by a function g, $g \geq 0$.

Points in Ω and on the boundary Γ are denoted by $\boldsymbol{x} = (x_1, x_2, \cdots, x_N)$ and s_i, $1 \leq i \leq N$, respectively. We confine our attention to infinitesimal deformations of the body and assume that the body has a linearly elastic behavior characterized by the generalized Hooke's law, $\sigma_{ij}(\boldsymbol{u}) = E_{ijkl}u_{k,l}$ in Ω $1 \leq i, j, k, l \leq N$ where the elasticities satisfy standard symmetry and ellipticity conditions.

The frictional contact behavior on the contact surface Γ_C is assumed to be governed by the compliant surface model studied by Oden and Martins [9] in which the stress satisfies the power law,

$$\begin{aligned} \sigma_n &= -c_n a^{m_n} \\ \boldsymbol{\sigma}_T &= c_T a^{m_T} \boldsymbol{\tau}, \quad \text{if sliding occurs} \end{aligned} \tag{1}$$

where c_n, m_n, c_T and m_T are material interface parameters, $\boldsymbol{\tau}$ is a unit vector tangent to the contact surface, and a is the penetration approach of the contact surface. In [9], extensive arguments are given in support of such phenomenological laws based on experimental, theoretical, and computational results.

Assuming sufficient smoothness for all the functions and data introduced in the above relations and equating the penetration approach to the normal distance between compliant surfaces, $a = (u_n - g)_+$, $(\cdot)_+$ denoting the positive part of the argument, with $\boldsymbol{u}_n$ the normal component of the displacement field $\boldsymbol{u}$, we arrive at the following system of equations and inequalities governing the displacement field $\boldsymbol{u}$:

$$\begin{aligned} \sigma_{ij}(\boldsymbol{u}),_j + f_i &= 0 \quad \text{in} \quad \Omega \\ u_i &= 0 \quad \text{on} \quad \Gamma_D \\ \sigma_{ij}(\boldsymbol{u}) n_j &= t_i \quad \text{on} \quad \Gamma_F \end{aligned} \tag{2}$$

on Γ_C

$$\sigma_n = -c_n(u_n - g)_+^{m_n}$$

when $\boldsymbol{u}_T = \boldsymbol{0}$

$$|\boldsymbol{\sigma}_T| < c_T(u_n - g)_+^{m_T}$$

when $\boldsymbol{u}_T \neq \boldsymbol{0}$

$$\begin{aligned} |\boldsymbol{\sigma}_T| &= c_T(u_n - g)_+^{m_T} \\ \boldsymbol{u}_T &= -\lambda \boldsymbol{\sigma}_T, \quad \text{for } \lambda \geq 0 \end{aligned} \tag{3}$$

In order to establish a weak form of this problem, we introduce the space V of admissible displacements defined by $V = \{\boldsymbol{v} \in [H^1(\Omega)]^N : \boldsymbol{v} = \boldsymbol{0} \text{ on } \Gamma_D\}$. We also assume that $f_i \in L^2(\Omega), t_i \in L^2(\Gamma_F), g \in H^{1/2}(\Gamma_C)$,and in order to make boundary integrals on Γ_C well defined, we put restrictions on the coefficients $c_n(\boldsymbol{s}), c_T(\boldsymbol{s}), m_n$ and m_T: $c_n(\boldsymbol{s}), c_T(\boldsymbol{s}) \in L^\infty(\Gamma)$ and $1 \leq m_n, m_T < \infty$, when $N = 2$ or $1 \leq m_n, m_T \leq 3$, when $N = 3$. These restrictions on m_n and m_T basically come from the embedding theorem which states that for $\boldsymbol{v} \in [H^1(\Omega)]^N, \gamma(\boldsymbol{v}) \in [L^q(\Gamma)]^N$ with $2 \leq q \leq \infty$ for $N = 2$, and with $2 \leq q \leq 4$ for $N = 3$.

Then the weak form of the frictional contact problem is

Find $\boldsymbol{u} \in V$ such that $\forall\, \boldsymbol{v} \in V$

$$\begin{aligned} a(\boldsymbol{u}, \boldsymbol{v} - \boldsymbol{u}) &+ j_n(\boldsymbol{u}, \boldsymbol{v} - \boldsymbol{u}) + j_T(\boldsymbol{u}, \boldsymbol{v}) \\ &- j_T(\boldsymbol{u}, \boldsymbol{u}) \geq F(\boldsymbol{v} - \boldsymbol{u}) \end{aligned} \tag{4}$$

where

$$\begin{aligned} a(\boldsymbol{u}, \boldsymbol{v}) &= \int_\Omega E_{ijkl} u_{i,j} v_{k,l}\, dx \\ j_n(\boldsymbol{u}, \boldsymbol{v}) &= \int_{\Gamma_C} c_n(\boldsymbol{s})(u_n - g)_+^{m_n} v_n\, ds \\ j_T(\boldsymbol{u}, \boldsymbol{v}) &= \int_{\Gamma_C} c_T(\boldsymbol{s})(u_n - g)_+^{m_T} |\boldsymbol{v}_T|\, ds \\ F(\boldsymbol{v}) &= \int_\Omega f_i v_i\, dx + \int_{\Gamma_F} t_i v_i\, ds \end{aligned} \tag{5}$$

The equivalence of (4) with problem (2)–(3) is established by standard methods.

3 *A POSTERIORI* ERROR ESTIMATION

We now introduce an *a posteriori* error estimator as appeared in [8] and confine our attention to two-dimensional problems for simplicity, which is not necessary.

Let $\boldsymbol{u}$ and $\boldsymbol{u}^h$ be the exact and approximate solutions, respectively. The finite element approximation error $\boldsymbol{e}$ is defined as $\boldsymbol{e} \in V$, $\boldsymbol{e} = \boldsymbol{u} - \boldsymbol{u}^h$. By using the bilinear form $a(\cdot,\cdot)$, we also define the energy norm as

$$\|\boldsymbol{v}\|_E = \sqrt{a(\boldsymbol{v}, \boldsymbol{v})} \tag{6}$$

We define local solution spaces V_K on a finite element Ω_K in a partition $\mathcal{P}$ so that $V(\mathcal{P}) = \prod_{\Omega_K \in \mathcal{P}} V_K$. With this local spaces V_K, we restrict the bilinear form $a(\cdot,\cdot)$, the linear functional $F(\cdot)$, and the nonlinear terms $j_n(\cdot,\cdot)$, $J(\cdot)$, and $T(\cdot)$ to a finite element Ω_K.

We introduce the averaging function $\alpha_{KL}^{(i)}(\boldsymbol{s})$ on the interelement boundaries Γ_{KL} shared by elements Ω_K and Ω_L such as (see [2])

$$\begin{aligned} \alpha_{KL}^{(i)} &: \Gamma_{KL} \to \mathbb{R}, \; k = 1, 2 \\ \alpha_{KL}^{(i)}(\boldsymbol{s}) &+ \alpha_{LK}^{(i)}(\boldsymbol{s}) = 1 \qquad \boldsymbol{s} \in \Gamma_{KL} \end{aligned} \tag{7}$$

In terms of this averaging function $\alpha_{KL}^{(i)}(\boldsymbol{s})$, we define an averaged normal stress on Γ_{KL} as

$$\begin{aligned} \left\langle \sigma_{ij}(\boldsymbol{u}^h) n_{Kj} \right\rangle_{1-\alpha} = \; &\alpha_{LK}^{(i)}(\boldsymbol{s}) \sigma_{ij}(\boldsymbol{u}_K^h) n_{Kj} \\ &+ \alpha_{KL}^{(i)}(\boldsymbol{s}) \sigma_{ij}(\boldsymbol{u}_L^h) n_{Kj} \end{aligned} \tag{8}$$

where i is not summed, $\boldsymbol{n}_K$ is an outward unit normal vector on $\partial\Omega_K$ and $\boldsymbol{u}_K = \boldsymbol{u}\big|_{\Omega_K}$ is restriction of $\boldsymbol{u}^h$ to Ω_K. The jump of $\boldsymbol{v} \in V(\mathcal{P})$ on the interelement boundaries Γ_{KL} is denoted by $[\![\boldsymbol{v}]\!]_i$.

We have

$$\begin{aligned} \|e\|_E^2 \leq \; &a_K(\hat{\boldsymbol{u}}_K - \boldsymbol{u}^h, \hat{\boldsymbol{u}}_K - \boldsymbol{u}^h) - 2\,[J_K(\hat{\boldsymbol{u}}_K) \\ &- J_K(\boldsymbol{u}^h) - j_{nK}(\hat{\boldsymbol{u}}_K, \hat{\boldsymbol{u}}_K - \boldsymbol{u}^h)] \end{aligned} \tag{9}$$

4 QUASISTATIC CONTACT PROBLEMS

We consider the motion of a linear-elastic body which is assumed to be very slow in comparison with the speed of elastic waves in the body. Time and derivatives with respect to time are denoted by t and $\dot{()}$, respectively, and we consider a time interval $t \in [0, T]$.

The frictional contact behavior on the contact surface Γ_C is assumed to be governed by the compliant surface model studied by Oden and Martins [9]

$$\sigma_n = -c_n(u_n - g)_+^{m_n}$$

when $u_n \leq g$,

$$\boldsymbol{\sigma}_T(\boldsymbol{u}) = 0$$

when $u_n > g$,

$$\begin{cases} |\sigma_T(u)| & < & c_T(u_n - g)_+^{m_T} \Longrightarrow \dot{u}_T = 0 \\ |\sigma_T(u)| & = & c_T(u_n - g)_+^{m_T} \\ & & \Longrightarrow \exists\ \lambda \geq 0 \text{ such that} \\ & & \dot{u}_T = -\lambda\sigma_T(u) \end{cases} \tag{10}$$

Note that in *dynamic* contact conditions, the tangential displacement, u_T, in the *static* case is replaced by the tangential velocity, $\dot{u}_T$.

Then the weak form of the *dynamic* frictional contact problem is

Find $u(t): (0,T] \rightarrow V$ such that

$$\begin{aligned} &\langle\rho\ddot{u}(t), v - \dot{u}\rangle + a(u(t), v - \dot{u}) \\ &+j_n(u(t), v - \dot{u}) + j_T(u(t), v) \\ &-j_T(u(t), \dot{u}) \geq F(v - \dot{u}) \qquad \forall\ v \in V \end{aligned} \tag{11}$$

with the initial conditions $u(0) = u_0$, $\dot{u}(0) = u_1$ where $\langle u, v\rangle = \int_\Omega u_i v_i dx$ The formal equivalence of (11) with the classical problem is established by standard methods [3, 7].

Now consider cases in which the inertia term is small enough to be neglected; for example, when the density of the material is small or the acceleration of the whole body is small. Then the inertia term can be omitted and we have the *quasistatic* formulation:

Find $u \in V$, $\forall\ v \in V$ such that

$$\begin{aligned} a(u(t), v - \dot{u}) \quad &+j_n(u(t), v - \dot{u}) + j_T(u(t), v) \\ &-j_T(u(t), \dot{u}) \geq F(v - \dot{u}) \end{aligned} \tag{12}$$

with the initial conditions $u(0) = u_0$.

A time discretized approximation of problem (12) can also derived as in [4, 7]. Owing to space limitations, we omit details and refer the readers to [8].

5 NUMERICAL EXPERIMENTS

5.1 Static contact problems

We consider a two-dimensional plane strain problem involving an infinitely long linearly elastic cylinder in contact with a rigid flat foundation. The configuration of the problem is illustrated in Fig. 1. Material properties are taken to be as: Young's modulus $E = 1.4 \times 10^3$ ($10^5 kg\ cm^{-1}s^{-2}$), Poisson's ratio $\nu = 0.25$, and $c_n = 1.0 \times 10^5$ ($10^5 kg\ cm^{-4}$), $c_T = 0.3 \times 10^5$ ($10^5 kg\ cm^{-3}s^{-1}$), $m_n = m_T = 2$. No body forces are applied, and external pressure is exerted on the upper boundary

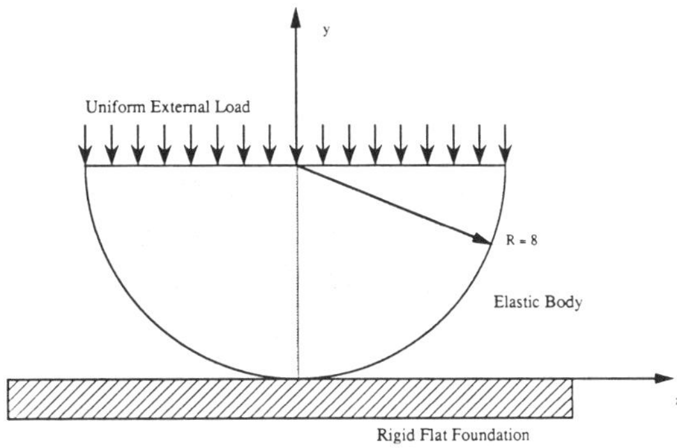

Figure 1: Configuration of the model problem

of the body. From the symmetry of the problem, only half of the body is considered. The regularization parameter is taken to be 10^{-1} and Newton's method is used to solve the resulting nonlinear problems. Since exact solutions are not available, we judge the performance of the error estimations by calculating the error based on the reference solutions, presumed to be very accurate, which are obtained by the same finite elements with element order $p = 8$. Table 1 shows the global and local errors and effectivity indices of estimated errors which are calculated by two different methods. A three step scheme (the "Texas 3-step" method of [10]) was adapted to the present class of problems; details are in [8]. The scheme calls for specifying a target error index, $\eta_T = \frac{\theta}{\|e\|_E}$ and an intermediate error index η_I, where θ is the estimated global error and $\|e\|_E$ is an energy norm of an (approximate) global error. Two different cases of the intermediate and target errors η_I and η_T were calculated. Test 1 is performed with $\eta_I = 0.12, \eta_T = 0.06, 0.04$ and in Fig. 2,

Table 1: Global and local errors and effectivity indices

nel.	true error	equilibrated	1/2	eff. (equil)	eff. (1/2)
1	0.0848165	0.0881246	0.0837193	1.04	0.99
2	0.0804220	0.1222878	0.1126967	1.52	1.40
3	0.0189538	0.0193209	0.0309098	1.02	1.63
4	0.0081559	0.0096104	0.0143641	1.18	1.76
5	0.0181394	0.0259803	0.0273508	1.43	1.51
6	0.0182997	0.0280273	0.0298395	1.53	1.63
7	0.0276181	0.0366286	0.0424145	1.33	1.54
8	0.0199138	0.0313325	0.0415174	1.57	2.08
9	0.0142488	0.0256926	0.0269879	1.80	1.89
10	0.0240408	0.0437543	0.0417718	1.82	1.74
11	0.0182348	0.0258899	0.0310386	1.42	1.70
12	0.0186214	0.0220363	0.0229575	1.18	1.23
13	0.0280526	0.0443224	0.0438456	1.58	1.56
14	0.0241867	0.0288076	0.0335939	1.19	1.39
15	0.0168413	0.0201029	0.0196873	1.19	1.17
total	0.1379361	0.1841161	0.1828515	1.33	1.33

the convergence of the solutions is compared with that of uniform h or p elements, with respect to number of degrees of freedom. As shown in the figure, rates of convergence are almost as steep as that of uniform p element case. The resultant final meshes are shown in Fig. 3 and 4 for $\eta_T = 0.06$ and $\eta_T = 0.04$, respectively. We expect that as a optimal mesh for this problem, low p and small h should appear around the point where the contact begins and higher p and bigger h as the element are located farther from the contact point. Hence, as seen in Fig. 3 and 4, the final meshes might not be "optimal", but the final errors are well distributed and controlled. Final target errors are calculated with equilibration for $\eta_I = 0.16, \eta_T = 0.04$ case and distribution of estimated local errors is shown in Fig. 5. From the figures it can be observed that final errors are relatively well distributed.

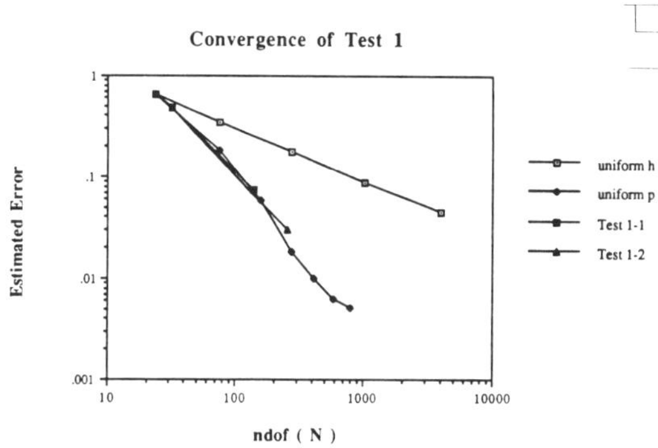

Figure 2: Convergence of the three-step method

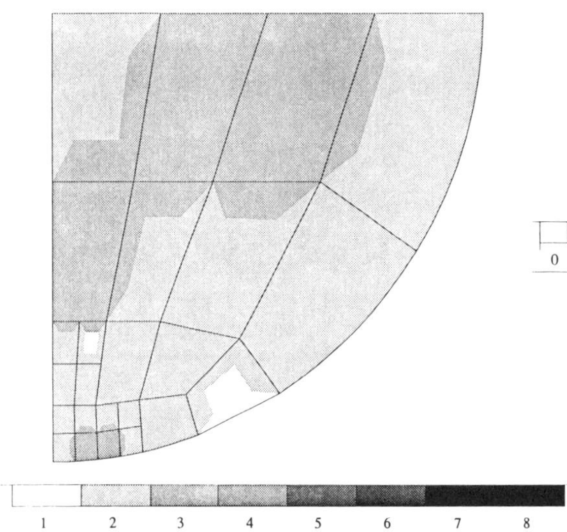

Figure 3: Final configuration of mesh ($\eta_I = 0.16, \eta_T = 0.04$)

Figure 4: Final configuration of mesh ($\eta_I = 0.16, \eta_T = 0.02$)

Figure 5: Distribution of local errors ($\eta_I = 0.16, \eta_T = 0.04$)

Test 2 is performed with $\eta_I = 0.12, \eta_T = 0.06, 0.04$ and in Fig. 6, the convergence of the solutions are compared with those of uniform h or p elements, with respect to number of degrees of freedom. The resultant final meshes are shown in Fig. 7 and 8, respectively. Final target errors are

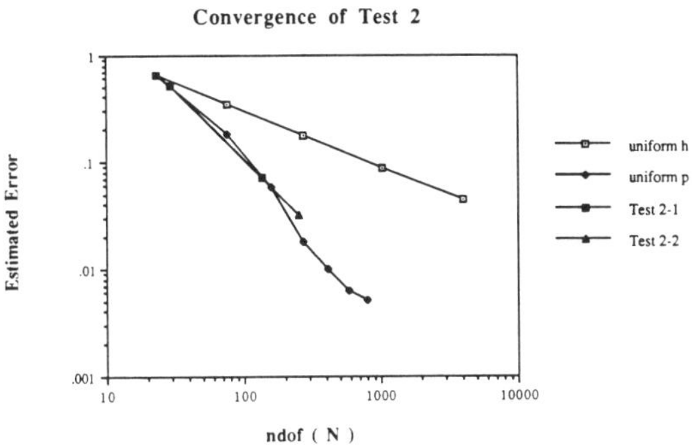

Figure 6: Convergence of the three-step method

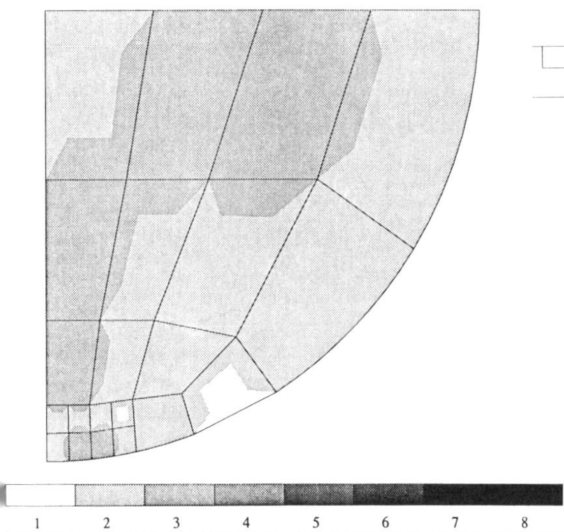

Figure 7: Final configuration of mesh ($\eta_I = 0.12, \eta_T = 0.06$)

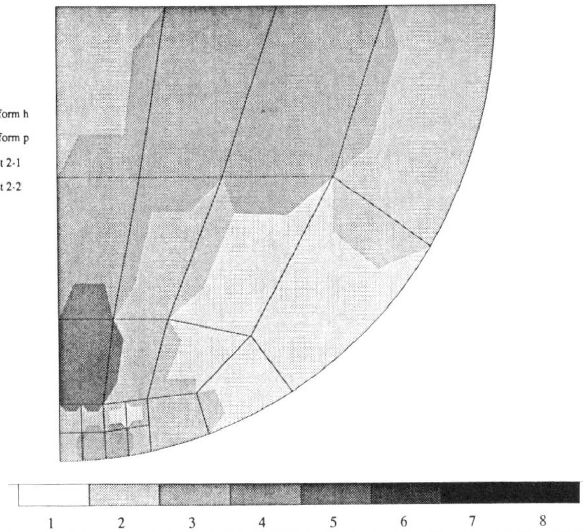

Figure 8: Final configuration of mesh ($\eta_I = 0.12, \eta_T = 0.04$)

calculated for $\eta_I = 0.12, \eta_T = 0.06$ case and distribution of estimated local errors is shown in Fig. 9. From those figures we observe the same results as Test 1.

5.2 Quasistatic contact problems

We introduced the *quasistatic* (incremental) problem and the regularization of the non-differentiable term of the variation inequality formulation. By substituting variables and scaling, we recast the regularized *quasistatic* problem into the same form as the *static* case in [7] and use the same error estimators including friction in the solution method. We consider the same problem mentioned above with the mass density $\rho = 7 \times 10^{-9}$ ($10^5 kg\ cm^{-3}$). Two different types of load histories are considered: ramp and sinusoidal, as shown in Figs. 10 and 11. The time step is taken as 1(sec) and the range of time is from 0 to 10(sec). The initial conditions used were the following: the cylinder rests on the rigid flat foundation without any external force and it is in *static* equilibrium position as $u(\boldsymbol{x}, 0) = 0$ and $\dot{u}(\boldsymbol{x}, 0) = 0$, $\boldsymbol{x} \in \Omega$.

The problem is solved using the *quasistatic* theory as in [7]. This problem is considered to be *quasistatic*, since the load is applied very slowly compared with the speed of elastic waves ($\approx 10^6\ cm\, s^{-1}$). The Newton-Raphson method was used to solve the resultant nonlinear system of equations at each time step and backward differences ($\theta = 1$) were used to discretize variables in time.

The three-step method is used as adaptive strategy at each time step. We choose to estimate errors with equilibration, which was verified to give tighter upper bound to errors than without equilibration. We follow the same steps as in the static case and apply the three-step method by taking the previous time step as the initial state of the current time. In other words, the final solutions and configuration of mesh of the previous time are used as the initial conditions and mesh of the current time step. But whenever elements are unrefined or unriched, information from the previous time steps are lost, which may lead incorrect initial conditions for current step. Hence in our numerical experiments, we do not consider unrefinements and keep all degree of freedom from the previous time steps, which may lead to create more degrees

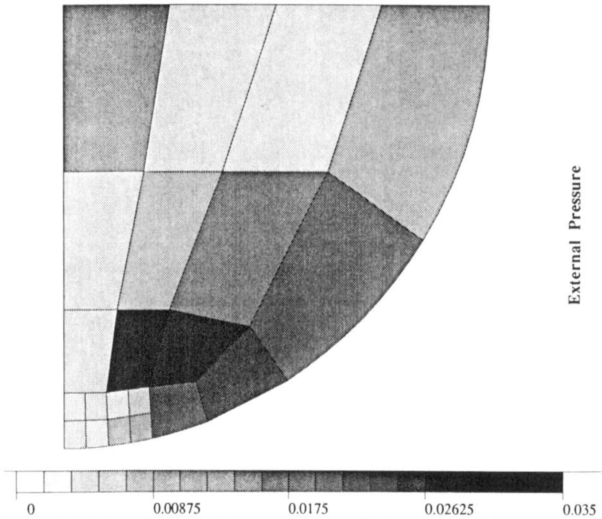

Figure 9: Distribution of local errors ($\eta_I = 0.12, \eta_T = 0.06$)

of freedom every time step. But we observed that after some time steps, we do not need more degrees of freedom to produce a given target error level. We also observed that since the previously adapted mesh is used at next time step, the initial error level is already low enough and h or p refinement, one of which is sufficient to satisfy the given error level, could be skipped at some time steps.

The final adapted mesh and distribution of local errors after five time steps are shown in Fig. 12 and 13 for load type 1. Most of time the intermediate parameter γ was taken as two to five, which is smaller than that suggested by [10], because the error levels are small enough not to need much h-type refinements. After time step 5, since the external force is decreasing no more new degrees of freedom was needed to tolerate the target error level. On the contrary, h or p unrefinements are suggested by the three-step method, which we avoid on purpose for the reasons stated earlier. Hence after time step 5, the same mesh as at time step 5 was used for the rest of time steps.

We could observe from the estimated errors that after each time step, local errors are distributed evenly over elements. Again, since no external force is present after time step 10, zero solution is obtained.

As a conclusion, we succeeded to solve the static or quasistatic contact problems with friction by using the three-step adaptive strategy. But some questions are not clearly answered about the choice of h or p refinements; i.e., choice of intermediate parameter γ.

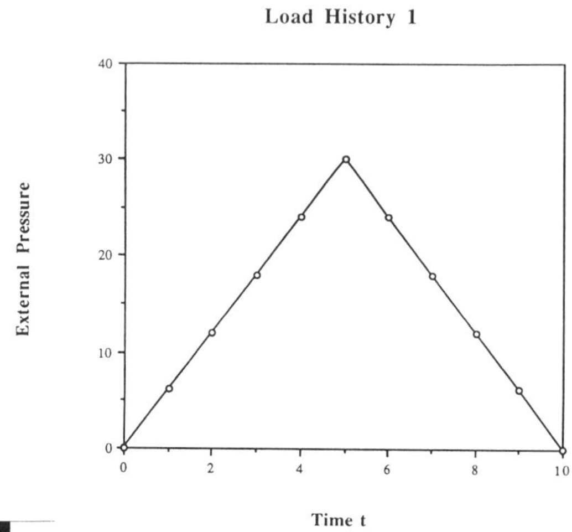

Figure 10: Time history of load type 1

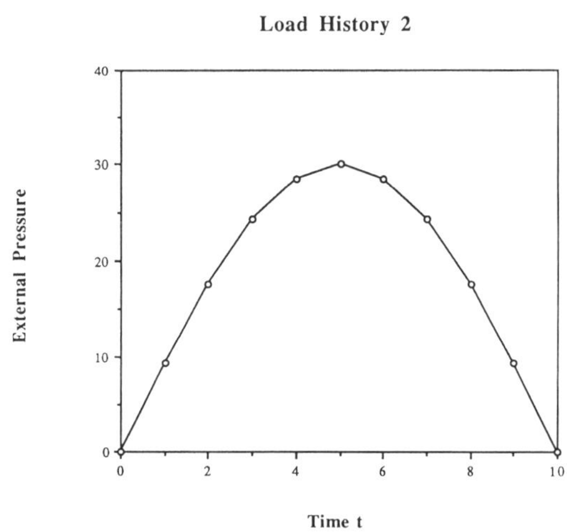

Figure 11: Time history of load type 2

Acknowledgments. The support of the U. S. Air Force Office of Scientific Research under Contract F49620-91-C-0011 is gratefully acknowledged. The partial support of one author, C. Y. Lee, under a grant from the Alcoa Foundation is also acknowledged.

References

[1] Ainsworth, M., and Oden, J. T., A Unified Approach to *A Posteriori* Error Estimation Using Element Residual Methods, *Numer. Math.* (to appear)

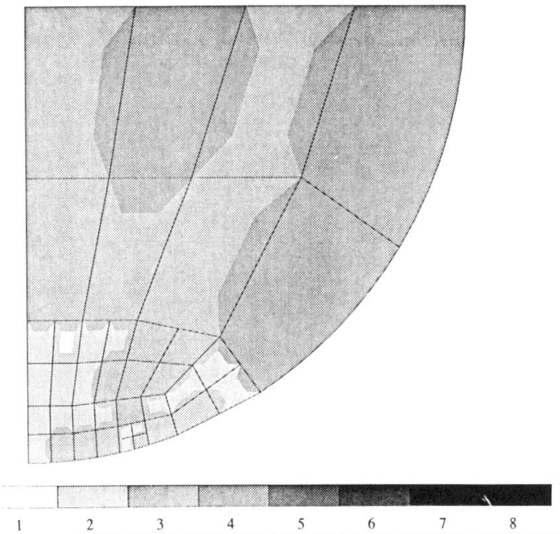

Figure 12: Final configuration of mesh (load type 1)

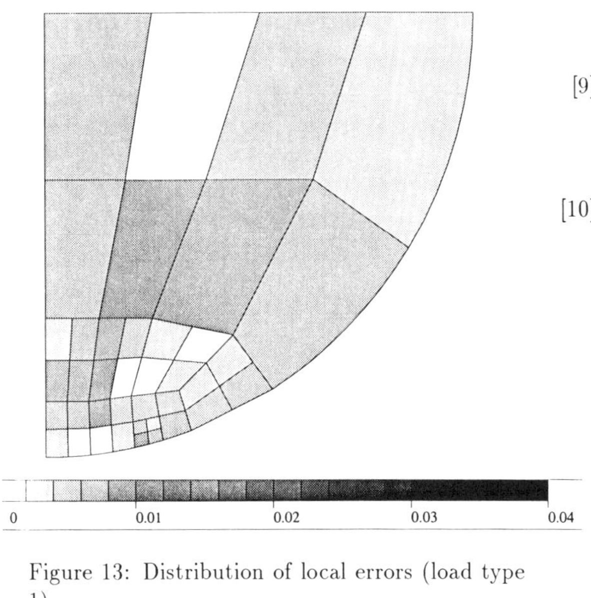

Figure 13: Distribution of local errors (load type 1)

[2] Ainsworth, M., and Oden, J. T., *A Posteriori* Error Estimations for Second–Order Elliptic Systems, Part 2. An Optimal Order Process for Calculating Self–Equilibrating Fluxes, *Computers and Mathematics with Applications* (to appear)

[3] Kikuchi, N., and Oden, J. T., Contact Problems in Elasticity, *SIAM*, Philadelphia, 1988

[4] Klarbring, A., Mikelic, A., and Shillor, M., Frictional Contact Problems With Normal Compliance, *Int. J. Engng. Sci.*, **26**, pp. 811–832, 1988

[5] Klarbring, A., Mikelic, A., and Shillor, M., On Friction Problems With Normal Compliance, *Nonlinear Anal. TMA*, **13**, pp. 935–955, 1989

[6] Lee, C. Y., and Oden, J. T., *A Priori* Error Estimation of *hp*-Finite Element Approximations of Frictional Contact Problems With Normal Compliances, *International Journal of Engineering Science* (to appear)

[7] Lee, C. Y., and Oden, J. T., Theory and Approximation of Quasistatic Frictional Contact Problems, *Comp. Meth. App. Mech. Eng.* (to appear)

[8] Lee, C.Y. and Oden, J.T., A Posteriori Error Estimation of hp-Finite Element Approximations of Frictional Contact Problems, (preprint)

[9] Oden, J. T., and Martins, J. A. C., Models and Computational Methods for Dynamic Friction Phenomena, *Comp. Meth. App. Mech. Eng.*, **52**, pp. 527–634, 1985

[10] Oden, J. T., Patra, A. and Feng, Y., An hp Adaptive Strategy, *Adaptive Multilevel and Hierarchical Computational Strategies*, A.K. Noor (ed.), **AMD-Vol. 157**, ASME Publications, pp. 23–46, 1992

Impact of Computational Mechanics on Engineering Problems, Pulmano & Murti (eds)

 ISBN 90 5410 324 8

Finite element analysis of steel fibre reinforced cement composites

P. Paramasivam
National University of Singapore, Singapore

ABSTRACT: A three dimensional finite element analysis based on micromechanics approach has been employed to determine the linear and non-linear behaviour of steel fibre reinforced cement composites under uniaxial loading. Nine numerical models of the composites are presented by varying the parameters such as volume fraction, aspect ratio and geometric arrangement of the fibres. The results obtained from the finite element analysis are compared with the experimental results and discussed.

1 INTRODUCTION

It has been well established that fibre reinforced cement composites provide better crack arrest mechanism, increased cracking strength, toughness, improved resistance to thermal shock and higher fatigue strength[1,2]. Prediction of the mechanical properties and micromechanical understanding of this material can be very useful in the design and selection of a composite material for any particular application. Extensive experimental and theoretical investigations have been carried out by many researchers [3-5] to predict the mechanical properties of cement composites with continuous and discontinuous (short) fibres, in terms of the mechanical and geometrical properties of the constituent material using the law of mixtures and spacing concepts. These concepts can predict only the first crack and ultimate strength of the composites, but it is extremely difficult to predict the composite behaviour after elastic limit and also the interaction between the constituent materials. Therefore finite element analysis has been employed to study the linear and non-linear behaviour of steel fibre composites under uniaxial loading.

The finite element is a powerful numerical tool for composite micromechanics analysis. It has been extensively used to determine the behaviour of ductile matrix-fibre composites with fibre volume content of 20% to 80% such as polymer or resin based composites [6,7]. Most of these analyses were based on axisymmetric and plane strain models. A fully three dimensional finite element analysis was used by Curiskis [8,9] to study the micromechanics behaviour of uni directional ductile matrix composites with fibre volume of 20% to 80% and aspect ratio of 10-20. Later this method was extended to brittle-matrix composites, such as cement paste reinforced with steel fibres, under uniaxial loading by Paramasivam et al. [10]. In their investigation, validity of the technique for low volume fraction of fibres with aspect ratio of upto 53 was examined and the numerical results were compared with limited experimental results for only two volume fractions with one aspect ratio of the fibre. In this investigation, fibre composites with different parameters such as volume fraction, aspect ratio and geometric arrangement have been consiered and numerical results are compared with the experimental results and discussed.

2 FINITE ELEMENT METHOD

The finite element programs entitled SPLIT1 and SPLIT2 which were developed by Valliappan and Curiskis [8] have been employed in this study. The constituent materials of the composite (fibre and composite) are represented by 20 node isoparametric elements. The advantage of this concept is that it achieves the analysis efficiently by allowing an arbitrary discretization of the elements of different materials even with curved boundaries. The constituents are assumed to be isotropic and perfect bond is assumed at the interface between the fibre and matrix. The elastic-plastic behaviour of the materials are considered for the analysis. The plasticity analysis is based on Von Mises yield criterion with the associated flow rule. The initial stress method [11,12] has been used in the iterative technique for the elastic-plastic analysis.

3 COMPOSITE ANALYSIS

In the real problem of steel fibre reinforced cement composites, the fibres are randomly oriented three dimensionally as shown in Fig. 1(a). Generally the mechanical properties of such composites are determined by idealising the fibres in unidirection using the various expressions for the orientation factors [4]. In the present study, the discrete short fibres are arranged in a regular three dimensional triply-periodic array geometry using the orientation factor. The geometric idealisation of the composites is shown in Fig. 1. The fibres are assumed to be identical cylinders with constant length, cross-section and geometry with perfect square packing in the transverse plane. In the longitudinal direction, the fibres are considered either in row and column or staggered arrangement as shown in Figs. 1(b) and 1(c) respectively. The staggered arrangement is probably more like an actual uni-directional composite. Because of the symmetry, a typical repeating unit such as that indicated by the dashed lines in Fig. 1 can be isolated for detailed micro-mechanics analysis using three dimensional finite element method. The main aspects and details of the formulation of the boundary conditions of the finite element are presented in reference [8,10,13].

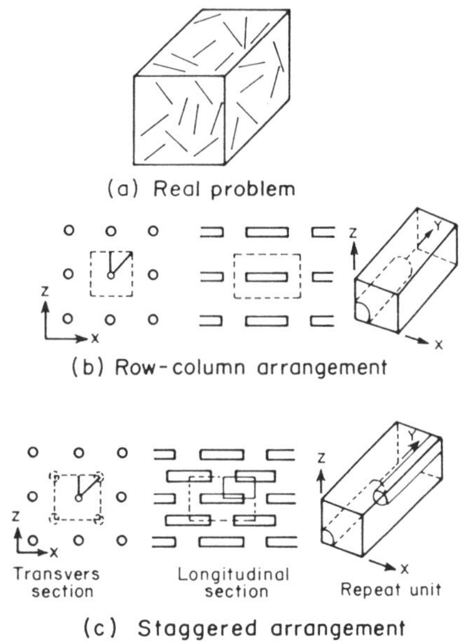

Fig. 1 Idealised geometry of composites

4 NUMERICAL EXAMPLES

The finite element programmes were used to analyse the mechanical properties of nine different cases of steel fibre reinforced composites as shown in Table. 1. The first four cases were adopted to investigate the influence of fibre volume fraction on characteristic properties of the composite by keeping the aspect ratio (l_f/D) constant at 53.6. The next three cases were used by varying the aspect ratio at constant volume fraction of 4% and the last two cases were used to investigate the effect of geomertric arrangements on the properties of the composites. The material properties used in these examples are the same as obtained by Nathan et al. [3] from their experimental

Table 1 Geometric parameters for numerical models

No.	V_f%	L_1/D	L_1 (mm)	L_2 (mm)	a (mm)	Model
1	5	53.6	30	4.0	1.43	row-column
2	4	53.6	30	4.0	1.61	row-column
3	3	53.6	30	4.0	1.85	row-column
4	2	53.6	30	4.0	2.77	row-column
5	4	35.7	20	2.67	1.61	row-column
6	4	53.6	30	4.0	1.61	row-column
7	4	71.4	40	5.33	1.61	row-column
8	2	53.6	30	4.0	4.54	staggered
9	3	53.6	30	4.0	3.70	staggered

investigation and are given in Table. 2.

For the finite element modelling, only one-eighth of the cross section and one quarter of the longitudinal section have been considered. The orientation factor of 0.47 proposed by Nathan et al. [3] is used for unidirectional idealisation. The finite element divisions of row-column and staggered arrangements are shown in Fig. 2. The geometric parameters relevant to all these cases are given in Table. 1. The mesh with 81 elements associated with 437 nodal points has been employed for the finite element analysis. The composite properties of elastic modulus and Poisson's ratio are tabulated in Tables 3 and 4 for the different volume fraction with constant aspect ratio of 53.6 and for different aspect ratios with constant volume fraction of 4% respectively. It has been seen that the modulus of elasticity of composites are higher for a larger volume fraction and smaller aspect ratios of the fibres. The modulus of elasticity for the staggered arrangement is slightly higher than the row-column arrangement. The values of the Poisson's ratio are not affected.

Typical contours of longitudinal stress in the composite are as shown in Figs. 3 and 4 for the longitudinal section as well as the cross section in the case of row column arrangement with V_f = 4% for aspect ratios of 53.6 and 35.7 respectively. The stresses shown have been obtained for an applied stress of unity. The stress contours plotted for the cross sections in Figs. 3 and 4 shows the existance of triaxial stress state. It can be seen from the longitudinal section along sections A-A and B-B as shown in Figs. 3 and 4 that stress concentration exists near the fibre end as expected and the stress distribution away from the fibre end is almost uniform and the value is close to the applied load. A comparison of these stress values between the row-column and

Table 2 Properties of constituent materials (From Ref. 4)

	Cement Paste	Steel Fibre
Modulus of Elasticity (N/mm^2)	15900	196500
Poisson's ratio	0.2	0.4
Ultimate tensile strength (N/mm^2)	1.45	374
Crushing Strength (N/mm^2)	36.16	

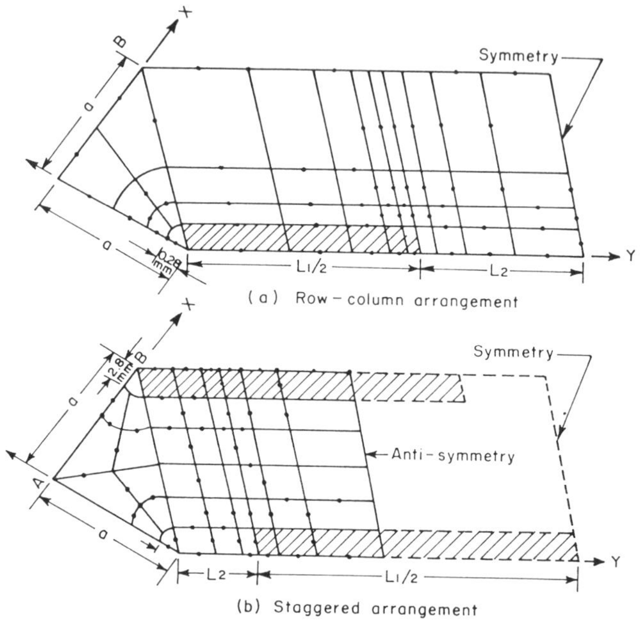

Fig. 2 Finite element idealisation

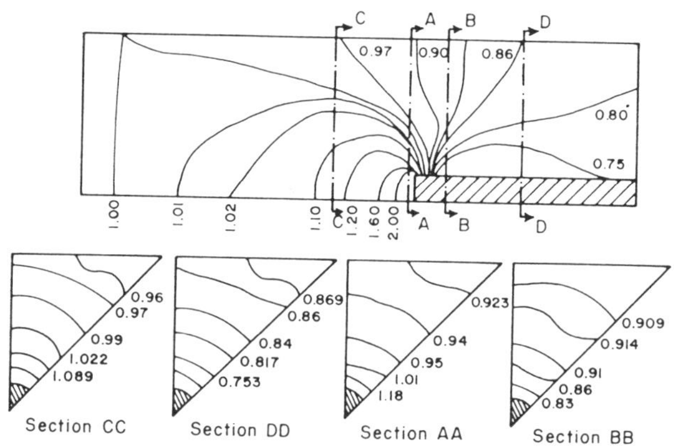

Fig. 3 Stress Contours - row and column arrangement ($V_f = 4\%$, $L_1/D = 53.6$)

staggered arrangement for the same value of volume fraction and aspect ratio are shown in Figs. 5(a) and 5(b) indicates that the stress concentration in the row-column case is higher than that for the staggered arrangement.

The linear and non-linear behaviour of fibre composites of the nine cases were carried out. The predicted stress-strain curves for the composites with various volume fractions and aspect ratio of the fibres, are plotted in Figs. 6(a) to 6(f). These results are compared with experimental results by Nathan et al. [3]. A comparison between the composite properties obtained from the finite element micromechanics analysis and those from the experimental investigation of Nathan et al [3] is given in

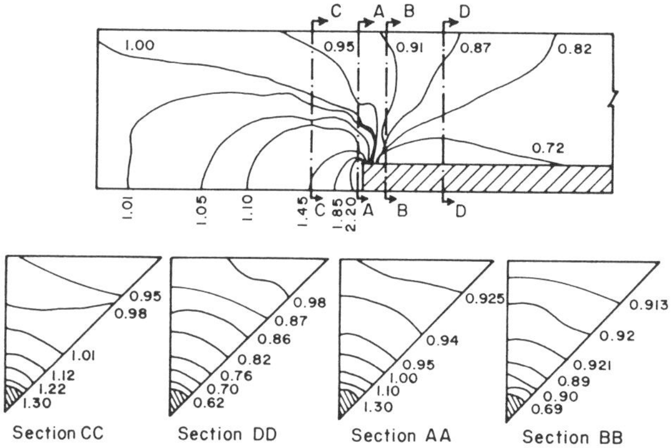

Fig. 4 Stress contours - row and column arrangement (V_f = 4%, L_1/D = 35.7)

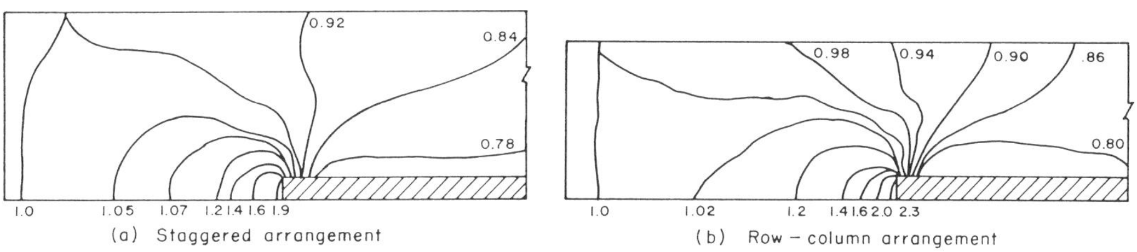

Fig. 5 Stress contours for different geometric arrangement (V_f = 3%, l_1/D = 53.6)

Table 3 Composite properties for different volume fractions and geometric arrangements.

V_f	L_1/D	Modulus of elasticity (N/mm^2)		Poisson's ratio	
		Row-col	Stagg	Row-col	Stagg
2	53.6	17 502	17 896	0.2 013	0.2 013
3	53.6	18 260	18 362	0.2 019	0.2 019
4	53.6	18 983	-	0.2 025	-
5	53.6	19 809	-	0.2 030	-

Table 4 Composite properties for different aspect ratios

V_f	L_1/D	Modulus of elasticity (N/mm^2)	Poisson's ratio
4	35.7	19 114	0.2 024
4	53.6	18 983	0.2 025
4	71.4	18 897	0.2 024

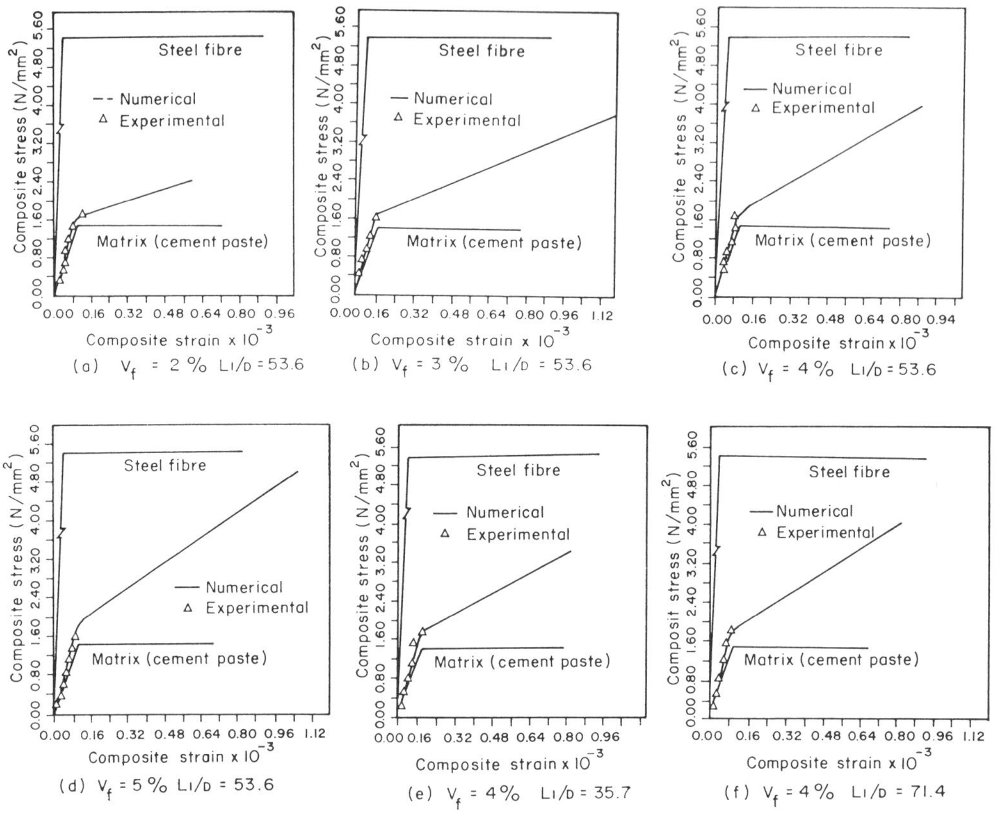

Fig. 6 Comparison of numerical and experimental stress-strain curves

Table 5 Comparison between numerical and experimental values of first crack and ultimate strength for different volume fractions with $L_l/D=53.6$

V_f%	Stress at first crack (N/mm^2) Numerical Row-col.	Stagg.	Expt.	Ultimate stress (N/mm^2) Numerical Row-col.	Stagg.	Expt.
2	1.60	1.59	1.58	2.05	1.84	1.79
3	1.68	1.76	1.84	2.79	2.72	2.61
4	1.72	-	2.07	2.98		3.85
5	1.90	-	2.23	4.94	-	5.25

Tables 5 and 6 for different volume fractions with an aspect ratio of 53.6 and different aspect ratios with a volume fraction of 4% respectively. It can be seen from the Tables 5 and 6 that the numerical results favour comparably with the experimental results. The composites with staggered arrangement give slightly closer value to the experimental values compared to the row-column arrangement.

A typical variation of the longitudinal stress

Table 6 Comparison between numerical and experimental values of first crack and ultimate strength for different aspect ratios with $V_f=4\%$

L_1/D	Stress at first crack (N/mm^2)		Ultimate stress (N/mm^2)	
	Numerical	Experimental	Numerical	Experimental
35.7	1.70	1.85	2.39	2.49
53.7	1.72	2.07	2.98	3.85
71.4	1.80	2.17	4.72	4.55

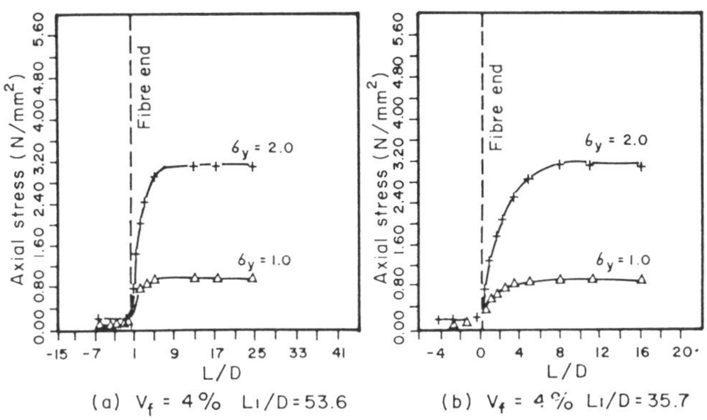

Fig. 7 Longitudinal stress near fibre axis

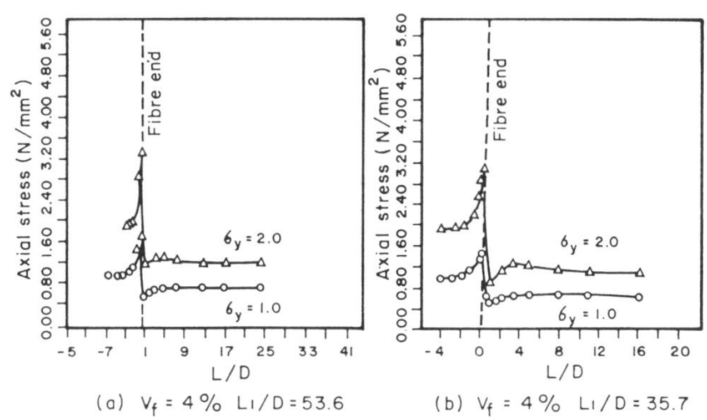

Fig. 8 Matrix axial stress near interface

near the fibre axis for the composite with $V_f=4\%$ and different aspect ratios is given in Figs. 7(a) and 7(b) respectively. Two curves corresponding to axial stress before yielding ($\sigma_y=2.0$) are plotted. It exhibits the nature of load transfer from the matrix to the fibre at the matrix interface. The stress infront of the fibre is almost constant at the applied stress until just slightly infront of the fibre end, where it increases sharply and attains the maximum value about 8 times the aspect ratio and remains constant there after.

Similar variation for the other cases are also observed. Figs. 8(a) and 8(b) show the typical variation of longitudinal stress in the matrix near the interface of fibre and matrix for a composite with $V_f = 4\%$ and different aspect ratios of 53.6 and 35.7 respectively. In these figures two curves are plotted - one just before yielding ($\sigma_y = 1.0$) and another after yielding ($\sigma_y = 2.0$). It can be seen that the longitudinal stress increases sharply infront of the fibre to a maximum value and then drops sharply to a lower value of about 1 time the aspect ratio behind the fibre end axis. After this, the stress starts rising again and reaches another peak at amtrix yielding, at about 8 times the aspect ratio and remains constant thereafter. These observations are similar to those reported by Agarwal et al [7].

5 CONCLUSION

A three dimensional finite element model using the micromechanics approach has been adopted to study the linear and non-linear behaviour of steel fibre reinforced composites under uniaxial loading. Numerical examples for which the experimental results are available, have been illustrated to asses the validity of the approach. From the results obtained it can be seen that the comparison is reasonably good. The study shows usefulness of the method to understand the behaviour of composites and the interaction between matrix and fibre.

6 REFERENCES

1. Batson, G. 1973. State-of-the-art report on fibre reinforced concrete. Journal of the ACI. 11 : 729-744.

2. Johnson, C.D. 1973. Steel fibre reinforced mortar and concrete : A review of mechanical properties. ACI Publication SP-44, 127-142. Detroit : ACI.

3. Nathan, G.K, Paramasivam, P, and Lee, S.L., 1977. Tensile behaviour of fibre reinforced cement paste. Journal of ferrocement 7(2) 59-77.

4. Ganeshalingam, R., Paramasivam, P., and Nathan, G.K. 1981. An evaluation of theories and a design method of fibre cement composites. International Journal of Cement composites and Lightweight concrete 3 103-114.

5. Swamy, R.N., Mangat, P.S., and Rao, C.V.S.K. 1973. The mechanics of fibre reinforcement of cement matrices. ACI Publication SP-44, 1-28. Detroit : ACI.

6. Adams, D.F 1974. Practical problems associated with the application of the finite element method to composite micromechanical analysis. Fibre Science and Technology 7 : 111-122.

7. Agarwal, B.D., and Broutman, L.J. 1974. Three dimensional finite element analysis of spherical particle composites. Fibre Science and Technology. 7:63-74.

8. Curiskis, J.I 1978. A study of the micromechanics of fibrous-reinforced composite materials using finite element techniques, Ph.D. Thesis, University of New South Wales, Sydney.

9. Curiskis, J.I 1980. On the determination of the non-linear longitudinal mechanical behaviour of of short fibre composites by a three dimensional finite element procedure. Proceedings Symposium of the Third International Conference on Composite Materials 1 : 796-811, Paris.

10. Paramasivam, P, Curiskis, J.I, and Valliappan, S. 1984. Micromechanics Analysis of Fibre Reinforced Composites. Fibre Science and Technology 20 : 99-120.

11. Zienkiewicz, O.C., Valliappan, S., and King, I.P. 1969. Elasto-plastic solution of engineering problems "Initial stress method" finite element approach. International Journal of Numerical Methods in Engin 1:75- 100.

12. Zienkiewicz, O.C. 1977. The finite element method, 3rd edn., London : McGraw Hill Book Co.

13. Curiskis, J.I., and Valliappan, S. 1978. A solution algorithm for linear constraint equations in finite element analysis. Computers and Structures 8 : 117-124.

Impact of Computational Mechanics on Engineering Problems, Pulmano & Murti (eds)
© 1993 Balkema, Rotterdam, ISBN 90 5410 324 8

Measures of fuzziness in FE geotechnical solutions

T. D. Pham
University of New South Wales, N.S.W., Australia

ABSTRACT: This paper deals with imprecision of the input geotechnical parameters required in FE (finite element) analysis. This uncertainty is considered to be non-random or fuzzy. Therefore, fuzzy sets are introduced in the formulation of the elasticity matrix. The possibility distributions of finite element results are investigated using fuzzy entropies, and fuzzy measures are adopted to derive the error rate of subjective estimates.

1 INTRODUCTION

Soil properties are not completely known in practice; hence the sense of exactness is doubtful in the analysis of geotechnical structures. In this context, the uncertainty of geotechnical information is addressed not in view of probability principles but within the conceptual framework of fuzzy-set theory.

Since the statistical data of soils are hardly obtained and regardless of the magnitude of testing programs, engineering judgment certainly plays an important role in the design process. Expert estimates of soil parameters are assessed and fuzzified in the deterministic finite element analysis of an elastic soil medium loaded at the surface. With fuzzy input, the results will be presented in terms of possibility distributions which show the lower and upper bounds of displacements and stresses at certain membership grades.

The fuzzy behavior of the soil medium is further investigated by considering the fuzzy entropies of the possibility distributions. Finally, Sugeno's fuzzy measure is used to derive the error rate of subjective estimates.

2 FUZZY SETS

The notion of fuzzy sets was first introduced by Zadeh (1965) to describe the imprecise real-life classes which may not be solved using the clear-cut logic of classical mathematics. A fuzzy set is a class of objects in which there is a continuum of grades of membership. Such a set is characterized by a membership function which maps each object to a grade of membership with a value ranging from zero to one in a gradual manner.

2.1 *Notation of a fuzzy set*:

Let U = {x} be a space of objects. A fuzzy set A in U is the union of the ordered pairs and can be expressed as

$$A = \int_U \mu_A(x)/x \qquad (1)$$

where the integral sign denotes the union of the fuzzy singletons $\mu_A(x)/x$. $\mu_A(x)$ is the membership grade of x in A. The "/" sign is used to separate $\mu_A(x)$ from x for convenient reading. If A is discrete then Eq.(1) may be written as

$$A = \mu_1/x_1 + \ldots + \mu_n/x_n \qquad (2)$$

in which μ_i is the membership grade of x_i in A, i = 1,n.

2.2 *Complement of a fuzzy set*

The complement of A is denoted by Ā and defined by

$$\mu_{\bar{A}}(x) = 1 - \mu_A(x) \tag{3}$$

2.3 *α-level set*

An α-level set of A is a non-fuzzy set denoted by A_α which consists all elements in U whose membership grade is greater or equal to α:

$$A_\alpha = \{\mu_A(x)/x,\ \mu_A(x) \geq \alpha\} \tag{4}$$

2.4 *Convex fuzzy set*

A is convex if and only if

$$\mu_A[\eta x_1 + (1-\eta)x_2] \geq Min[\mu_A(x_1), \mu_A(x_2)] \tag{5}$$

for all x_1 and x_2 in U and all η in [0,1].

2.5 *Operations on fuzzy numbers*

If two fuzzy numbers N and M can be defined as

$$N = [n_1\ ,\ n_2] \tag{6a}$$
$$M = [m_1\ ,\ m_2] \tag{6b}$$

where (n_1 and n_2) and (m_1 and m_2) denote the left and the right values of N and M respectively; then the addition, subtraction, multiplication and division of two fuzzy numbers N and M and at an α-level (Kaufmann & Gupta 1985) can be defined as

$$N_\alpha(+)M_\alpha = [n_1^{(\alpha)} + m_1^{(\alpha)},\ n_2^{(\alpha)} + m_2^{(\alpha)}] \tag{7}$$

$$N_\alpha(-)M_\alpha = [n_1^{(\alpha)} - m_2^{(\alpha)},\ n_2^{(\alpha)} - m_1^{(\alpha)}] \tag{8}$$

$$N_\alpha(\times)M_\alpha = [n_1^{(\alpha)} \times m_1^{(\alpha)},\ n_2^{(\alpha)} \times m_2^{(\alpha)}] \tag{9}$$

$$N_\alpha(:)M_\alpha = [n_1^{(\alpha)} : m_2^{(\alpha)},\ n_2^{(\alpha)} : m_1^{(\alpha)}] \tag{10}$$

where the symbol (.) stands for the fuzzy arithmetic operator.

3 MEMBERSHIP FUNCTIONS OF GEOTECHNICAL PARAMETERS

Elastic modulus and Poisson's ratio are modeled as convex fuzzy sets whose membership functions have been proposed as either triangular or pi functions (Valliappan & Pham 1993 b). The construction of membership functions of engineering parameters with objective and subjective information has also been studied by Valliappan and Pham (1993 a). In this paper, the elastic modulus and the Poisson's ratio of soil are characterized by the convex fuzzy sets with triangular membership functions which can be defined by

$$\mu(x) = \begin{cases} 0 & for\ x \leq L' \\ \dfrac{x-L'}{P-L'} & for\ L' \leq x \leq P \\ \dfrac{H'-x}{H'-P} & for\ P \leq x \leq H' \\ 0 & for\ x \geq H' \end{cases} \tag{11}$$

and

$$L' = \begin{cases} P-2(P-L) & for\ P \geq 2(P-L) \\ 0 & for\ P \leq 2(P-L) \end{cases} \tag{12}$$

$$H' = P + 2(H-P) \tag{13}$$

where L, P and H are expert estimates of low, most likely and high values respectively.

4 FUZZIFICATION OF THE ELASTICITY MATRIX

In finite element analysis, the load-displacement relation is written as

$$[K]\ \{\delta\} = \{P\} \tag{14}$$

where [K] is the stiffness matrix, {δ} is the nodal displacement vector, and {P} is the nodal force vector.

In two-dimensional finite element analysis, the stiffness matrix of an element is expressed as

$$[k] = \iint [B]^T[D][B] \; t \; dx \; dy \qquad (15)$$

where [B] is the strain-displacement matrix, [D] is the elasticity matrix and t is the element thickness.

For plane-strain idealization in an isotropic material, [D] will be fuzzified to give the lower and upper bounds of displacements using the operations of fuzzy numbers at an α-level as follows.

$$[D]_L^\alpha = \frac{E}{(1+\nu_L^\alpha)(1-2\nu_R^\alpha)} \begin{bmatrix} 1-\nu_L^\alpha & \nu_R^\alpha & 0 \\ \nu_R^\alpha & 1-\nu_L^\alpha & 0 \\ 0 & 0 & \frac{1-2\nu_R^\alpha}{2} \end{bmatrix} \qquad (16)$$

$$[D]_R^\alpha = \frac{E}{(1+\nu_R^\alpha)(1-2\nu_L^\alpha)} \begin{bmatrix} 1-\nu_R^\alpha & \nu_L^\alpha & 0 \\ \nu_L^\alpha & 1-\nu_R^\alpha & 0 \\ 0 & 0 & \frac{1-2\nu_L^\alpha}{2} \end{bmatrix} \qquad (17)$$

where L and R stand for the left and the right values of a fuzzy number respectively. E and ν are elastic modulus and Poisson's ratio respectively. And with the understanding that the elements D_{33} in the above two matrices in Eqs. (16) & (17) are derived to yield fuzzy shear moduli.

5 FUZZY ENTROPY

The entropy of a fuzzy set A denoted by d(A) which was introduced by De Luca and Termini (1972) is the measure of the degree of fuzziness of A and satisfies at least the following properties.

p1: $d(\mu_A)$ must be 0 if and only if μ_A takes the value 0 or 1 (crisp set).

p2: $d(\mu_A)$ must assume the maximum value if and only if μ_A assumes always the value 0.5 (maximally fuzzy).

p3: $d(\mu_A)$ must be greater or equal to $d(\mu^*_A)$ where μ^*_A is sharper than μ_A, such that $\mu^*_A \geq \mu_A$ if $\mu_A \geq 0.5$ and $\mu^*_A \leq \mu_A$ if $\mu_A \leq 0.5$.

This fuzzy entropy is defined by

$$d(A) = K \sum_{h=1}^{N} S(\mu(x_h)) \qquad (18)$$

where K is a constant and $S(\mu(x_h))$ is expressed using Shannon's function:

$$S(\mu(x_h)) = -\mu(x_h) \ln \mu(x_h) - (1-\mu(x_h)) \ln (1-\mu(x_h)) \qquad (19)$$

with the assumption that 0 ln 0 = 0.

The normalized version of d(A) is obtained by the functional

$$d_n(A) = \frac{1}{N} \sum_{h=1}^{N} S(\mu(x_h)) \qquad (20)$$

where $S(\mu(x_h))$ is now expressed in logarithm in base 2.

Another measure of fuzziness is also suggested in terms of the Minkowski class of metric distances (Klir & Folger 1988):

$$d_p(A) = |X|^{1/p} - D_p(A, \bar{A}) \qquad (21)$$

where |X| denotes the cardinality of the universal set X (x ε X), p ε [1,∞] and

$$D_p(A, \bar{A}) = \left[\sum_{x \in X} \delta_A^p(x)\right]^{1/p} \qquad (22)$$

with

$$\delta_A(x) = |\mu_A(x) - \mu_{\bar{A}}(x)| \qquad (23)$$

It is noted that when p = 1 or 2, Eq.(21) involves the Hamming distance or the Euclidean distance respectively. The measure of fuzziness $d_p(A)$ can be normalized as

$$d_{n,p}(A) = 1 - \frac{D_p(A, \bar{A})}{|X|^{1/p}} \qquad (24)$$

6 FUZZY MEASURE

The concept of fuzzy measure was introduced by Sugeno (1977). The definition of Sugeno's fuzzy measure is different from that of the measure of fuzziness as described in section 5. In contrast with probability measures which are used as scales for randomness, fuzzy measures are subjective scales for fuzziness and useful in a fuzzy environment where human subjectivity plays an important role.

Let X be an arbitrary set and $\mathfrak{B}$ a Borel field of X. A set function g defined on $\mathfrak{B}$ is a fuzzy measure if it has the following axioms.

A1 (boundary conditions): $g(\varnothing)=0$, $g(X)=1$;
A2 (monotonicity): If A, B ε $\mathfrak{B}$ and $A \subseteq B$, then $g(A) \leq g(B)$;
A3 (continuity): If F_n ε $\mathfrak{B}$ for $1 \leq n \leq \infty$ and a sequence $\{F_n\}$ is monotone, then $\lim g_{n\to\infty}(F_n) = g_{n\to\infty}(\lim F_n)$.

A λ-rule which satisfies, in addition to axioms A1 to A3, the following relation: for all A, B ε X and $A \cap B = \varnothing$ is proposed to obtain the measure of the union of two disjoint subsets:

$$g(A \cup B)=g(A)+g(B)+\lambda g(A)g(B),\ \lambda>-1 \qquad (25)$$

Eq.(25) is also called a g_λ-fuzzy measure, and for $\lambda=0$ the g_λ-fuzzy measure becomes a probability measure.

Now let $X = \{x_1, ..., x_n\}$ be a finite set, and $g^i = g(\{x^i\})$ for i=1,n be the values of g over the set of singletons of X. g^i will also be called a fuzzy density function of g_λ. Assume that $A = \{x_{i1}, ..., x_{im}\} \subseteq X$, then g(A) can be expressed as (Tahani & Keller 1990)

$$g(A)=\sum_{j=1}^{m} g^{ij}+\lambda\sum_{j=1}^{m-1}\sum_{k=j+1}^{m} g^{ij}g^{ik}+...+\lambda^{m-1}g^{il} \ ...\ g^{im} = \frac{1}{\lambda}\left[\prod_{x_i\in A}(1+\lambda g^{i})-1\right],\ \lambda\neq 0 \qquad (26)$$

Recall that g(X)=1 and if the fuzzy densities g^i are known, then the g_λ-fuzzy measure can be constructed by solving the equation:

$$\lambda+1 = \prod_{i=1}^{n}(1+\lambda g^{i}) \qquad (27)$$

7 g_λ-FUZZY MEASURE IN FUZZY FEM

A fuzzy measure g_λ is used as a mathematical model for subjective measures (Sugeno 1977) by assigning a value to each crisp set (or the union of subsets) of the universal set to signify the grade of importance that a particular element belongs to the set. Here let us show how fuzzy measures can be applied in considering the uncertainty involving in finite element analysis.

The three main factors which govern the reliability of the fuzzy finite element method (FEM) of an elastic soil medium may be listed as:

1. Reliability in expert estimate of E and its shape of membership function.
2. Reliability in expert estimate of ν and its shape of membership function.
3. Reliability in mesh discretization.

If other factors can be negligible and the weights of contribution to the accuracy of FE solutions for the above three factors are assigned in such a way:

$$w(E) + w(\nu) + w(M) = 1 \qquad (28)$$

where
w(E): weight of contribution relating to E.
w(ν): weight of contribution relating to ν.
w(M): weight of contribution relating to mesh discretization.

Let R = {E, ν, M} and $\wp(R)$ = {∅, {E}, {ν}, {M}, {E,ν}, {E,M}, {ν,M}, {E,ν,M}} be the power set of R.

If the numerical value of each weight can be determined either subjectively or objectively, then our interest is to establish the fuzzy measures, which imply the "grades of efficiency" to the solution accuracy, on $\wp(R)$, ie. $g(\varnothing)$, $g_\lambda(\{E\})$, $g_\lambda(\{\nu\})$, $g_\lambda(\{M\})$, $g_\lambda(\{E,\nu\})$, $g_\lambda(\{E,M\})$, $g_\lambda(\{\nu,M\})$ and $g_\lambda(\{E,\nu,M\})$. And these are called the grades of ideal efficiency. However, in real applications, the grades of actual efficiency are usually less than those of the ideal owing to the errors in the subjective estimates of material properties and in the mesh discretization, ie. $w_E + w_\nu + w_M < 1$. From this standpoint, the error rate of subjective estimates denoted by ξ is defined by the following equation.

$$\xi = \frac{\sum_{i=1}^{r} w(F_i) \left(g(F_i)_{old} - g(F_i)_{new} \right)}{\sum_{i=1}^{r} w(F_i)\, g(F_i)_{old}} \tag{29}$$

where
$F_i \subseteq \wp(R)$ excluding $\varnothing$ and $\{E,\nu,M\}$,
$r=|\wp(R)|-2=2^q-2$, $q=|R|$ (q=3 in this case),
$w(F_i)$ is the weight of a subset F in $\wp(R)$.
$g(F_i)_{old}$ is the grade of ideal efficiency of F_i,
$g(F_i)_{new}$ is the grade of actual efficiency of F_i,
and $g(F_i)_{old} \geq g(F_i)_{new}$.

The value of ξ is set in the limit: $0 \leq \xi \leq 1$. $\xi = 0$ means that the estimates of E, ν and M are totally reliable, while the increase of ξ implies the magnitude of the error affecting the result accuracy with reference to the estimates of E, ν and M.

If the subjective weights $w_j = w(A_j)\ \varepsilon[0,1]$ are given for all subsets A_j of a finite set X, then the fuzzy densities g^i can be determined by the following algorithm (Wierzchon 1983).

1. Compute λ by solving the equation

$$\prod_{j=1}^{m} (1+\lambda w_j)^{|A_j|} = (1+\lambda)^u \tag{30}$$

where $k=|X|$, $m=2^k-2$, $u=2^{k-2}(k+1)-k$

2. Calculate z_i:

$$z_i = \log_{(\lambda+1)} \prod_{j=1}^{m} (1+\lambda w_j d_{ij}), \quad i=1,k \tag{31}$$

where $d_{ij}=1$ if $x_i \varepsilon A_j$ and $d_{ij}=0$ otherwise.

3. Compute p_i:

$$p_i = 2^{2-k}(z_i+1)-1, \quad i=1,k \tag{32}$$

4. Calculate g^i:

$$g^i = \frac{(\lambda+1)^{p_i} - 1}{\lambda} \tag{33}$$

8 NUMERICAL EXAMPLE

Consider the half symmetrical finite element mesh of a soil medium which consists of two layers with a soft soil being on top of a stiffer layer as shown in Fig.1. The estimates of soil properties in the form of low (L), most likely (P) and high values (H) are as follows.

1. First layer (soft soil):
E_1 (kPa)= 1500 (L), 2000 (P), 2600 (H).
ν_1= 0.2 (L), 0.3 (P), 0.4 (H).
2. Second layer (stiff soil):
E_2 (kPa)= 8000 (L), 10000 (P), 13500 (H).
ν_2= 0.2 (L), 0.3 (P), 0.4 (H).

The strip load is assumed to be 240 kN/m.

The membership functions of E_1, ν_1, E_2 and ν_2 are constructed using Eqs.(11), (12) and (13) (see Figs.2, 3 and 4).

Fig.5 shows the possible vertical displacements at the center along the depth for the membership grades of 0.8, 0.9 and 1. The displacement variations are greater in the top (softer soil) than the second layer. Fig.6 shows the variations of the vertical stresses about the center along the soil depth. The vertical stresses in the top soil layer are not much affected by the variations of the fuzzy soil parameters. The vertical stress variations in the stiffer soil are also not considerable.

Fig.7 and Fig.8 show the normalized fuzzy entropies of the vertical displacements and vertical stresses respectively in the soil mass where the measures of fuzziness using the Euclidean distance (EN), the Hamming distance (HG) (Eq.(24)) and De Luca & Termini's entropy (D & T) (Eq.(20)) may be taken as the lower bound, median and upper bound respectively.

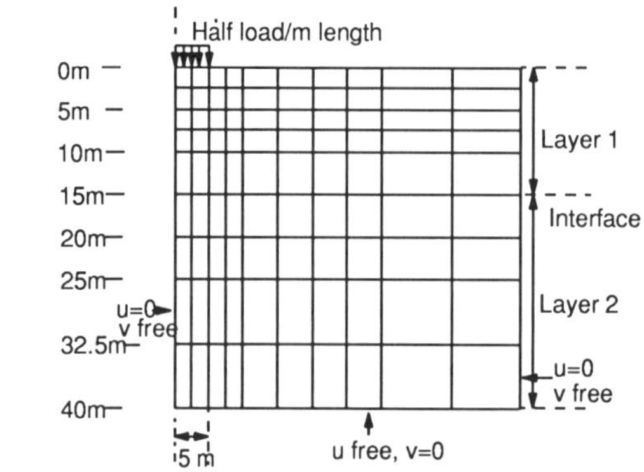

Fig.1 - A half symmetrical finite element mesh of a two-layered soil mass.

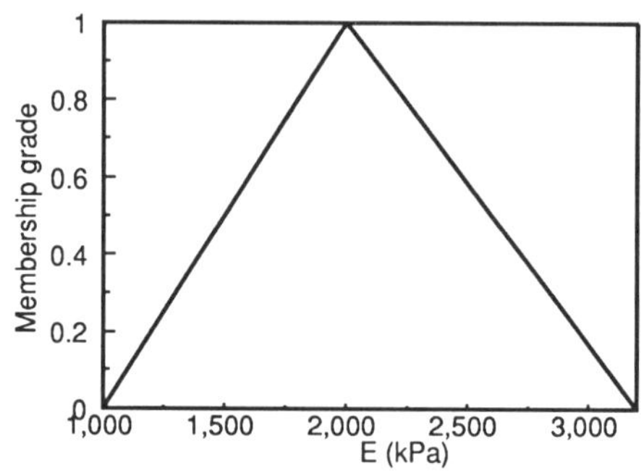

Fig.2 - Membership function of E for first soil layer

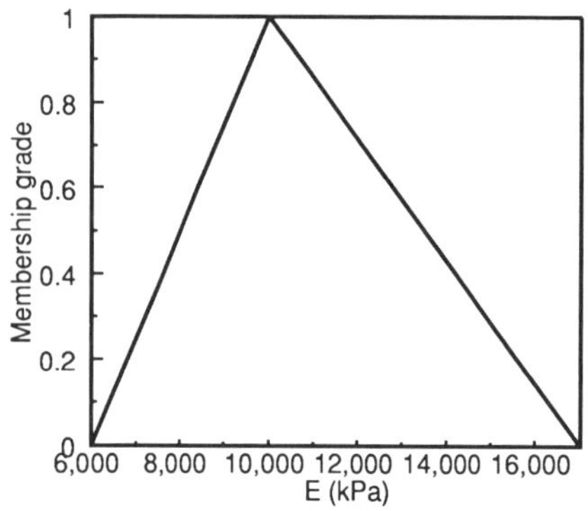

Fig.3 - Membership function of E for second soil layer

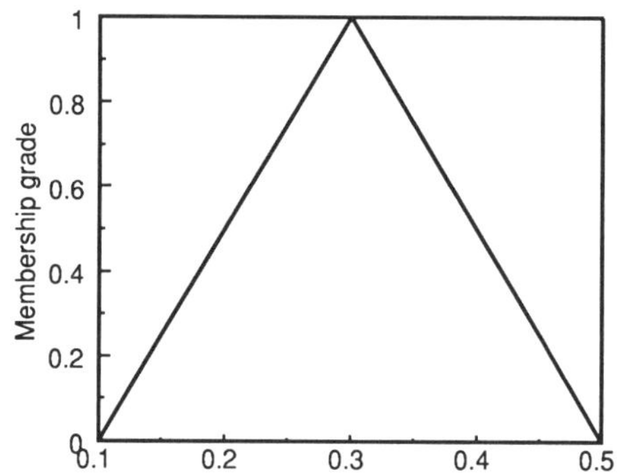

Fig.4 - Membership function of ν for both soil layers

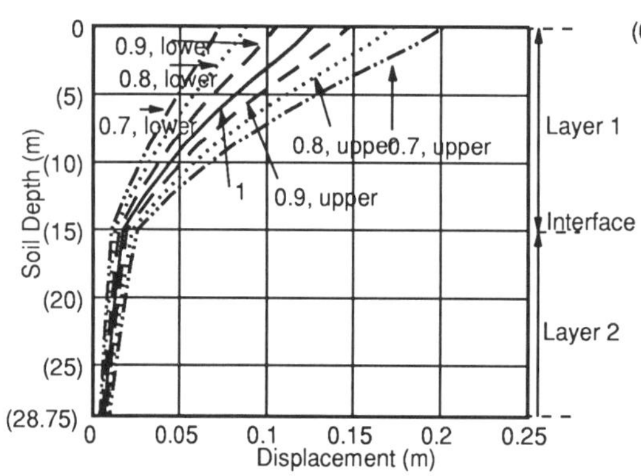

Fig. 5 - Vertical displacements below the center line of a strip load in a two-layered soil mass at different membership grades.

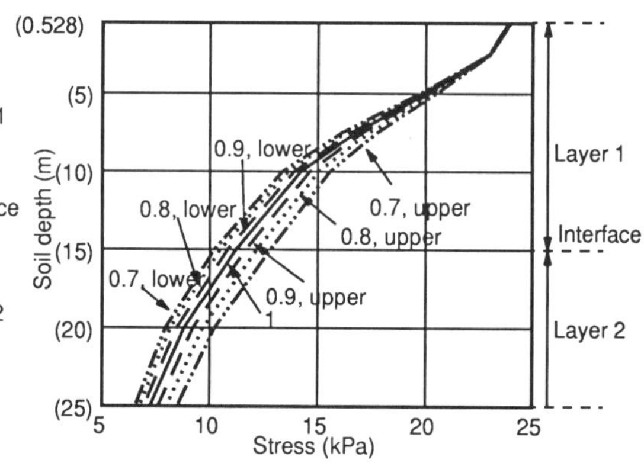

Fig.6 - Vertical stresses at Gaussian points below loading cente in a 2-layered soil mass at different membership grades.

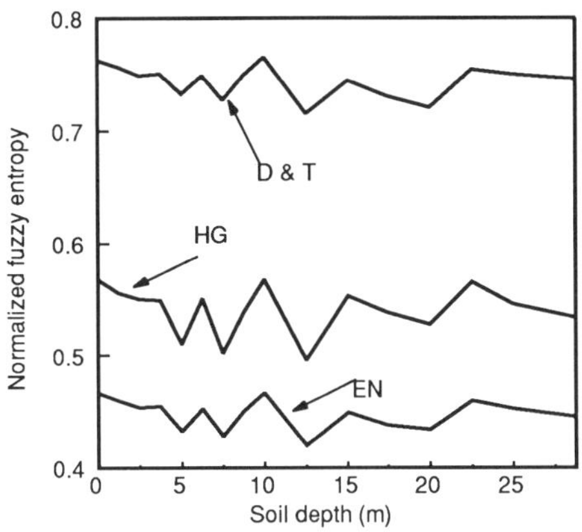

Fig.7 - Measures of fuzziness of vertical displacements

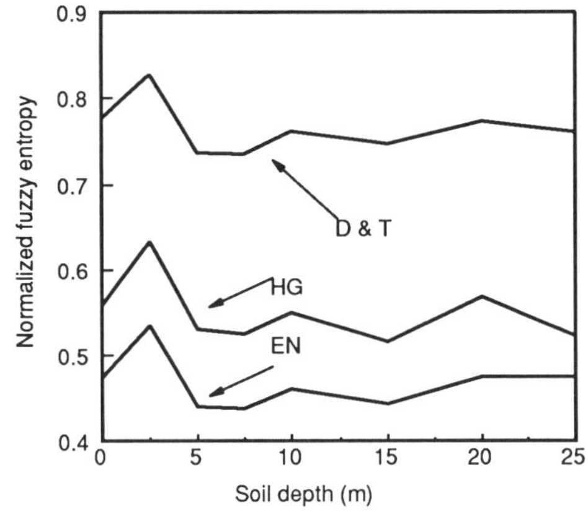

Fig.8 - Measures of fuzziness of vertical stresses

To illustrate how $g\lambda$-fuzzy measure is applied to calculate the error rate of subjective estimates in system reliability, we consider the following case.

For the problem of foundation settlement, the determination of the displacements is of most importance. Thus, the geotechnical engineer would rank the contributions of E, ν and M to the accuracy of the displacements obtained from the finite element analysis with different weights as w(E) = 0.5, w(ν) = 0.2 and w(M) = 0.3. If all the weights are assumed to be equal to the fuzzy densities, then $g^1 + g^2 + g^3 = 1$; therefore, λ is zero. The g_0-fuzzy measures of $\wp(R)$ are determined from Eq.(25) and tabulated in Table 1.

Now according to the degree of experience in estimating the soil parameters, the intrinsic uncertainty of the membership-function shape, and the degree of the mesh accuracy; the engineer decides to reduce the fuzzy densities to the new values: $g^1 = 0.3$, $g^2 = 0.1$ and $g^3 = 0.2$. Using Eq.(27) a new λ can be obtained by solving the following equation.

$$0.006\lambda^2 + 0.11\lambda - 0.4 = 0$$

Taking the root being greater than -1 in the above equation, we have λ = 3.109.

Again using Eq.(25), the new fuzzy measures $g_{3.109}$ of all subsets A_j are calculated and shown in Table 2.

Table 1. g_0-fuzzy measures

Subset A_j	$g_0(A_j)$
∅	0
{E}	0.5
{ν}	0.2
{M}	0.3
{E,ν}	0.7
{E,M}	0.8
{ν,M}	0.5
{E,ν,M}	1.0

Table 2. $g_{3.109}$-fuzzy measures

Subset A_j	$g_{3.109}(A_j)$
∅	0
{E}	0.3
{ν}	0.1
{M}	0.2
{E,ν}	0.493
{E,M}	0.687
{ν,M}	0.362
{E,ν,M}	1.0

Finally the error rate of the subjective estimates ξ = 0.258 which is calculated from Eq.(29).

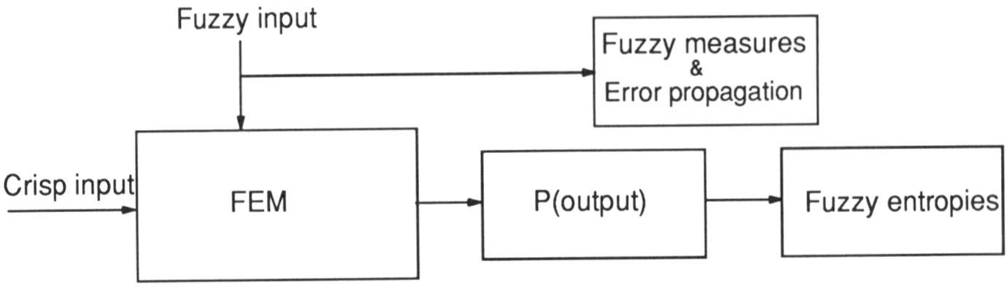

Fig.9 - Diagramatic representation of a FFEA

9 CONCLUSION

In the preceeding sections of this paper, the mathematical aspects of fuzzy-set theory have been applied to the finite element analysis where subjective information concerns. Fuzzy mathematical techniques have also been implemented in elasto-plastic finite element method by Valliappan & Pham (1993 c). In addition to the basic formulation of the fuzzy elastic matrix in the finite element method, we have attempted to elaborate the framework of the fuzzy finite element method by exploring the concepts of fuzzy entropies and Sugeno's fuzzy measures to study the possibility distributions and estimate the error propagation in the fuzzy-deterministic system. A diagramatic representation of a fuzzy finite element analysis (FFEA) is shown in Fig.9 where P(output) denotes the possibility distributions of displacements or stresses.

In the light of this pilot study, the fuzzy finite element method would be pervasive in the treatment of uncertainty in civil engineering designs. In fact, large ill-defined geotechnical systems can be tackled effectively at a reasonable cost of computation by means of the procedures as described in this paper.

ACKNOWLEDGEMENT

This work, under the direction of Professor Valliappan, forms a part of the ongoing project supported by the Australian Research Council.

REFERENCES

De Luca, A. & S. Termini 1972. A definition of a nonprobabilistic entropy in the setting of fuzzy sets theory. *Information and Control* 20: 301-312.

Kaufmann, A. & M.M. Gupta 1985. *Introduction to fuzzy arithmetic*. New York: Van Nostrand Reinhold.

Klir, G.J. & T.A. Folger 1988. *Fuzzy sets, uncertainty, and information*. Englewood Cliffs, NJ: Prentice-Hall.

Sugeno, M. 1977. Fuzzy measures and fuzzy integrals: A survey. In M.M. Gupta, G.N. Saridis & B.R. Gaines (eds.), *Fuzzy automata and decision processes*. Amsterdam, North-Holland.

Tahani, H. & J.M. Keller 1990. Information fusion in computer vision using fuzzy integral. *IEEE Trans. Systems, Man, and Cybernetics* 20: 733-741.

Valliappan, S. & T.D. Pham (1993 a). Constructing the membership function of a fuzzy set with objective and subjective information. *Microcomputers in Civil Engineering* 8: 75-82.

Valliappan, S. & T.D. Pham (1993 b). Fuzzy finite element analysis of a foundation on an elastic soil medium. Submitted to *Int. J. Numerical and Analytical Methods in Geomechanics*.

Valliappan, S. & T.D. Pham (1993 c). Elasto-plastic finite element analysis using fuzzy logic. Submitted to *Int. J. Numerical Methods in Engineering*.

Wierzchon, S.T. 1983. An algorithm for identification of fuzzy measure. *Fuzzy Sets and Systems* 9: 69-78.

Zadeh, L.A. 1965. Fuzzy sets. *Information and Control* 8: 338-353.

Impact of Computational Mechanics on Engineering Problems, Pulmano & Murti (eds)
 ISBN 90 5410 324 8

Computations of vertical soil stresses underneath foundations with curved boundary using parametric mapping technique

V.A. Pulmano, V. Murti & S. L. S. Chiu
University of New South Wales, N.S.W., Australia

ABSTRACT: A simple computational technique is proposed to evaluate the vertical soil stresses underneath foundations with curved boundary. By parametric mapping of the local and Cartesian coordinates which define an eight-node square domain and an eight-node domain with curved boundary, respectively, the integral expression of the Boussinesq solution over a loaded domain with curved boundary is transformed into an expression which can be easily integrated using the Gaussian integration scheme. Numerical example is presented to demonstrate the performance of the proposed technique.

1 INTRODUCTION

By assuming the soil as an elastic, isotropic and homogeneous semi-infinite continuum, the vertical stress at any point in the soil mass may be obtained by integration of the Boussinesq solution for a vertical, point surface force over the loaded region. However, the analytical evaluation of the resulting integral is in general quite complicated so that only analytical solutions for loaded regions such as circular, rectangular and specially oriented triangular areas are available, and are reproduced in the form of charts and/or tables for easy reference (e.g., Holtz et al. (1981), Poulos and Davies (1974)). For areas of arbitrary shape under uniformly distributed pressure, the approximate values of the vertical soil stresses may be obtained by using the well-known graphical method which utilizes an influence chart developed by Newmark (1942).

In recent years, computationally oriented techniques have been published elsewhere (e.g., Murti et al. (1990), Murti and Wang (1991), Li and Lee (1991), Pulmano et al. (1991), Li and Pulmano (1992)) to calculate the vertical soil stresses underneath arbitrarily shaped foundations. However, in all these techniques, the boundary of the loaded area is consists of a series of straight line segments, thereby introducing an approximation to any curved boundary of the loaded area.

In this paper, a technique is proposed to evaluate the integral expression of the Boussinesq solution over a loaded basic area/domain with four curved sides. This is accomplished by simple transformation of coordinates of an eight-node square domain and an eight-node domain with curved boundary by the use of parametric mapping technique which is similar to the mapping technique used in finite element method in dealing with isoparametric finite elements. This transformation allows the resulting Boussinesq integral expression to be evaluated by using the Gaussian integration scheme.

2 PARAMETRIC MAPPING

Consider a basic domain/area with four curved sides under a uniformly distributed load, q as shown in Fig. 1. This basic area, shown in Fig. 2a, is defined by the Cartesian coordinates of its eight nodes. The vertical soil stress, σ_z, at any point $P(x_P, y_P, z_P)$ due to the pressure acting on the loaded area is obtained by integration of the Boussinesq solution, and is given by

$$\sigma_z = (1.5z^3q/\pi) \int\int (1/R^5)\, dx\, dy \qquad (1)$$

in which $R^2=(x_P-x)^2+y_P-y)^2+(z_P)^2$. In the proposed technique, a simple transformation of coordinates is first performed before integration of Eq. 1 is carried out. This involves the mapping of

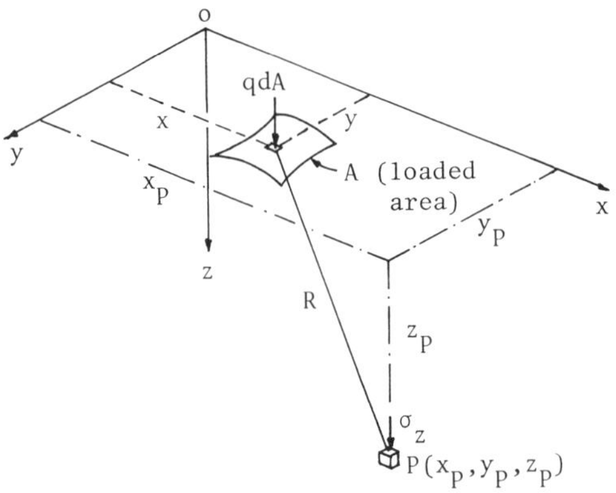

Fig. 1 Vertical soil pressure σ_z at point P underneath a loaded area with curved sides

local rs- and Cartesian xy-coordinates which define the eight-node square domain (Fig. 2b) and the eight-node loaded domain with four curved sides (Fig. 2a), respectively. With reference to Fig. 2, consider the mapping of the Cartesian coordinate x_i using the relation,

$$x = [f(r,s)]\{a\} \tag{2}$$

in which

$$[f(r,s)] = [1 \;\; r \;\; s \;\; r^2 \;\; rs \;\; s^2 \;\; r^2s \;\; rs^2]$$

and $\{a\}$=vector of undetermined coefficients. Substitution of nodal coordinates into Eq. 2 gives,

$$\{x^*\} = [C]\{a\} \tag{3}$$

in which

$$\{x^*\} = \{x_i\} \quad . \; . \; . \quad (i=1,8) \tag{4}$$

$$[C] = \begin{bmatrix} f(-1,-1) \\ f(\;0,-1) \\ f(+1,-1) \\ f(+1,\;0) \\ f(+1,+1) \\ f(\;0,+1) \\ f(-1,+1) \\ f(-1,\;0) \end{bmatrix} \tag{5}$$

Solving for vector $\{a\}$ in Eq. 2 gives,

$$\{a\} = [C^{-1}]\{x^*\} = \{\bar{x}\} \tag{6}$$

Substitution of Eq. 6 into Eq. 2 gives,

$$x = [f]\{\bar{x}\} \tag{7}$$

Similarly,

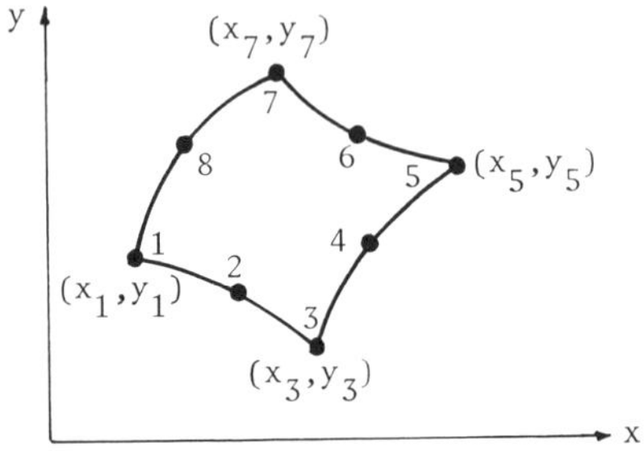

(a) Cartesian coordinates

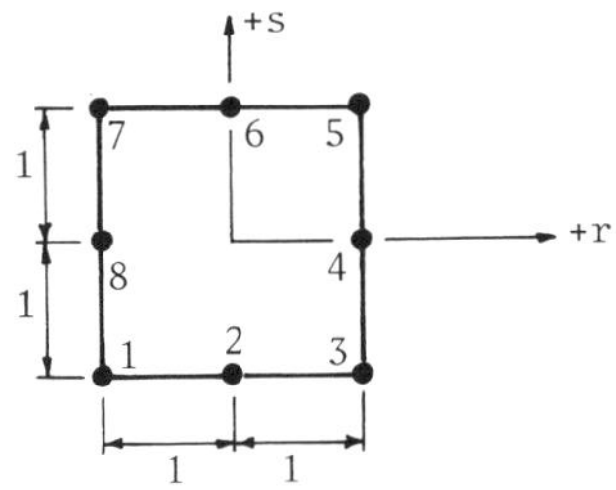

(b) Local coordinates

Fig. 2 Mapping of coordinates

$$y = [f]\{\bar{y}\} \tag{8}$$

in which

$$\{\bar{y}\} = [C^{-1}]\{y^*\} \tag{9}$$

$$\{y^*\} = \{y_i\} \quad . \; . \; . \quad (i=1,8) \tag{10}$$

In terms of the rs-coordinates, the elemental area dA is equal to $|J|$drds, in which $|J|$ is the determinant of the Jacobian matrix J, and is given by,

$$[J] = \begin{bmatrix} x_{,r} & y_{,r} \\ x_{,s} & y_{,s} \end{bmatrix} = \begin{bmatrix} J_{11} & J_{12} \\ J_{21} & J_{22} \end{bmatrix} \tag{11}$$

in which comma denotes differentiation and

$$\left.\begin{aligned} J_{11} &= [f_{,r}]\{\bar{x}\} \quad ; \quad J_{21} = [f_{,s}]\{\bar{x}\} \\ J_{12} &= [f_{,r}]\{\bar{y}\} \quad ; \quad J_{22} = [f_{,s}]\{\bar{y}\} \\ [f_{,r}] &= [0 \;\; 1 \;\; 0 \;\; 2r \;\; s \;\; 0 \;\; 2rs \;\; s^2] \\ [f_{,s}] &= [0 \;\; 0 \;\; 1 \;\; 0 \;\; r \;\; 2s \;\; r^2 \;\; 2rs] \end{aligned}\right\} \tag{12}$$

By utilizing the transformation relations given by Eqs. 7 to 12, Eq. 1 reduces to,

$$\sigma_z = (1.5 z_P{}^3 q/\pi) \int_{-1}^{1}\int_{-1}^{1} (1/R_P{}^5)\,|J|\,drds \tag{13}$$

in which

$$R_P = \sqrt{(x_P-[f]\{x\})^2 + (y_P-[f]\{y\})^2 + z_P^2}$$

Equation 12 gives the expression for σ_z in local rs-coordinates, and can be integrated quite easily by using the Gaussian integration scheme. When the loaded area/foundation has to be subdivided into several basic domains, the total vertical soil stress at any point P is found by summing the contributions of each loaded basic domain/area, that is,

$$\sigma_z = \sum_{i=1}^{N} (\sigma_z)_i \qquad (14)$$

in which $(\sigma_z)_i$ is the vertical soil stress at point P due to the loaded i^{th}-domain and N is the total number of basic areas comprising the loaded area.

3 ILLUSTRATIVE EXAMPLE

To demonstrate the performance of the proposed computational technique, the evaluation of the vertical soil stresses at various locations underneath a uniformly loaded circular area with radius, r = 1 m, was considered. A uniformly loaded circular area was chosen so as to allow comparison of results with existing analytical solutions. The vertical soil stresses were evaluated at two locations, P1(0,0,5) and P2(0,0,10), by considering the discretization of the circular area into (a) one basic domain for the whole circular area (Case 1) and (b) four basic domains (Case 2) as shown in Fig.3. For Case 2, because of symmetry only the contribution of one basic domain (that is, one quadrant) needs to be evaluated, and the final value of the vertical soil stress can be obtained by adding the contributions of the other three quadrants, (i.e., multiply σ_z obtained for one quadrant by four). The computed values of the vertical soil

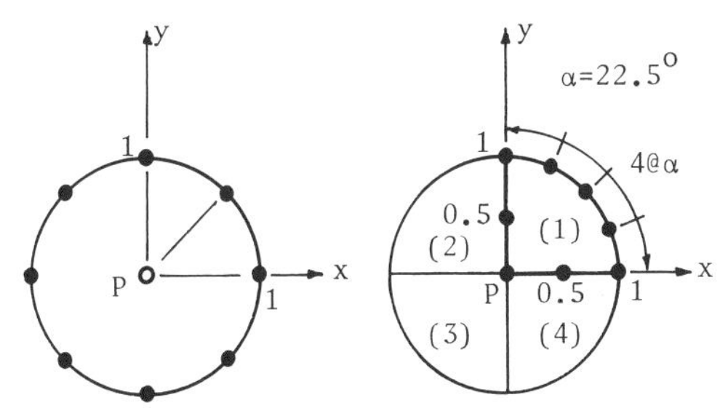

Fig. 3 Discretization of loaded circular area

stresses for Cases 1 and 2 are summarized in Table 1. It can be observed that, except for the result obtained in Case 1 (using one basic area) at location P1, the computed and analytical solutions are identical. In general, more accurate solutions can be obtained as the discretization of the loaded area is refined.

Table 1. Vertical soil stresses under a uniformly loaded circular area.

Depth	Vertical soil stress, σ_z/q		
z (m)	Case 1 (1 domain)	Case 2 (4 domains)	Timoshenko & Goodier,1951
5	0.0585	0.0571	0.0571
10	0.0148	0.0148	0.0148

4 CONCLUSIONS

A simple computational technique is proposed to calculate the vertical soil stresses underneath a loaded area with curved boundaries. It involves the transformation of the integral expression of the Boussinesq solution over a loaded basic area with four curved sides into an expression that can be readily integrated using the Gaussian integration scheme. The is done by parametric mapping of local and Cartesian coordinates which define the eight-node square domain and the eight-node domain with four curved sides, respectively. The transformation procedure is similar to the mapping technique used in finite element method involving isoparametric finite elements. Finally, the proposed technique is simple, easy to follow and can easily be programmed on microcomputers.

REFERENCES

Holtz,R.D. and Kovacs,W.D. 1981. An introduction to geotechnical engineering. Prentice-Hall, Englewood Cliffs, N.J.

Poulos,H.G. and Davies,E.H. 1974. Elastic solutions for solid and rock mechanics, John Wiley, New York.

Newmark,N.M. 1942. Influence charts for computation of stresses in elastic foundations. Univ. of Illinois Eng. Expt. Sta. Bull. No. 338.

Murti, S., Wang, Y.C., Sastra, H. and Chandrasekaran, V. 1990. Vertical stress distribution due to an arbitrarily shape foundation using triangulation and an indirect method. Computers and Structures, 36, 4: 667-680.

Murti,V. and Wang,Y.C. 1991. Vertical stress distribution due to an arbitrarily shaped foundation using triangulation and an analytical expression for a triangular loaded region. Int. Jnl. Numer. and Analytical Meth. Geomechanics, 15: 51-60.

Li,K.S. and Lee,I.K. 1991. Discussion of Murti and Wang, 1991. Int. Jnl. Numer. and Analytical Meth. Geomechanics.

Pulmano,V.A., Murti,V. and Chiu,S.L.S. 1991. Vertical soil stresses under foundations of arbitrary shape and loading using parametric mapping. Computational Mechanics, Edited by Y.K.Cheung, J.H.Lee & A.Y.T. Leung, Balkema, Rotterdam.

Li,K.S. and Pulmano,V.A. 1992. Technique for calculating vertical soil stresses under foundations of arbitrary shape and loading distribution. Computational Meths. in Engrg.-Advances and Applications, World Scientific Publishing Co.

Timoshenko,S.P. and Goodier,J.N. 1951. Theory of elasticity. McGraw-Hill Book Co., New York.

Impact of Computational Mechanics on Engineering Problems, Pulmano & Murti (eds)
 ISBN 90 5410 324 8

Field investigation of mechanical behaviour of a shallow tunnel due to ground surface excavation

S. Sakurai
Kobe University, Japan

N. Shimizu
Yamaguchi University, Japan

ABSTRACT: The ground surface above a shallow tunnel was excavated due to the operation of quarry. The ground consists of sand and gravel materials. The deformational behaviour of the tunnel was measured by convergence meters and extensometers. Strain of steel ribs was also measured during the surface excavation. In order to investigate the arching effect of shallow tunnels, a trench parallel to the tunnel axis was excavated on the ground surface above the tunnel. This paper describes all the details of this field experiment, and discusses the mechanism of developing the earth pressure acting on shallow tunnels.

1 INTRODUCTION

Engineers often encounter a case in which the ground surface above existing tunnels is excavated due to the construction of roads and the reclamation of residential areas. If tunnels are located in a shallow depth, the following questions may arise regarding the effect of excavation on the existing tunnels.

(1) How much rebound of tunnels occurs?

(2) What is the change in tunnel earth pressure?

(3) Does any damage of tunnel linings occur or not?

Although these are very important engineering problems, they have not been fully clarified yet. Of course, it is possible to adopt a numerical analysis method, such as the finite element method, to predict the deformational behaviour of the tunnels. However, it is not so easy to analyze a problem, such as the rebound of tunnels due to surface excavation, even though the finite element method is used. The most difficult task for analyzing the rebound of tunnels, particularly in two-dimensional finite element analysis, is how to determine the domain of the finite element model. This is particularly the case in the depth of the lower boundary of the model . The amount of the rebound of the tunnels depends entirely on how much depth of the model is taken into account. The deeper the depth considered, the larger the rebound to be calculated.

In engineering practice, therefore, the engineers must determine the lower boundary of the model on the basis of their experiences in similar cases. We have had a chance to measure the deformational behaviour of a shallow tunnel due to ground surface excavation. In this paper, the results of the field experiments are described and a discussion is given on the mechanism of developing the earth pressure acting on tunnel linings.

2 OUTLINE OF FIELD EXPERIMENTS

The tunnel of concern is an access tunnel in a quarry, located in the ground consisting of sand and gravel materials. The tunnel diameter is approximately 3m. The overburden for the tunnel is about 21m. The tunnel was constructed by the rock bolts and shotcrete method. A cross-section of the tunnel is shown in Fig. 1, where support structures are also shown. The plan view of the tunnel with the contour lines of the ground surface is shown in Fig. 2. The ground surface is excavated by bulldozers in quarry operations, layer by later, until the layer of the tunnel floor is reached, as indicated in Fig. 3. The total depth of excavation is approximately 25m.

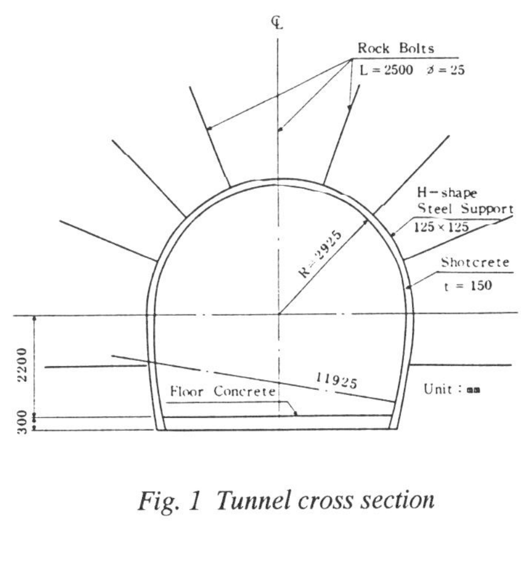

Fig. 1 Tunnel cross section

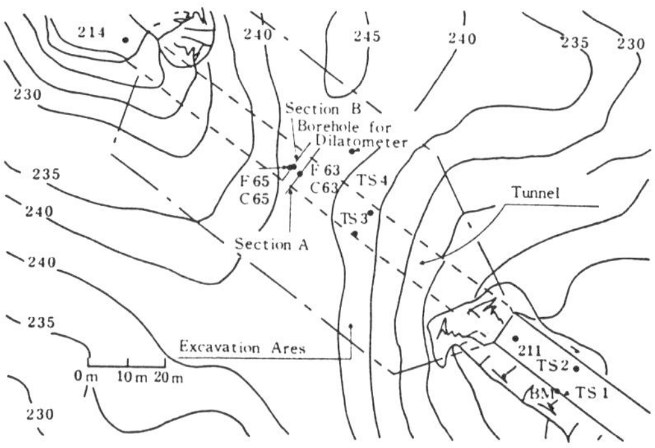

Fig. 2 Plan view of the tunnel with the contour lines of ground surface

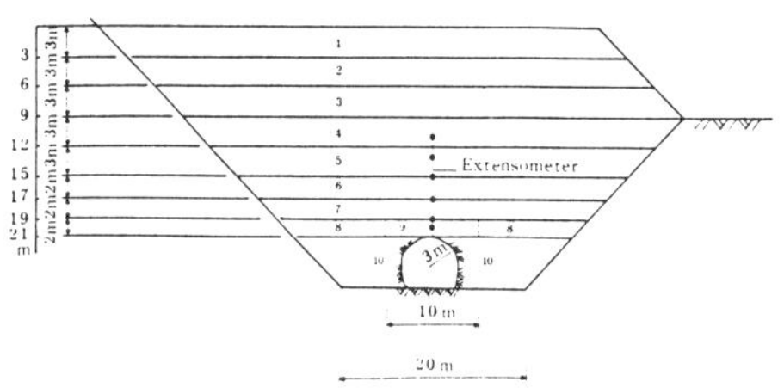

(a) Transverse cross section

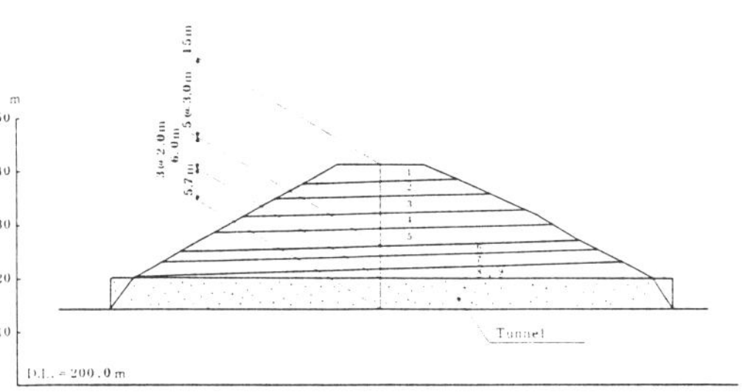

(b) Longitudinal cross section

Fig. 3 Excavation procedures of ground surface above the tunnel

Field measurements such as convergence and extensometer measurements are carried out during surface excavation. Strain of steel ribs is measured by strain gages. In order to investigate the rebound of the tunnel the high precision levelling survey is also carried out. The measuring sections of tunnel and measurement points are shown in Figs. 2 and 3. The locations of convergence measurements and extensometer measurements are indicated in Fig. 4. The locations of strain gage measurements are also shown in Fig. 4.

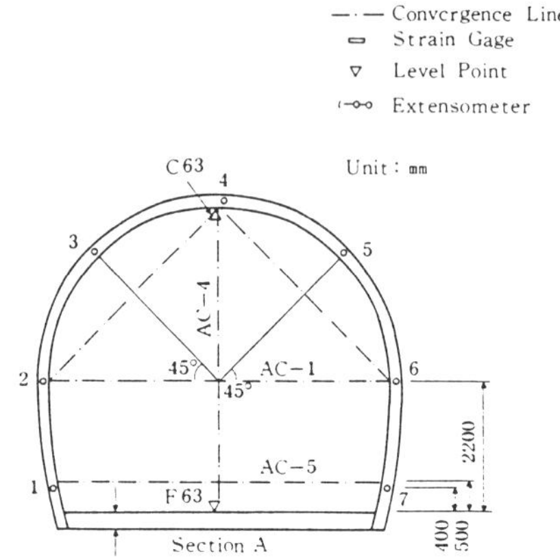

(a) Measurement lines of convergence (Section A)

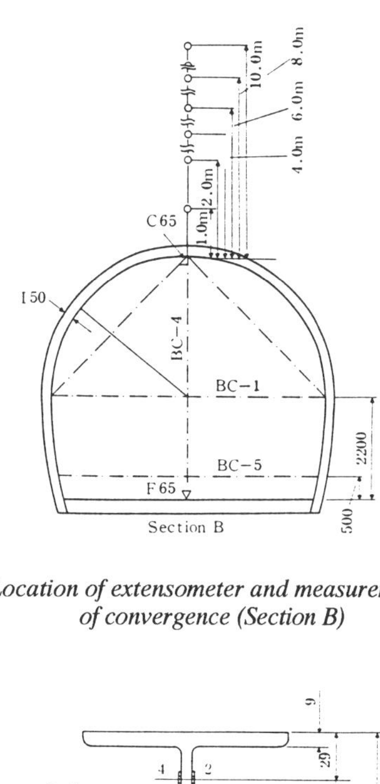

(b) Location of extensometer and measurement lines of convergence (Section B)

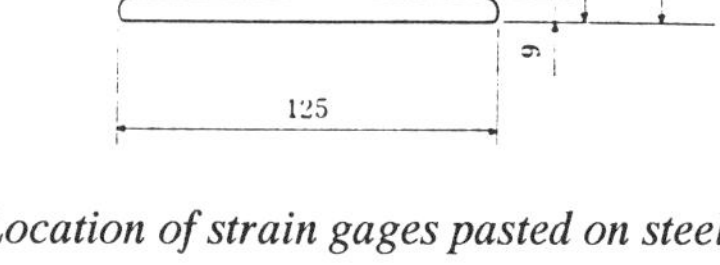

(c) Location of strain gages pasted on steel ribs

Fig. 4 Location of measuring points

3 MEASUREMENT RESULTS AND DISCUSSIONS

3.1 Extensometer Measurements

The measurement results of the extensometer installed at the tunnel crown are shown in Fig. 5. It is obvious from this figure that all the measurement points tend to move upward as the ground surface excavation progresses. It should be noted, however, that the displacements shown in this figure are relative displacements in considering the tunnel crown as a reference point. Therefore, the absolute values of displacements at each measuring point must be greater than those given here,. In Fig. 5 there are two values of displacements corresponding to the depth of 15m. This is due to the fact that no excavation activities were performed during a certain period of time. Thus, the difference between the two values is displacements caused by the creep effect.

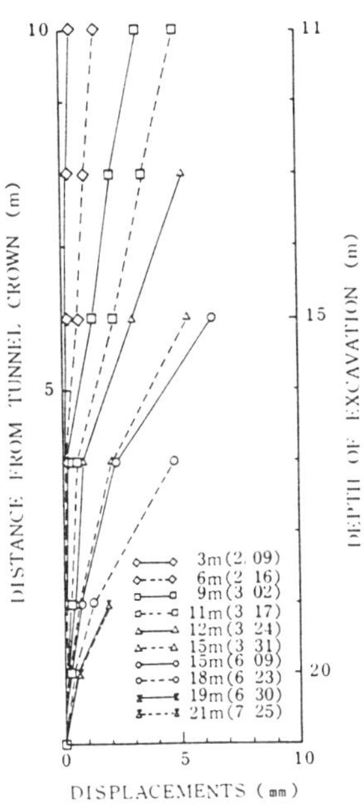

Fig. 5 Results of extensometer measurements

By using the displacements measured by the extensometer, the modulus of deformation of the ground is back analyzed. The results range 200-300 MPa, depending on the range of excavation.

The borehole pressuremeter tests are conducted before starting the surface excavation, and the results indicate that the modulus of deformation is 25-40 MPa for loading, and 200-250 MPa for unloading. It might be of interest to know that the modulus of deformation obtained by back analysis of measured displacements in the surface excavation is almost the same value as the one determined under the unloading of borehole pressuremeter tests.

The convergence and crown settlement for the existing tunnel were measured when it was constructed. All these data are available, and used for a back analysis to determine the modulus of deformation, and we then obtain the value of 200 MPa. This indicates that an almost identical value of the modulus of deformation is obtained for both cases, that is, the cases of ground surface excavation and tunnel excavation.

3.2 Absolute Value of Displacements

Extensometer measurements only provide a relative displacement between two measurement points. In order to measure the absolute value of displacements, the high precision levelling survey is conducted, which can measure the vertical component of absolute displacement. The location of measurement points is indicated in Figs. 2 and 4 where TS1 and TS2 denote the bench marks located outside of the tunnel. In the figure TS3 and TS4 are measurement points located in the left and right sides of tunnel floor, respectively. They are approximately 20 m away from the measuring sections A and B. The measuring points F63 and C63 are situated on the centre of the tunnel floor and on the crown, respectively, in measuring section A, while F65 and C65 are also on the center of tunnel floor and on the crown, respectively, in measuring section B.

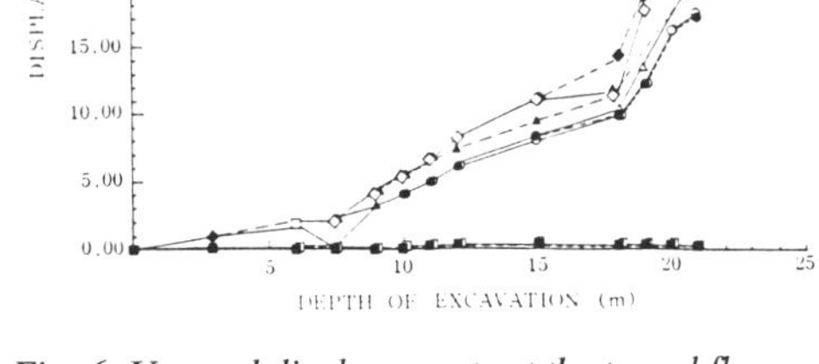

Fig. 6 Upward displacements at the tunnel floor and crown measured by high precision levelling survey

The results of the levelling survey are shown in Fig. 6. It is seen from this figure that the rebound of the tunnel at sections A and B (the distance between sections A and B is 1.5m), and the rebound at the measuring points TS3 and TS4 are almost identical to each other. This implies that the tunnel moves upwards uniformly in the whole excavated area.

3.3 Influence Depth

It may be of interest to discuss how much depth of the ground moves upwards by the surface excavation. Though this is an important engineering problem, it has not been fully investigated yet. The depth, called "influence depth," is particularly important when predicting rebound of the ground by the two-dimensional finite element method. The influence depth is defined as the depth of the ground where upward displacements are observed

When calculating the rebound of the tunnel due to surface excavations by means of the two-dimensional finite element method, the results of vertical displacements depend entirely on the depth of the lower boundary of the finite element model. The deeper the depth of the model, the larger the rebound obtained. Therefore, in two-dimensional finite element analysis the lower boundary of the model must be carefully evaluated by taking into account the influence depth of the ground.

Nevertheless, the influence depth can hardly be determined theoretically, but it usually can be determined on the basis of an engineer's experiences. In order to increase the reliability of the analysis, the influence depth must be investigated carefully on the basis of the field measurement results.

Assuming that the ground underneath tunnels deforms vertically in a one-dimensional fashion, we adopt the plane strain condition in horizontal directions. Hence, the influence depth D is calculated by the following equation:

$$D=\beta(E/p)\delta \ , \ \beta=(1-\nu)/(1+\nu)/(1-2\nu) \qquad (1)$$

where δ denotes vertical displacement (upwards) at the tunnel floor. E and ν and modulus of deformation and Poisson's ratio, respectively. Released pressure due to excavation is denoted by p ($p=\gamma d$, γ: unit weight, d: depth of excavation).

In the tunnel studied here the influence depth is calculated by Eq.(1). The result shows that the influence depth at the final stage of excavation becomes approximately 10-12m, where we assume E=200MPa, $\nu=0.3$, $\gamma=2.1$ t/m^3.

Observing Fig. 6 in more detail, we realize that the upward displacement at the tunnel crown is greater than that of the floor. This implies that the tunnel deforms slightly in an oval shape with a long axis in a vertical direction. The difference in the displacements of the crown and the floor is about 14mm, and it is almost identical in both measurement sections A and B. The results of the convergence measurements also show the tunnel deformed slightly into an oval shape. Fig 6 shows that the upward displacements of the tunnel increase monotonously with the progress of excavation. This indicates that the earth pressure acting on the tunnel lining tends to decrease. In other words, the effect of surface excavation on the tunnels is always unloading for the earth pressure.

4 TRENCH EXCAVATION

When excavating the ground surface above the tunnel, earth pressure acting on the tunnel lining tends to decrease, or rather, to unload. Thus, the tunnel rebounds and the shape of the tunnel deforms slightly from circular to oval with the long axis in the vertical direction.

In a shallow tunnel, however, it might be thought that the earth pressure starts to increase as the overburden becomes smaller than a certain depth and the ground arch existing around the tunnel vanishes. However, this is not the case. Fig. 6 demonstrates that the earth pressure decreases even in the case of a very small overburden of 2m. In other words, the ground arch always exists around the tunnel even with such a small overburden.

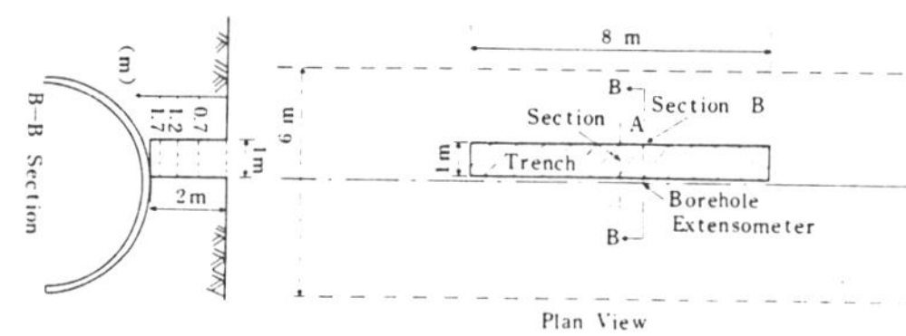

Fig. 7 Configuration of trench excavation on the ground surface above the tunnel

In order to investigate the mechanism of the ground arch appearing around a shallow tunnel, a trench 2m deep and 1m wide was excavated on the ground surface, parallel to the tunnel axis, as shown in Fig. 7. If the whole weight of the overburden on the tunnel acts as a load on the tunnel lining, the excavation of the trench must cause an unloading effect on the earth pressure because of a reduction in the load acting on the lining.On the contrary, it would be possible to assume that the trench makes the ground arch vanish if the ground arch exists around the tunnels with small overburdens. In this case, therefore, the earth pressure acting on the lining should increase.

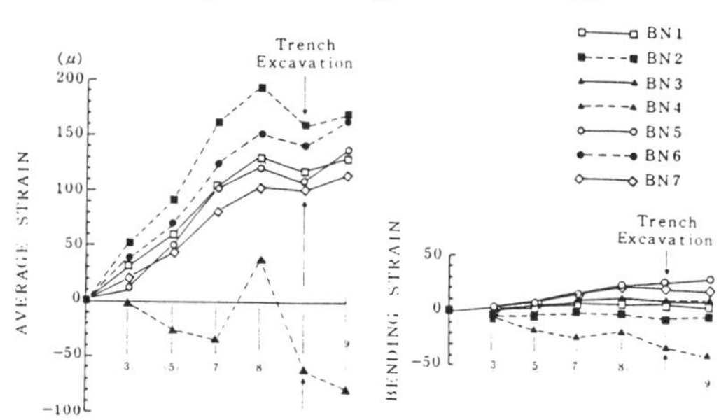

Fig. 8 Strain of steel ribs caused by trench excavation

The strain of the steel ribs due to surface excavation is measured, and the results are shown in Fig. 8. Convergence of the tunnel is also shown in Fig. 9 with the progress of a layered excavation. The numbers indicated in the axis of abscissa of the figure correspond to each layer of the excavation as shown in Fig. 3.

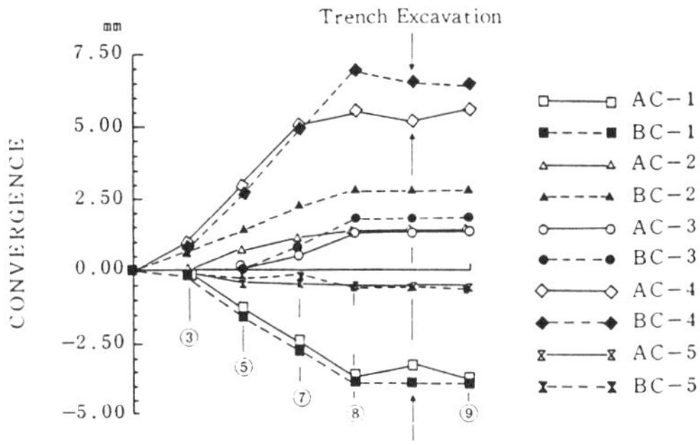

Fig. 9 Convergence measurements during trench excavation

The trench excavation is carried out in between the excavation of layers 8 and 9. These figures show that strain and convergence increase or decrease monotonously with the progress of excavations. However, when excavating the trench, both strain and convergence tend to change to the opposite direction as before.This implies that the earth pressure tends to increase with the excavation of the trench.

This may be proof that the ground arch existing above the crown has failed due to the excavation. It may be concluded that the ground arch exists around the tunnels, even when the overburden is only 2m above the tunnel crown.

5 CONCLUSIONS

1) The behaviour of tunnel is just as though it were under unloading conditions during the progress of surface excavation. Thus, the tunnel shape deforms from circular to an oval shape with the long axis in the vertical direction.

2) The modulus of deformation obtained through back analysis of measured displacements of the surface excavation is almost identical to the one obtained under an unloading condition for borehole pressure meter tests.

3) The influence depth defined by the depth of the rebound zone is approximately 10m underneath the tunnel.

4) No damage appears on the lining when the overburden of 21m was removed after the excavation operation.

5) The ground arch exists around tunnels even when the overburden has become only 2m. The effect of surface excavation is always unloading on the earth pressure.

ACKNOWLEDGEMENTS

Support from the Development Bureau of Kobe City Office is gratefully acknowledged.

Impact of Computational Mechanics on Engineering Problems, Pulmano & Murti (eds)
© 1993 Balkema, Rotterdam, ISBN 90 5410 324 8

Complex number expression of 3D deviatoric tensor

M. Satake
Tohoku Gakuin University, Japan

ABSTRACT: This paper proposes a new expression of 3D deviatoric tensor. It is shown that, in 3D case, a deviatoric tensor is expressed by a complex number named *complex deviation*. The properties and the geometrical representation of the complex deviation are explained, taking the stress tensor as an example.

1 INTRODUCTION

As is wellknown, a symmetric tensor is decomposed into two parts; isotropic part and deviatoric one. In 2D case, we have two scalars, *mean (principal) value* and *deviation*, which correspond to the above two parts respectively. However, in 3D case, a scalar corresponding to the deviatoric part has not been kown. This paper shows that such a scalar is also obtainable in 3D case and it becomes a complex number. The derivation of this scalar, *complex deviation*, and its important properties and geometrical representation are explained.

2 DECOMPOSITION OF A 2D SYMMETRIC TENSOR

The decomposition of a symmetric tensor is very useful in continuum mechanics. First we explain the decomposition in 2D case by taking the stress tensor as an example.

A 2D stress tensor $\boldsymbol{\sigma}$ is decomposed as

$$\boldsymbol{\sigma} = p\,\boldsymbol{I} + q\,\boldsymbol{J}, \tag{2.1}$$

where

$$\boldsymbol{I} = \begin{pmatrix} 1 & \\ & 1 \end{pmatrix}, \qquad \boldsymbol{J} = \begin{pmatrix} 1 & \\ & -1 \end{pmatrix} \tag{2.2}$$

with respect to the principal axes. $\boldsymbol{I}$ is the unit tensor and $\boldsymbol{J}$ is a tensor called the *unit deviatoric tensor*, respectively, and

$$\left.\begin{aligned} p &= \frac{1}{2}\boldsymbol{\sigma}\cdot\cdot\boldsymbol{I} = \frac{1}{2}(\sigma_1 + \sigma_2) \\ q &= \frac{1}{2}\boldsymbol{\sigma}\cdot\cdot\boldsymbol{J} = \frac{1}{2}(\sigma_1 - \sigma_2) \end{aligned}\right\} \tag{2.3}$$

where $\sigma_1 \geq \sigma_2$ are two principal stresses. p and q are called the *mean stress* and the *deviatoric stress*, respectively.

It is noted that this decomposition is based on the following orthogonality condition:

$$\boldsymbol{I}\cdot\cdot\boldsymbol{I} = \boldsymbol{J}\cdot\cdot\boldsymbol{J} = 2\,, \quad \boldsymbol{I}\cdot\cdot\boldsymbol{J} = 0 \tag{2.4}$$

As is wellknown, the decomposition of a symmetric tensor is very important in the energy analysis. Table 1 shows the decomposition of a 2D stress and strain increment tensors, where the coaxiality is assumed, and the corresponding decomposition of the work increment $dW = \boldsymbol{\sigma}\cdot\cdot d\boldsymbol{\varepsilon}$. It is noted that the work is only done along the solid lines and is not done along the dotted lines.

Introducing a parameter φ defined by

$$\sin\varphi = \frac{q}{p} \tag{2.5}$$

we can write as

$$\left.\begin{aligned} \sigma_1 &= p + q = p(1 + \sin\varphi) \\ \sigma_2 &= p - q = p(1 - \sin\varphi) \end{aligned}\right\} \tag{2.6}$$

$$\sqrt{\frac{\sigma_1}{\sigma_2}} = \sqrt{\frac{1+\sin\varphi}{1-\sin\varphi}} = \tan\left(\frac{\pi}{2} + \frac{\varphi}{4}\right). \tag{2.7}$$

φ is called the *mobilized angle of internal friction* in mechanics of granular materials, and it describes the anisotropy of the stress tensor.

Table 1. Decomposition of 2-D work increment.

$\boldsymbol{\sigma}$	=	$p\boldsymbol{I}$	+	$q\boldsymbol{J}$
$d\boldsymbol{\varepsilon}$	=	$\frac{d\varepsilon}{2}\boldsymbol{I}$	+	$\frac{d\gamma}{2}\boldsymbol{J}$
dW	=	$p\,d\varepsilon$	+	$q\,d\gamma$

3 3D CASE AND COMPLEX DEVIATION

To extend the 2D decomposition of a symmetric tensor to 3D, we introduce the following 3D unit deviatoric tensor:

$$\boldsymbol{J} = \begin{pmatrix} 1 & & \\ & \omega & \\ & & \omega^2 \end{pmatrix} \tag{3.1}$$

where

$$\omega = \frac{1}{2}(-1 + \sqrt{3}\, i) \tag{3.2}$$

is a root of equation $z^3 = 1$. Using the 3D unit tensor $\boldsymbol{I}$ and the 3D unit deviatoric tensor $\boldsymbol{J}$, we obtain the following decomposition of a 3D stress tensor:

$$\boldsymbol{\sigma} = p\,\boldsymbol{I} + q\,\bar{\boldsymbol{J}} + \bar{q}\,\boldsymbol{J} \tag{3.3}$$

where

$$\left.\begin{aligned} p &= \frac{1}{3}\boldsymbol{\sigma}\cdot\cdot\boldsymbol{I} = \frac{1}{3}(\sigma_1 + \sigma_2 + \sigma_3) \\ q &= \frac{1}{3}\boldsymbol{\sigma}\cdot\cdot\boldsymbol{J} = \frac{1}{3}(\sigma_1 + \omega\,\sigma_2 + \omega^2\,\sigma_3) \\ \bar{q} &= \frac{1}{3}\boldsymbol{\sigma}\cdot\cdot\bar{\boldsymbol{J}} = \frac{1}{3}(\sigma_1 + \omega^2\sigma_2 + \omega\,\sigma_3) \end{aligned}\right\} \tag{3.4}$$

where $\sigma_1 \geq \sigma_2 \geq \sigma_3$ are the principal stresses and - indicates conjugate complex number or a tensor which has conjugate complex numbers as its corresponding elements. q is called the *complex deviatoric stress* (*complex deviation*, in general case).
Quite similarly as in Eq.(2.4), we have the relations

$$\left.\begin{aligned} \boldsymbol{I}\cdot\cdot\boldsymbol{I} = \boldsymbol{J}\cdot\cdot\bar{\boldsymbol{J}} = 3 \\ \boldsymbol{I}\cdot\cdot\boldsymbol{J} = \boldsymbol{J}\cdot\cdot\boldsymbol{J} = 0 \end{aligned}\right\}. \tag{3.5}$$

As q is a complex number, we can write it as

$$q = |q|\, e^{i\alpha} \tag{3.6}$$

where $|q|$ and α denote the absolute value and argument of q, respectively.
$|q|$ and α are closely related to S_2 and S_3, the invariants of the deviatoric stress tensor $\boldsymbol{\sigma}'$, and μ, the Lode's parameter. The relationship is expressed as

$$\left.\begin{aligned} |q| &= \sqrt{\frac{S_2}{3}} \\ \cos 3\,\alpha &= \frac{1}{2\,|q|^3}\sqrt{S_3} \end{aligned}\right\} \tag{3.7}$$

$$\alpha = \tan^{-1}\frac{\mu}{\sqrt{3}} + \frac{\pi}{6} \tag{3.8}$$

The deviatoric principal stresses σ_1', σ_2', and σ_3' are written as

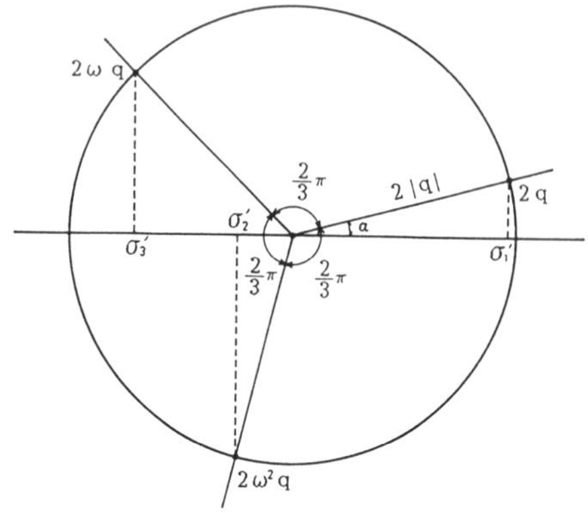

Fig. 1. Illustration of $2q$, $2\omega q$, $2\omega^2 q$ in the Gaussian plane, which gives the three principal deviatoric stresses.

$$\left.\begin{aligned} \sigma_1' &= 2\,|q|\cos\alpha \\ \sigma_2' &= 2\,|q|\cos(\alpha - \frac{2}{3}\pi) \\ \sigma_3' &= 2\,|q|\cos(\alpha + \frac{2}{3}\pi) \end{aligned}\right\} \tag{3.9}$$

This relation is illustrated in Fig.1.
Introducing a parameter φ defined by

$$\sin\varphi = \frac{|q|}{p} \tag{3.10}$$

we can write as

$$\left.\begin{aligned} \sigma_1 &= p\,(1 + 2\sin\varphi\cos\alpha) \\ \sigma_2 &= p\left\{1 + 2\sin\varphi\cos\left(\alpha - \frac{2}{3}\pi\right)\right\} \\ \sigma_3 &= p\left\{1 + 2\sin\varphi\cos\left(\alpha + \frac{2}{3}\pi\right)\right\} \end{aligned}\right\}. \tag{3.11}$$

Similarly as in 2D case, we have the decomposition of a work increment in 3D, as is shown in Table 2. Here it is also noted that stress tensor $\boldsymbol{\sigma}$ and strain increment tensor $d\boldsymbol{\varepsilon}$ are decomposed into two parts respectively and the work is done only between the corresponding terms, which is indicated by the solid lines in the table.

Table 2. Decomposition of 3-D work increment.

$\boldsymbol{\sigma}$	=	$p\boldsymbol{I}$	+	$(q\,\boldsymbol{J}$	+	$\bar{q}\,\boldsymbol{J})$
$d\boldsymbol{\varepsilon}$	=	$\frac{d\varepsilon}{3}\boldsymbol{I}$	+	$\frac{1}{3}(d\gamma\,\bar{\boldsymbol{J}}$	+	$\overline{d\gamma}\,\boldsymbol{J})$
dW	=	$p\,d\varepsilon$	+	$(\bar{q}\,d\gamma$	+	$q\,\overline{d\gamma})$

4 GEOMETRICAL REPRESENTATION

In the analysis of stress states, the principal stress space is used very conveniently. In the principal stress space a stress with principal stresses σ_i is represented by a point with coordinates σ_i.

Fig.2 shows a 2D principal stress space. Point Q represents a stress with principal stresses (σ_1,σ_2). As is shown, vector $\overrightarrow{OQ}$ is decomposed as

$$\overrightarrow{OQ} = \overrightarrow{OP} + \overrightarrow{OQ'} \tag{4.1}$$

which is the geometrical representation of Eq.(2.1). $\overrightarrow{OP}$ lies on the space diagonal ($\sigma_1 = \sigma_2 = p$) and $\overrightarrow{OQ'}$ lies on π -line ($\sigma_1 + \sigma_2 = 0$). Here we note that

$$\overline{OQ'} = \sqrt{2}\, q = \frac{1}{\sqrt{2}}(\sigma_1 - \sigma_2) \tag{4.2}$$

From Eq.(4.1), It is easy to see that

$$(\sqrt{2}\, p)^2 + (\sqrt{2}\, q)^2 = \sigma_1^{\,2} + \sigma_2^{\,2} \tag{4.3}$$

Fig.3 shows a 3D principal stress space. Point Q represents a stress with principal stresses $(\sigma_1,\sigma_2,\sigma_3)$. Similarly as in 2D, $\overrightarrow{OQ}$ is decomposed as Eq.(4.1). As is shown in Fig.3, $\overrightarrow{OP}$ lies on the space diagonal ($\sigma_1 = \sigma_2 = \sigma_3 = p$) and $\overrightarrow{OQ'}$ lies on π -plane ($\sigma_1 + \sigma_2 + \sigma_3 = 0$).

Fig.4 shows π -plane together with three coordinate axes $\sqrt{\frac{2}{3}}\sigma_1$, $\sqrt{\frac{2}{3}}\sigma_2$, $\sqrt{\frac{2}{3}}\sigma_3$, which make angle of $\frac{2}{3}\pi$ each other and are the projection of three coordinate axes $\sigma_1,\sigma_2,\sigma_3$ in the principal stress space respectively. π -plane is also considered as a Gaussian plane and we can introduce coordinate axes 1 and i ,as is shown in Fig.4, and the above-mentioned three coordinate axes correspond to the axes of 1 , ω , ω^2 , respectively. It is easily seen that

$$\overrightarrow{OQ'} = \sqrt{\frac{2}{3}}(\sigma_1 + \omega\,\sigma_2 + \omega^2\,\sigma_3) = \sqrt{6}\, q \tag{4.4}$$

Thus we find that the complex deviatoric stress q lies on $\overrightarrow{OQ'}$. We have also

$$\overline{OQ'} = \sqrt{6}\,|\,q\,| = \frac{1}{\sqrt{3}}\sqrt{(\sigma_1-\sigma_2)^2+(\sigma_2-\sigma_3)^2+(\sigma_3-\sigma_1)^2} \tag{4.5}$$

Eqs. (4.4) and (4.5) in 3D correspond to Eq.(4.2) in 2D. In 3D case, we have from Eq.(4.1)

$$(\sqrt{3}\, p)^2 + (\sqrt{6}\,|\,q\,|)^2 = \sigma_1^{\,2} + \sigma_2^{\,2} + \sigma_3^{\,2} \tag{4.6}$$

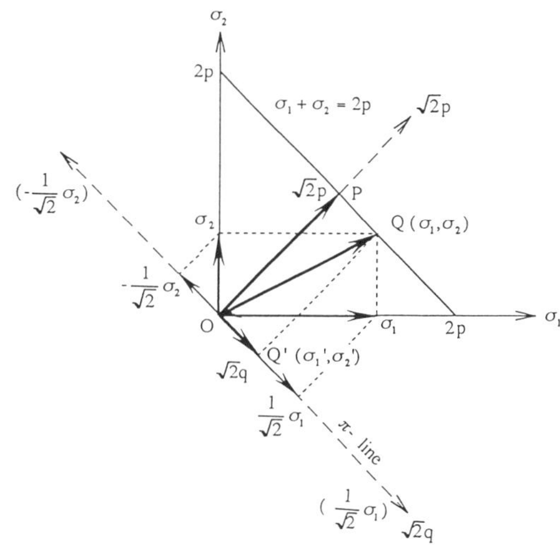

Fig.2 2D decomposition of a principal stress vector

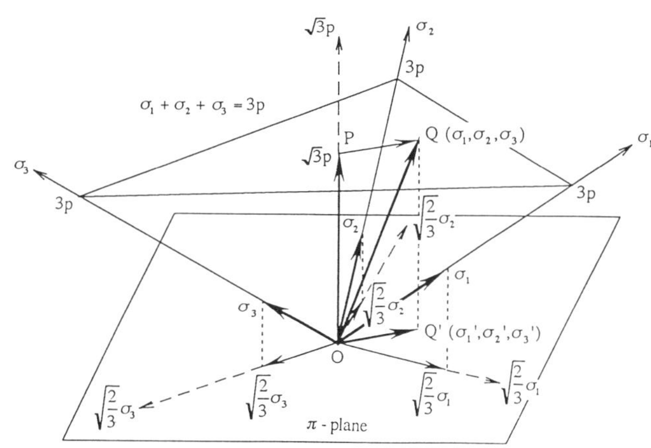

Fig.3 3D decomposition of a principal stress vector

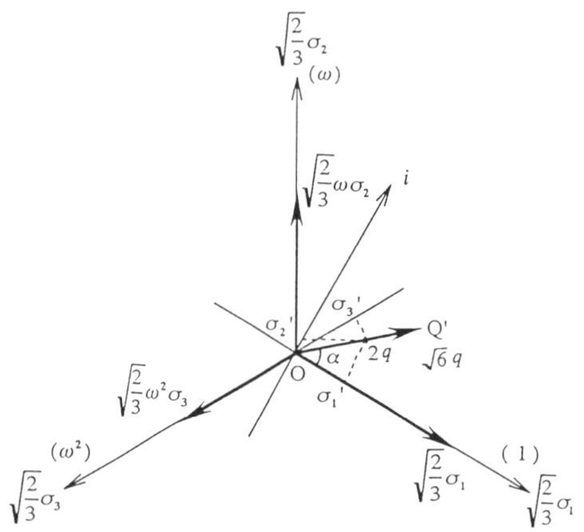

Fig.4 π-plane and complex deviatoric stress q ($2q$ and $\sqrt{6}q$ are illustrated)

5 CONCLUSION

This paper presented a complex number expression of a 3D deviatoric tensor, *complex deviation*, and explained its properties and geometrical representation. In using this expression, a 3D deviatoric tensor is simply expressed by only one complex number. So it is hoped that this expression will be effectively applied in analysis of continuum mechanics.

REFERENCES

Satake, M. 1989. Expression of three-dimensional deviatoric tensor through complex number ω. Theoretical and Applied Mechanics, Vol.37, University of Tokyo Press: 329-337.

Impact of Computational Mechanics on Engineering Problems, Pulmano & Murti (eds)
© 1993 Balkema, Rotterdam, ISBN 90 5410 324 8

Pollutant transport in deformable semi-saturated geomaterials

B.A.Schrefler, L.Simoni & X.Y.Zhan
University of Padua, Italy

ABSTRACT: The paper deals with a coupled mathematical model for the analysis of transport of contaminant substances in unsaturated deformable porous media. The different behaviour among miscible and immiscible pollutants is accounted for. The resulting equations are discretized by the finite element method and a subset of these is applied for the solution of a test problem, which was investigated also experimentally, concerning transport in a column of sand.

1 INTRODUCTION

In recent years water flow and contaminant transport in the unsaturated zone has received particular attention due to the importance of environmental problems involving pollution. A review of the conceptual understanding of the basic process and of various mathematical models can be found in Nielsen et al. (1986). A quantitative solution of the problem is at present the aim of several researches. This includes the development and utilization of numerical models, the observation of actual field scale behaviour and the performance of laboratory experiments, to obtain the necessary constitutive relationships. Furthermore an experimental assessment of the importance of the different contributions involved in the phenomenon is necessary. It is important, however, to remark that these investigative techniques should be viewed not as independent efforts, but rather as aspects of an overall investigative methodology. The necessity of such an effort comes not only from the highly complicated mechanical aspects involved in transport, but also from the behaviour of various contaminants, ranging from complete immiscibility in the fluid phases (dense liquids or NAPL) to highly soluble pollutants.

In the present paper, a rather general set of governing equations is presented. They are then are discretized in space by the finite element method and in time by a single step algorithm. On this basis a computer code has been developed. Its validity was checked by simulating a series of experimental tests performed by Matticchio et al. (1991), for miscible contaminant. The case of NAPL is analyzed in Zhan (1993).

2 THEORETICAL MODEL

The unsaturated (vadose) zone, where contaminant transport occurs, is place of different mechanical phenomena, say infiltration, evaporation, groundwater recharge, soil moisture storage, soil deformation and erosion. These effects depend also on the temperature field, which is however at present disregarded.

In order to obtain a sufficiently general mathematical model, the soil is assumed as a multiphase system where the interstitial voids of the solid matrix are filled with water and with an inert gaseous phase and with the contaminant. This may be miscible or immiscible with the fluid phases present within the solid skeleton.

The mathematical model for fluids and contaminant transport, accounting also for the deformation of the solid matrix, is obtained within the framework of averaging theories. This approach analyzes the problem at two different scales. The balance equations of the continuum mechanics are assumed valid within each phase (microscopic level). Then they are averaged over a representative volume element (R.E.V.), which is large with regard to pore dimensions, but small compared to the dimension of the sample, and expressed in terms of mean values of the state variables. The basic ideas for such a transformation are shown e.g. in Schrefler et al. (1990). The pollutant in this context is only a new component of the mixture. The governing equations are here shortly recalled, introducing the terms of phase changes and the typical transport mechanism of the soluble component (i.e. the diffusion effect).

2.1 Mass conservation equations

The mass balance equation for the generic phase can be written as (Hassanizadeh and Gray, 1979)

$$\frac{\partial(\varepsilon_\pi \rho_\pi)}{\partial t}+\nabla\cdot(\varepsilon_\pi \rho_\pi \underline{v}^\pi)=\rho_\pi e^\pi(\rho) \tag{1}$$

where ε_π is the volume fraction and ρ_π the intrinsic density of the π-phase, $\underline{v}^\pi$ the intrinsic mass weighted phase velocity and $e^\pi(\rho)$ the local mass rate of change of phase per unit volume. The last term could differ from zero for all components, also for the solid. For components interested in mass exchanges due to phase change of the component β the following constraint holds

$$\sum_\beta \rho_\pi \, e^\pi(\rho) = 0 \tag{2}$$

Velocity $\underline{v}^\pi$ in equation (1) represents the mass weighted averaged value over the R.E.V. and is also called barycentric velocity of the mixture. The deviation of a constituent velocity from this one is the diffusion velocity $\tilde{\underline{v}}^\pi$. It can be proved (Allen et al. 1988) that

$$\sum_\pi \omega^\pi \, \tilde{\underline{v}}^\pi = 0 \tag{3}$$

where ω^π is the mass fraction of constituent π, say the mass of π per unit mass of mixture. For the related diffusive flux $\underline{i}^\pi$ the simplest constitutive assumption is Fick's law, which for a multicomponent fluid system takes the form

$$\underline{i}^\pi = - \sum_\beta D^{\beta\pi} \, \nabla \, \omega^\beta \tag{4}$$

where $D^{\beta\pi}$ is the coefficient of diffusion that expresses the contribution to the flux of the π component due to a unit gradient of the β-component.

By accounting for eq. (4), eq. (1) can be written as

$$\frac{\partial(\varepsilon_\pi \rho_\pi)}{\partial t}+\nabla\cdot(\varepsilon_\pi \rho_\pi \underline{v}^\pi)+\nabla\cdot\underline{i}^\pi=\rho_\pi e^\pi(\rho) \tag{5}$$

which is known as advection-diffusion transport equation.

Equation (5) must be written for the solid phase, for water, for gas (mixture of dry air and vapour) and for a immiscible contaminant (NAPL), denoted by the indices s, w ,g and n respectively. For all these components phase changes and diffusion are usually disregarded.

For a soluble pollutant propagating in the fluid phases (e.g. water and gas), under the hypothesis that no chemical reactions, no interphase exchange take place and no source or sink are present within the domain, eq. (5) results in two mass balance equations, which states the conservation of the solute in the water and in the gas component. These equations can be written in terms of mass per unit volume of the fluid component C_w (concentration in the water) and C_g^w (concentration in the gas) as

$$\frac{\partial(\varepsilon_w C_w)}{\partial t}+\nabla\cdot(\varepsilon_w C_w \underline{v}^w)=\nabla\cdot(\varepsilon_w D^w \nabla C_w) \tag{6}$$

and

$$\frac{\partial(\varepsilon_g C_g)}{\partial t}+\nabla\cdot(\varepsilon_g C_g \underline{v}^g)=\nabla\cdot(\varepsilon_g D^g \nabla C_g) \tag{7}$$

D^π being the effective diffusion coefficient of solute in the π fluid. In these equations the concentrations can be assumed as further state variables.

2.2 Linear momentum balance equations

We assume that transport of miscible contaminant is passive, i.e. it has negligible effects on the mechanical properties of the mixture and its components. Further only slow phenomena and small displacements are considered and no momentum supply is related to phase changes. Under these hypotheses, the linear momentum balance equation for each fluid phase, NAPL component included, yields generalized Darcy's law for the velocity of the phase relative to the solid $\underline{v}^{d\pi}$ (Schrefler et al. 1990)

$$\underline{v}^{d\pi}=\underline{k}_{a\pi} \frac{k_{r\pi}}{\mu_\pi}\left(-\nabla p_\pi+\varepsilon_\pi \rho_\pi \underline{g}\right) \quad \pi=w,g,n \tag{8}$$

where $k_{a\pi}$ is the absolute permeability matrix of the medium, μ_π its dynamic viscosity and $k_{r\pi}$ (relative permeability) an experimental function of the degree of saturation S_π.

By summing the momentum balance equations for nonsoluble components the equilibrium condition for the whole mixture is obtained. Accounting for the averaged density of the mixture defined as

$$\rho = \varepsilon_s\rho_s+\varepsilon_g\rho_g+\varepsilon_w\rho_w+\varepsilon_n\rho_n \tag{9}$$

we have

$$\nabla\cdot\underline{\sigma} + \rho\,\underline{g} = 0 \tag{10}$$

where $\underline{\sigma}$ is the total stress vector in the multiphase system.

Owing to the knowledge of the darcian velocities, a further transformation of the balance equations can be made by splitting the velocity of each fluid phase as

$$\underline{v}^\pi = \underline{v}^s + \underline{v}^{d\pi}/\varepsilon_\pi \tag{11}$$

By summing solid mass balance with that of each fluid we obtain for water

$$\frac{\partial(\varepsilon_w\rho_w)}{\partial t}+\frac{\partial(\varepsilon_s\rho_s)}{\partial t}+\nabla\cdot(\varepsilon_w\rho_w(\underline{v}_s+\frac{\underline{v}^{d\pi}}{\varepsilon_\pi})$$

$$+\nabla\cdot(\varepsilon_s\rho_s\underline{v}^s)=0 \tag{12}$$

for gas

$$\frac{\partial(\varepsilon_g\rho_g)}{\partial t}+\frac{\partial(\varepsilon_s\rho_s)}{\partial t}+\nabla\cdot(\varepsilon_g\rho_g(\underline{v}_s+\frac{\underline{v}^{dg}}{\varepsilon_g})$$

$$+\nabla\cdot(\varepsilon\,\rho_s\underline{v}_s^s)=0 \tag{13}$$

and for NAPL

$$\frac{\partial(\varepsilon_n\rho_n)}{\partial t}+\frac{\partial(\varepsilon_s\rho_s)}{\partial t}+\nabla\cdot(\varepsilon_n\rho_n(\underline{v}_s+\frac{\underline{v}^{dn}}{\varepsilon_n})$$

$$+\nabla\cdot(\varepsilon\,\rho_s\underline{v}_s^s)=0 \tag{14}$$

2.3 Angular momentum balance

All phases of the system are assumed microscopically nonpolar. No macroscopic angular momentum or exchange of angular momentum is assumed owing to the fact that the microscopic partial stresses are symmetric. Moreover the presence of soluble contaminant is totally neutral. Hence this equation is is always satisfied.

2.4 Summary of the governing equations

Depending on the particular contaminant considered, the mathematical model need four or five equations. We can distinguish two cases:

a) in presence of NAPL, equations (10) and (12)-(14) fully describe mass transfer in deformable porous media;

b) for a miscible pollutant moving in a unsaturated medium, the problem is described by equations (6), (7), (10), (12) and (13).

In every case, mass balance equations can be simplified by assuming the solid velocity to be zero and introducing darcian velocity represented by eq. (8).

More complicated situations could be obtained by activating the phase change term among the components.

For each case, different choices of the basic field variables are possible. In this work, displacement of the solid skeleton, water and gas pressures are used, the variables associated with the contaminant depending on the particular problem.

The mathematical model must be supplemented by the initial and boundary conditions. The latter establish the value assumed by the primary field variables or the related fluxes. These take the form

$$\varepsilon_\pi\rho_\pi\underline{v}^{d\pi}\cdot\underline{n} - q_\pi = 0 \tag{15}$$

for fluid phases and for concentration

$$\left(\varepsilon_\pi C_\pi\underline{v}^{d\pi} - \varepsilon_\pi D^\pi\nabla C_\pi\right)\cdot\underline{n} - q_{c\pi}=0 \tag{16}$$

being $\underline{n}$ the outer normal to the boundary, q_π the imposed flux of the π fluid and $q_{c\pi}$ the flux of the pollutant.

3 CONSTITUTIVE RELATIONSHIPS

We remark the assumption that the contaminating substance does not change the mechanical and hydraulic properties of each component of the multiphase system and does not chemically react. The usual hypotheses for the macroscopic behaviour of the components are used in the present paper (Schrefler et al. 1990):

- elastic or elastoplastic solid phase with constitutive relationship in terms of the effective stress;

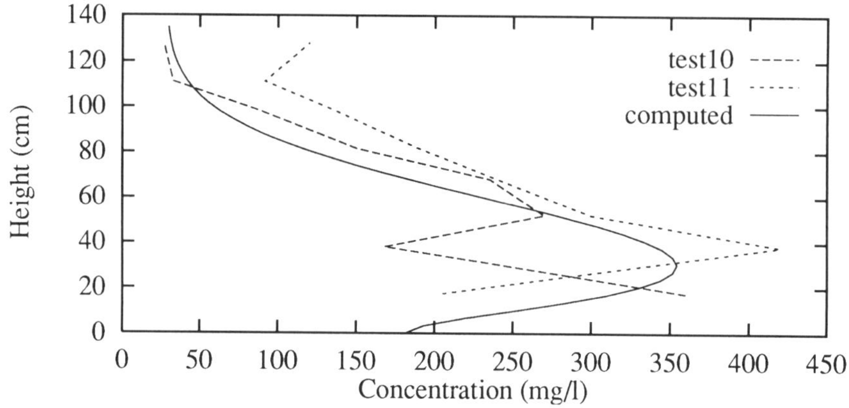

Figure 1. Comparisons of numerically and experimentally obtained concentration profiles (specific water inflow 0.40 cm/min)

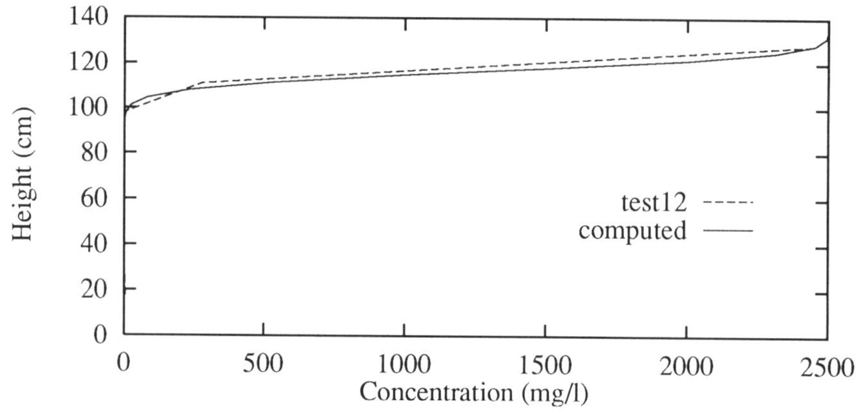

Figure 2. Comparisons of numerically and experimentally obtained concentration profiles (specific water inflow 0.054 cm/min)

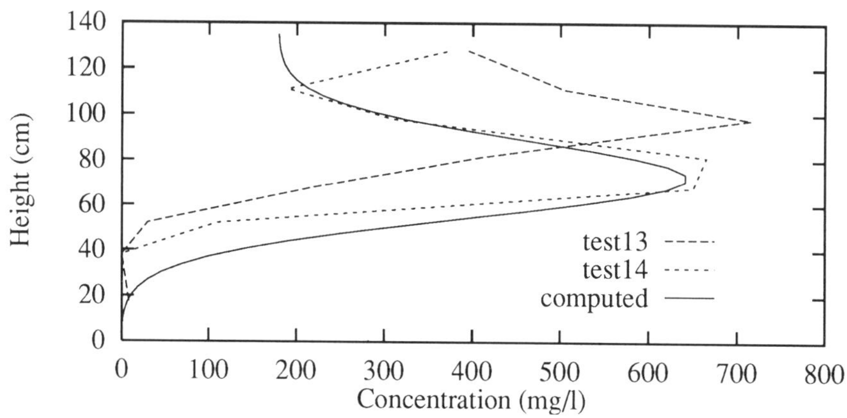

Figure 3. Comparisons of numerically and experimentally obtained concentration profiles (specific water inflow 0.23 cm/min)

- barotropic perfect fluid;
- saturation S_π of the π-phase is a known function of the capillary pressure p_c

$$S_\pi = S_\pi(p_c) \qquad (17)$$

- the relative permeabilities are known functions of the degree of saturation:

$$k_{r\pi} = k_{r\pi} (S_\pi). \qquad (18)$$

It is noteworthy to recall that capillary pressure must be defined for each couple of fluid phases present in the mixture.

4 DISCRETIZATION IN SPACE AND IN TIME

By applying Galerkin's procedure of weighted residuals, the governing equations are discretized in space and terms involving second spatial derivatives are transformed through Gauss' theorem. The unknown variables are expressed in terms of their nodal values as usual in finite element techniques, by approximated functions. In this way a semidiscretized nonlinear system of ordinary differential equations in time is obtained (Schrefler et al. 1990). In all cases this system is nonlinear, nonsymmetric and coupled, in the sense that no field can be independently solved. The complete solution is now obtained by means of time stepping algorithms (Zienkiewicz & Taylor 1991).

5 NUMERICAL APPLICATION

To check the mathematical model and its implementation in a computer code a laboratory infiltration experiment referenced in Matticchio et al. (1991) is chosen as example. The experiment was performed in a sand soil column 1.35 meters high with soil porosity of 0.439, having initially an average moisture content of 0.154, which correspond to a water saturation of 0.35. A stratum of pollutant, potassium nitrate, is placed on the top surface of the column. Water inflow at constant rate is maintained from t=0 and continued until the end of experiment. Water outflow with the pollutant through the bottom of the column is free and recorded during the test. Initially the concentration of potassium nitrate in the column is zero everywhere. As time goes on, the moisture front and a slug of the solute move downwards through the soil column. This may clearly be seen and measured at different time instants. Boundary conditions are given as function of time.

The potassium nitrate is soluble only in water, hence a subset of the previously presented governing equations is used for the numerical computation: incompressible soil and static air phase are assumed. This results in using the continuity equation of water (in unsaturated conditions) and the pollutant mass balance in terms of concentration (eq. 6). The equations are solved as coupled ones.

The absolute permeability of water is 0.08 cm/s, whereas the relative permeability and saturation as function of suction by Van Genuchten et al. (1977) are used in the simulation. The diffusion coefficient was set in Matticchio et al. (1991) as 4.6 cm^2/sec.

The spatial discretization is obtained by 40 equally sized elements. Neither flow nor pollutant transport occur across the lateral boundary. The analysis was performed with different water inflow rates, both experimentally and numerically.

The resulting profiles of pollutant concentration after 1 hour from the onset are shown in Figures 1-3 and are compared with the laboratory test results. The three cases differ only for the specific inflow rates, respectively 0.4 cm/min (tests 10 and 11), 0.056 cm/min (test 12) and 0.23 cm/min (tests 13-14).

Due to the difficulties in reading local values of concentration, they were measured as averaged values over a portion of the column 15 cm high. It must be recalled further that parameters such as permeability and diffusion coefficient were identified as mean values, without knowledge of their effective distribution in the column. Taking into account these facts, the results obtained through the numerical simulation can be considered in good agreement with the experimental ones.

6 CONCLUSIONS

A coupled numerical model for the analysis of mass and pollutant transfer in deformable partially saturated porous media is presented. Results obtained by this model compare well with those of a laboratory test. This shows that quantitative solutions can be obtained numerically. Great difficulty arises however in the choice of the parameters entering the stated governing equations.

Acknowledgements

This study was supported by MURST (40%) and CNR under contract number 89.03331.07.

REFERENCES

Allen, B.M., Herrera, I. & Pinder, G.F. 1988. Numerical modeling in science and engineering. Chichester: Wiley.

Bear, J. & Bachmat, Y. 1991. Introduction to modeling of transport phenomena in porous media. Dordrecht: Kluwer.

Hassanizadeh, M. & Gray, W.G. 1979. General conservation equations for multiphase systems: 2. Mass, momenta, energy and entropy equations. Advances Water Resources, **2**, 191-203.

Matticchio, B., D'Alpaos, L. & Lippe, E. 1991. Indagine sperimentale sul movimento di un inquinante in colonna di sabbia parzialmente satura. Report of Istituto di Idraulica, University of Padua.

Nielsen, D.R., van Genuchten, M.Th. & Biggar, J.W. 1986. Water flow and solute transport Process in the unsaturated zone. Water Resources Research, **22** 89S-108S.

Schrefler, B.A., Simoni, L., Li Xikui and Zienkiewicz O.C. 1990. Mechanics of partially saturated porous media. Numerical Methods and Constitutive Modelling in Geomechanics. CISM Lecture notes. Desai C.S. and Gioda G. (eds.), 169-209, Wien, Springer Verlag.

Van Genuchten, M.Th., Pinder, G.P. & Saukin, W.P. 1977. Modeling of leachate and soil interactions in an aquifer. Proc. Third Annual Mun. Solid Waste Res. Symp., EPA-600/9-77-026, 95-103.

Zienkiewicz, O.C. & Taylor R.L. 1991. The Finite Element Method, Vol. 2. London, McGraw-Hill.

Zhan, X.Y. 1993. Numerical simulation of heat and multiphase flow in deformable porous media, Ph.D. Thesis, University of Padua.

Impact of Computational Mechanics on Engineering Problems, Pulmano & Murti (eds)
© 1993 Balkema, Rotterdam, ISBN 90 5410 324 8

Design of rock bolts with numerical models

G. Swoboda
University of Innsbruck, Austria

M. Marenče
Geomechatronic Center Linz, Hagenberg i.M., Austria

S. Strohhäusl & H. Erten
Mayreder Consult, Linz, Austria

ABSTRACT: In order to calculate the real behaviour of the bolt in the rock mass two elements are suggested, one for the description of the rock bolts in continuous rock mass and one when the bolt intersects a joint. Application of this model is shown in a normal shear test and compared with experimental results and in a numerical modelling of excavation and stabilisation of a cavern.

1 INTRODUCTION

The strength of rock mass, especially jointed rock mass, is very often insufficient in tension and shear. It can be increased by rock bolts, one of the most effective and most economical means of support. The term rock bolt indicates any form of mechanical support inserted into the rock mass for the primary purpose of increasing its stiffness and strength with respect to tensile and shear loads. A rock bolt system consists of a cylindrical bar of high strength and a connecting medium between the bar and the surrounding ground.

The bearing capacity of rock bolts depends on the axial and shear resistance of the bolt but also on the rock mass. In the continuous or intact rock mass rock bolts act mostly by axial resistance; in the jointed rock mass, the bolt shear resistance becomes important where the bolt intersects the joint. While the axial resistance acts in the same way for all types of rock bolts, the shear resistance is only important for the bolts with a continuous contact between the rock mass and the bar, such as grouted and frictional bolts.

In numerical calculations, it is only possible to consider the real behaviour of the bolt when the bolt is modelled as a special discrete element. Because of the different bolt action in continuous rock mass and on joints, a model with two different elements is suggested.

2 BOLT IN CONTINUOUS ROCK MASS

Usually the rock bolts are in continuous rock mass simulated as truss or beam elements (see Laabmayr and Swoboda (1978)). These elements are not appropriate to describe rock bolts, especially grouted bolts, because they don't take into consideration the stiffness of the grout. The main disadvantage of the method is that it is not possible to take into consideration different displacements in the bolt and on the surface of the borehole (shear displacement which takes part in grout). To solve this problem some authors try to apply special interface elements between bolt, represented as one-dimensional elements, and rock mass. These interface elements were springs in nodes (see Wullschläger and Natau (1983)), or special interface elements. Such elements are normally two or three dimensional elements with characteristics of grouted material (see Spang (1988) and Siriwardane (1989)), or special joint elements as the axisymmetrical joint element by Ghamboussi (1973).

A further development of the rock bolt elements was an element which includes the stiffness of bolt and interface. Aydan et al. (1988) developed an element with four nodes, two connected to the rock and two to the bolt. This element can be coupled to regular finite element models with two nodes which represent the grout–rock coupling. The remaining two nodes on the bar are connected with the rock with the stiffness of grout (Figure 1).

Based on this element a modified rock bolt element was developed. In this model, the nodes on the steel bar and the grout–rock border have the same coordinates, but the real geometry is used in the stiffness matrix. This stiffness matrix is the sum of the axial stiffness of the bar and the stiffness of the grout annulus. Some assumptions are made for this element:

- the distribution of deformation in longitudinal (local x) direction of the bolt element is linear
- the distribution of deformation in radial direction, normal to the local axis of the bolt element is constant

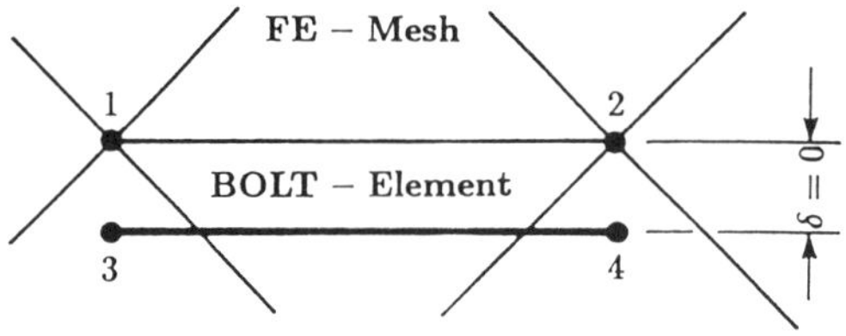

Figure 1: Four node bolt element. Coupling with Finite Element model

– the distribution of deformation in radial direction in the grout is based on a theoretical analytical solution for axisymmetric bodies. This assumption saves a discretization in the radial direction
– the normal stress in the longitudinal direction is transferred by the bolt, but in radial direction no stresses are taken into account
– the shear stress which is the result of a different longitudinal movement in the bolt and on the grout–rock border is taken by shear resistance of grout
– the shear stress which is the result of a transversal movement on the grout-rock border is taken by the shear resistance of the steel bar
– the transversal stiffness between the bar and the grout rock border, 'dowel' stiffness, is modelled as an additional spring between the bar and the grout-rock border nodes in transversal direction.

The bolt stiffness matrix $[K]$ under consideration of expressed assumptions can be written as (1) where

$$k_b = \frac{E_b A_b}{L}, \quad k_s = \frac{G_b A_b}{L}, \quad k_g = \pi G_g \frac{L}{3\ln(r_h/r_b)}$$

E_b and G_b are the elasticity modulus and the shear modulus of the bolt, G_g the shear modulus of the grout, A_b the area of the bolt, L the length of the element, r_b and r_h the radius of the bolt and the hole and k_d is the 'dowel' stiffness. The 'dowel stiffness' is important only when the bolt intersects the rock joint and it is equal to the shear stiffness of the Bolt Crossing element defined in the next section. If the 'dowel' is not defined in this way the value can also be calculated from work of Brady and Lorig (1988)

$$k_d = E_b I \left[\frac{E_g}{2E_b I (r_h/r_b - 1)} \right]^{3/4} \tag{2}$$

where E_g is the elasticity modulus of the grout and I a second moment of inertia of the bolt.
Since relative small displacements of the rock can produce very big forces in the bolt the bolt element requires to be modelled as an elasto-plastic element. The bolt element consists of two different parts, the bar and the grout, made of different materials, and the post yield definition for both materials have to follow their natural behaviour.

3 ROCK BOLT IN INTERSECTION WITH JOINT

A fully grouted, untensioned bolt in intersection with a joint has a great influence on the joint resistance to shear. The resistance of the bolted joint can be expressed as a sum of joint and bolt resistance. Bolt resistance was observed in laboratory and in situ tests and can be considered as the result of two effects: an increase in the bolt normal force and a 'dowel' effect. The increase in the bolt axial tensile force caused by relative movements of joint planes acts in two ways, one component normal to the joint furnishes additional shear resistance through a friction effect and one component parallel to the joint acts as a portion of the joint shear resistance. The 'dowel' effect describes resistance induced by shear and bending deformations of the bolt. The factors influencing the resistance of a bolted joint depend on the characteristics of the joint and many factors connected to the bolt such as the diameter of the bolt and the hole, steel quality, bolt extension, strength of the rock and the grout, and an angle between joint and bolt.
Many authors have worked on this problem in the last twenty years and their work can be divided into two groups: one dealing with only the maximal resistance and another concerning the behaviour of the bolted joint from the beginning to failure of the bolt.
Most theories in the first group (see Bjurström (1974) and Pels (1974)) concern an axial resistance based only on undeformed bolt geometry and on the assumption that the normal force in the bolt is equal to the yield force. Empirical formulas are

$$[K] = \begin{bmatrix} 2k_g & 0 & k_g & 0 & -2k_g & 0 & -k_g & 0 \\ 0 & k_s + k_d & 0 & -k_s & 0 & -k_d & 0 & 0 \\ k_g & 0 & 2k_g & 0 & -k_g & 0 & -2k_g & 0 \\ 0 & -k_s & 0 & k_s + k_d & 0 & 0 & 0 & -k_d \\ -2k_g & 0 & -k_g & 0 & k_b + 2k_g & 0 & -k_b + k_g & 0 \\ 0 & -k_d & 0 & 0 & 0 & k_d & 0 & 0 \\ -k_g & 0 & -2k_g & 0 & -k_b + k_g & 0 & k_b + 2k_g & 0 \\ 0 & 0 & 0 & -k_d & 0 & 0 & 0 & k_d \end{bmatrix} \tag{1}$$

used for the 'dowel' effect. Spang (1988) conducted many shear tests under different conditions and found the relation for the maximal resistance of the bolted joint, which includes most of the important parameters. All these theories consider only the maximal resistance of the bolted joint and do not discuss behaviour before failure.

In order to understand the behaviour of the bolted joint before failure, the complex combined system including the bolt, grout and rock mass around the bolt must be solved. The physical behaviour of this system must be described for shear movement and also for normal movement or for a decoupling of the joint. Panet (1987) suggested that these be calculated as separate problems for which elastic analytical solutions exist. This calculation correlates well with experimental results but only when the whole system remains elastic, which is only true for very small displacements. Schubert (1984) found an iterative procedure for shape and forces in the bolt crossing the joint. He included in his theory all geometrical and material parameters significant for the result. He included the strength of the grout and the rock as a 'penetration modulus', which he defined as the ratio of force to penetration depth as measured in a 20-mm diameter steel ball in rock material. From experimental results Yoshinaka et.al. (1988) tried to find a relationship between the displacement at the joint and the bolt resistance. This theory is based on the assumption of the circular deformed bolt shape with some parameters such as statistical values. Also, the bedding reaction forces are assumed to be uniformly distributed but independent on displacement, as is the case in reality. Brady and Lorig (1988) solved the problem of the bolted joint for the Finite Difference Method, but the same theory can also be included in the Finite Element Method. Resistance of the bolt in intersection with a joint is represented as an element with two springs, one parallel to the local axis of the element and one perpendicular to it. The elastic stiffness of the springs can be determined in experiments or with the given equations. The elastic stiffness produces very great forces in the element and plastic models must be used.

4 BOLT CROSSING JOINT ELEMENT

It is not possible to explain in closed form the interaction between the bolt and the joint due to the complexity of the problem. An iterative process is presented here, in which the intersection region is represented with a special element shown in Figure 2.

The Bolt Crossing Joint element (BCJ) is the element that connects the bolt elements (see Swoboda and Marenče (1991)) on both sides of discontinuity. The BCJ element is assumed as an element with two nodes, one on each side of the discontinuity and connecting the bar nodes of the bolt element. The other nodes of the bolt element on the grout–rock border are connected through an interface element representing the joint.

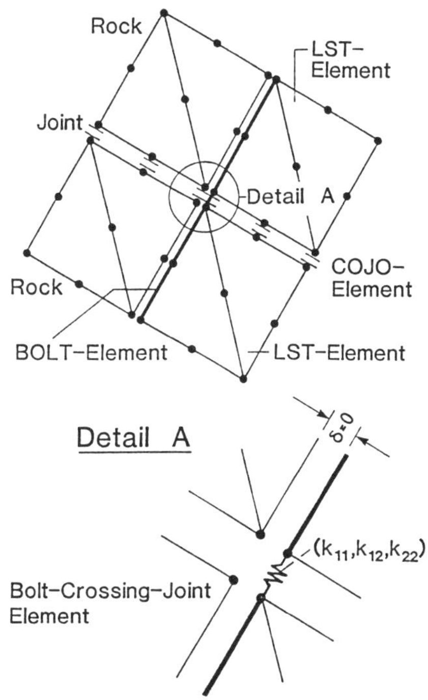

Figure 2: Bolt Crossing Joint element and connection with finite element mesh

The BCJ element is modelled as springs that describe bolt resistance according to movement at the joint. The stiffness matrix of the Bolt Crossing Joint element $[K]$ can be written as

$$[K] = \begin{bmatrix} k_{11} & k_{12} & -k_{11} & -k_{12} \\ k_{21} & k_{22} & -k_{21} & -k_{22} \\ -k_{11} & -k_{12} & k_{11} & k_{12} \\ -k_{21} & -k_{22} & k_{21} & k_{22} \end{bmatrix} \quad (3)$$

where k_{11} is the shear stiffness, k_{22} the normal stiffness, k_{12} represents a connection between a shear displacement and a normal force and k_{21} a connection between a normal displacement and a shear force. The stiffness values depend on many different parameters, and explicit expressions are very difficult to prescribe. Here, an additional finite element analysis is suggested for calculation of the stiffness matrix components.

The stiffness matrix components are calculated as a substructure in the separate calculation. In the substructure system and assuming that bolt deformation is symmetric on both sides of the joint, only the bolt on one side of the joint is taken into account. The bolt near the joint is calculated as a beam structure. The beams represent the bar, and the grout and rock around the bar are modelled as longitudinal and lateral continuously distributed non-linear springs. In this case, most of the influences can be taken into account such as the diameter of bolt and hole, steel quality, strength of rock and grout, angle between joint and bolt. Before breaking, the bolt can absorb displacement in the approximate range of the bolt diameter and the analysis is performed according to the theory of large displacements. The substructure stiffness matrix is represented as the sum of beam stiffness, spring stiffness and additional stiffness of the large displacement theory. A deformed bolt shape and forces in the bolt are calculated to determine real shear and normal displacement on the joint. The proposed analysis is an iterative procedure.

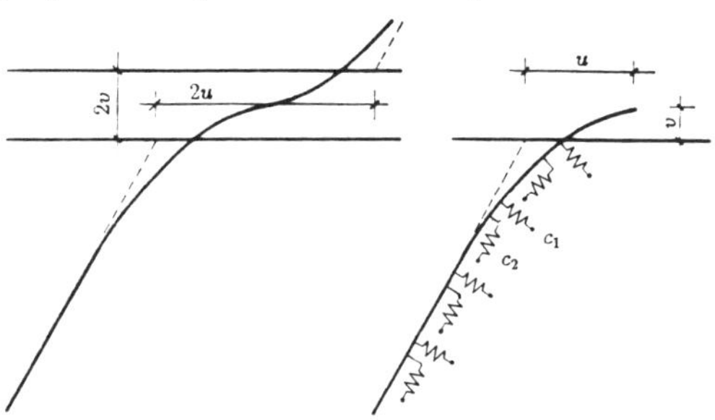

Figure 3: Substructure system

The stiffness of the substructure system is very sensitive to the displacement value and also to a combination of shear and normal displacements. The BCJ stiffness matrix terms are calculated on the defined stiffness of the substructure system by separately applying normal and shear displacement. The stiffness matrix of the BCJ element is not symmetric, but the difference between k_{12} and k_{21} is very small and in the numerical model a symmetricized stiffness matrix is used.

At the intersection with the rock joint, the bolt can absorb large shear and normal displacement before breaking, but only with plastic deformations of the bar and the surrounding material. The steel beam is modelled as an elastic-idealplastic material.

The surrounding material is modelled as non-linear springs in longitudinal and lateral direction. Spring stiffness is very sensitive to displacement and must be defined on the basis of it. The longitudinal spring stiffness can be defined on the basis of the bond–slip curve for bonds between bar and grout or it can be deducted from experiment results. The lateral spring stiffness is more complicated and can be determined through experiments or the Finite Element analysis in which the materials, especially grout and rock, must be very carefully modelled. Special consideration must be given to the region where plasticity occurs. This region is fully crushed and the material constants for the crushed material must be used.

5 SHEAR TEST EXAMPLE

5.1 Laboratory Shear Test

Shear tests were conducted, in Tunnelling Laboratory of Mayreder Consult in Linz, with fully grouted, untensioned rock bolts, in order to investigate the force-deformation-behaviour, which was used as a comparison basis with numerical calculations.

In each test, three precast concrete blocks were connected by an anchor, which was installed in pre drilled, central boreholes and grouted, according to regular procedures in tunnelling. After hardening of the grout, the side blocks were lifted, by means of a hydraulic device, while the middle block was forced to stay in its location. The shear force resulting from the (app. monotonous) shear displacement was measured.

The symmetrical experiment set-up was chosen to eliminate bending moments on the total system, and thus systematic errors. The ideal case of a pure shearing without friction was not possible to put into practice, because too big point loads would have to be applied on the concrete blocks. Thus the forces were introduced as surface loads by means of steel plates, taking into account, that this would produce bending moments on the joints. In order to reduce the friction between concrete blocks, Teflon plates of 4mm thickness were used (friction angle app. 5°). The deformation of the above mentioned steel plates as well as the

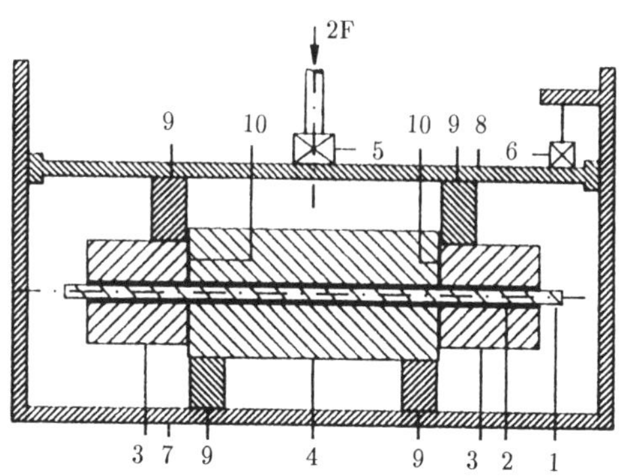

1 Ribbed steel bar of diameter 26 mm
2 Ready-mix-grout
3 Concrete block, 200*200*200 mm with a central hole of diameter 40 mm
4 Concrete block, 250*250*500 mm with a central hole of diameter 40 mm
5 Load cell
6 Inductive position transducer
7 Rigid steel plate
8 Movable steel plate
9 Block of steel plates for force distribution
10 Teflon plate for reduction of friction

Figure 4: Experiment set-up

weight of the concrete blocks were negligible. Only a constructive reinforcement was used for the concrete blocks. The grout had reached a 12 day strength, when the experiments were conducted. The force was induced by a hydraulic device, measured by a load cell and the displacement change was app. monotonous measured by an inductive position transducer. Further physical data were as follows. The concrete blocks had a strength to uniform pressure of 50 N/mm^2 and an E-modulus of 37 000 N/mm^2. The grout hat strength to uniform pressure of 72 N/mm^2 and an E-modulus of 20 000 N/mm^2. The steel bar had a tension strength of 750 N/mm^2 and an E-modulus of 211 000 N/mm^2. Five equal experiments were conducted. A maximal displacement of 20 mm was induced. At the end of the tests, the grout was observed to have been pressed through and the blocks cracked, near the joints. Figure 6 shows the measured force-displacement behaviour, characteristic for all experiments.

5.2 Numerical Shear Test

Based on the presented theory, the laboratory shear test is numerically modelled and the results are compared. Calculation is performed with the program package FINAL (Swoboda, 1990). In the numerical model the symmetrical experiment set-up is used and only a half of the system is modelled. The model is based on the assumption of plane stress. The concrete blocks are modelled with LST (Linear Strain Triangular) elements which in the test remain elastic. The joint between the blocks is modelled as an elasto-idealplastic joint with a friction angle of 5°. The bolt with it's geometry is modelled using BOLT elements and at the intersection with the joint with a BCJ element. The longitudinal spring stiffness curve for the substructure calculation is defined from pull-out tests and the lateral spring stiffness curve is defined from an experimental work of Poli (1988). The Finite Element mesh is shown in Figure 5.
The results of the laboratory test and numerical model are compared in Figure 6. The shear force - shear displacement diagram is shown for one half of the system. The numerical results are in very good agreement with the experiments till to the applied force on one side of 150 kN. At that point in the experiment the sheared block is splitted under the bolt. Changing of the lateral spring stiffness in the substructuring calculation of the BCJ element the splitting of the block can be also modelled.

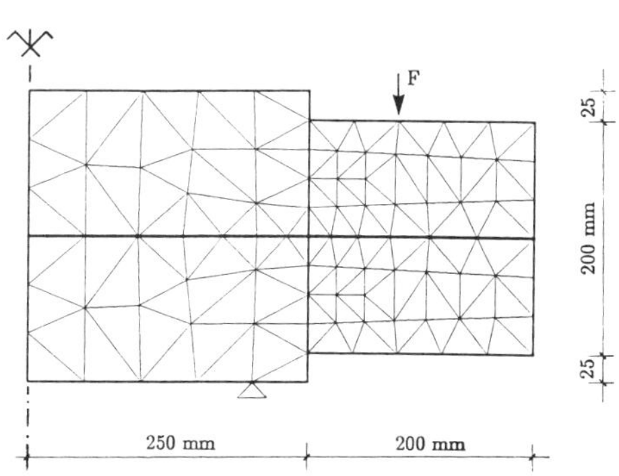

Figure 5: Shear test - finite element mesh

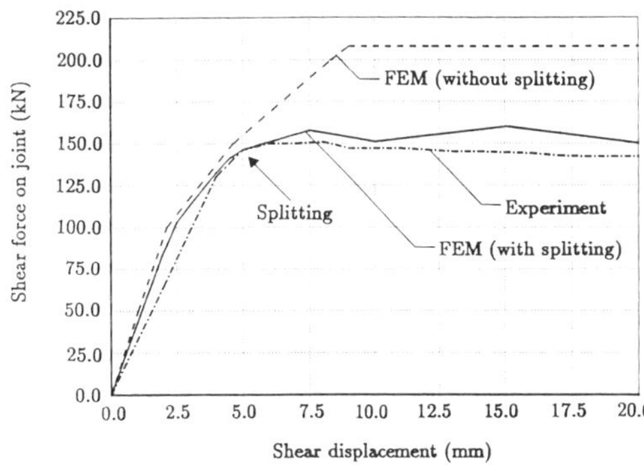

Figure 6: Measured and calculated shear force-shear displacement behaviour for one joint

6 BEHAVIOUR OF BOLTED JOINT

The effect of rock bolting on joint shear resistance can be described as an increase in joint cohesion without a change in the friction angle. This can be proved by a shear test placing different normal pressure on the joint. The increase in cohesion can be expressed as the additional joint shear resistance caused by the bolt divided by the joint area corresponding to one bolt.

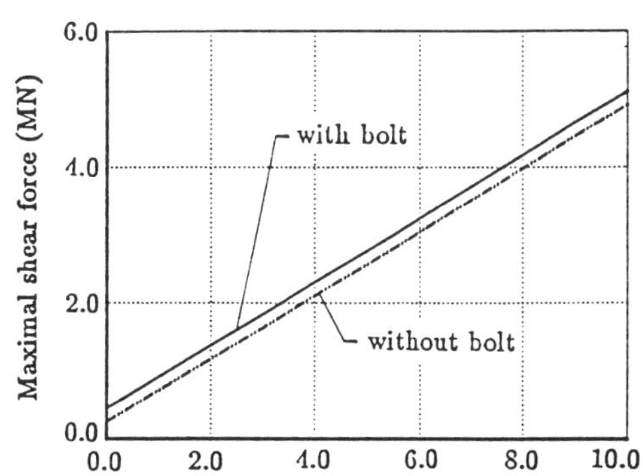

Figure 7: Increase in joint strength parameters caused by bolting

Behaviour of the bolted joint is very sensitive to the setting angle, as can be seen from Figure 8, where the maximal resistance of a bolted joint for a whole setting angle region is shown. The results are compared with different theories presented in the literature. The results obtained correlate well with other theories, except when the setting angle is approximately 90°. This is because all these theories only give consideration to the effect of the normal force and not to that of the 'dowel' effect. The equation defined by Spang (1988) took both effects into consideration because from many shear tests it was defined as a statistical equation. His equation is only valid for a setting angle between 60° and 90° and correlates well with the presented theory.

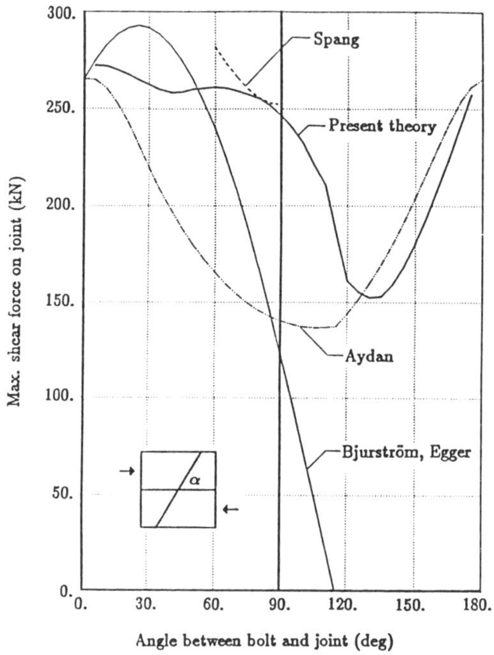

Figure 8: Influence of the setting angle on maximal bolted joint resistance

7 NUMERICAL MODELLING OF A CAVERN

The stability of an underground structure during and after excavation is the most important question for designers, because any kind of collapse can destroy large parts of finished tunnel, thereby causing major repair work and time loss. Preliminary calculations are therefore of great importance, but a calculation is only useful when the model correctly describes natural behaviour.

Implementation of rock bolts in the numerical model which simulates excavation and stabilisation of an underground opening is demonstrated on the cross section of one cavern (Schieber- und Montage-Kaverne) for a Hydroelectric power plant at Gerlos, Austria. The cavern is situated in Innsbruck's quartz-phylite (Innsbrucker Quarzphyllites) with up to 1-m-thick mylonite zones. The cavern's location is selected on the basis of geological predictions and drilled boreholes (Zenz and Kummerer (1991)). The geometry of the cavern with disturbed zones and primary support system is shown in Figure 9. The cavern is situated at a depth of 110 meters. The primary support system consists of a 20-cm-thick shotcrete lining and systematic bolting with 'short' grouted bolts 3, 4 and 6 m in length, and 'long' bolts 10 m in length. In order to stabilise the disturbed zones and fix the crane girders, three pre-stressed bolts were used. 'Short' bolts were set along the cavern at intervals of 1.5 m, and 'long' and pre-stressed bolts at intervals of 3 m. Displacements and their development in time were defined by convergence and extensometer measurements.

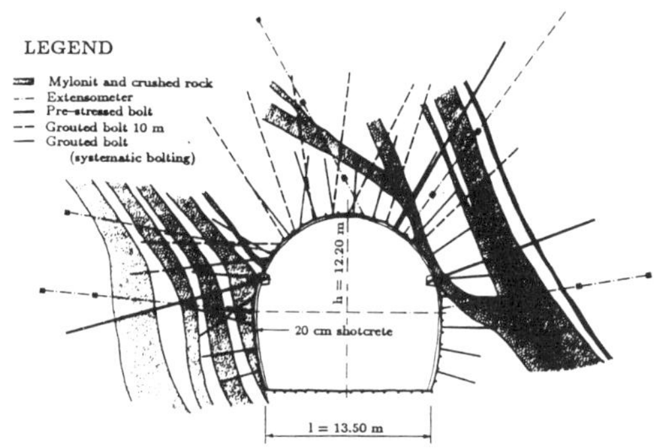

Figure 9: Geometry of the cavern with disturbed zones and primary support system

Excavation and stabilisation of the cavern was conducted in four steps:

- excavation of a side drift
- excavation of a crown
- excavation of a bench
- excavation of an invert

The numerical model follows the specified excavation steps and takes into account all important parameters. Calculation is performed with the finite element method using the program system Final (Swoboda (1990)) as a two-dimensional plane strain problem. The plane strain model thickness is not 1 m as usual, but 3 m, which is the longitudinal interval between rock bolt sets. This makes specification of a bolt, as a discrete element, much easier and has no influence on results.

The rock mass is modelled by linear strain triangular (LST) elements as a homogeneous isotropic elasto-plastic material. The rock mass is in reality an anisotropic material, but the isotropic model is used because of a lack of required parameters indispensable for anisotropic behaviour. The plastic behaviour is defined by the Mohr-Coulomb failure criteria with peak and residual strength parameters. Based on the geological conditions the rock mass is divided into four different materials. Materials I, II, and III specify the quartz-phylite zones

Table 1: Rock mass material properties

MATE-RIAL	ELASTIC PARAMETERS		STRENGTH PEAK		STRENGTH RESIDUAL	
	Elasticity modulus [MN/m²]	Poisson's number -	Cohesion [kN/m²]	Friction angle [°]	Cohesion [kN/m²]	Friction angle [°]
I	10 000.	0.18	500.	40.	250.	40.
II	1 800.	0.16	1000.	35.	500.	30.
III	1 300.	0.15	1000.	35.	500.	30.
IV	100.	0.30	2000.	30.	1000.	30.

with different degree of mylonitization. Material IV represents mylonite zones. Rock mass parameters used in calculation are specified in Table 1.
Shotcrete lining is modelled as beam elements (BEAM6) with a thickness of 20 cm. Development of shotcrete stiffness with time is, in the calculation, expressed by a modulus of elasticity dependent on shotcrete age; for young shotcrete 5 000 kN/m^2 and for old shotcrete 20 000 kN/m^2.
Bolts, grouted and pre–stressed, are modelled as BOLT elements with a bar diameter r_b=26 mm and a hole diameter r_h=40 mm. Pre–stressed bolts are tensioned by a force of 345 kN. The 'short' bolts which are spaced at longitudinal intervals of 1.5 m, are expressed twice in the model. The position of the rock bolts is shown in Figure 10.

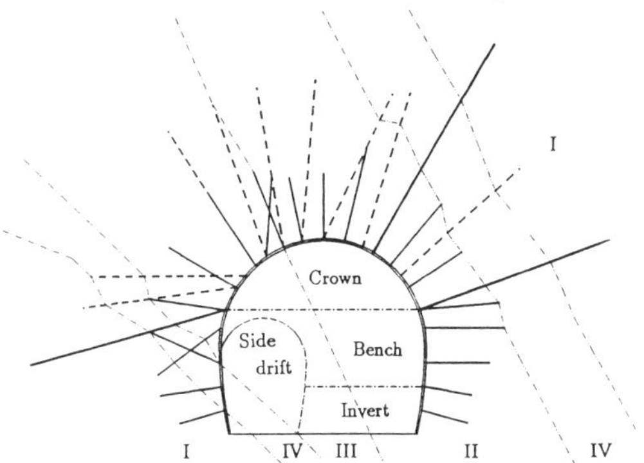

Figure 10: Disposition of rock bolts and excavation steps

The loading steps follow the excavation phases with indispensable partition. The initial state of stress is assumed from the geological conditions. Vertical initial state of stress is taken as a linear distribution in depth with an overburden of 110 m and a rock unit weight of 28 kN/m^3. The horizontal initial state of stress is taken as a portion of vertical stress with a coefficient of horizontal stress of 0.5.
The three–dimensional effect of excavation is taken into account by initial stress relief before excavation. In this way, displacements taking place before excavation are included in the calculation. The factor defining initial stress relief α is 0.3 (Swoboda, (1989)).
The bolts and shotcrete lining follow the excavation procedure. After finishing each excavation step, the plastic calculation is performed to partly release the over stressed rock regions. At the end, all over stressed rock regions are released. Compatibility between the numerical model and the 'in situ' conditions is achieved through the convergence measurements.
Some representative results, for the final situation, are shown here. Distribution of axial force in the 'short' bolts is shown in Figure 11. These bolts are mostly loaded by plastic displacements in the rock mass. This is proved when the distribution of axial forces in bolts is compared with the region in which plastic deformations occur. Some parts of 'short' bolts are loaded until the yield force.

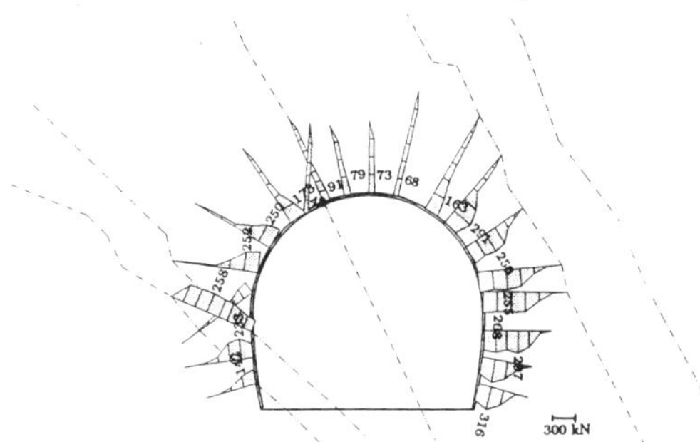

Figure 11: Axial force distribution in 'short' bolts for final loading step

The distribution of axial force in 'long' bolts shows the same tendency as in short bolts, except when they pass through weak mylonite zones (Figure 12). The bolts passing through or located only in the mylonite zone situated on the right side of the cavern, because of large displacements which occur in this zone, are also loaded at the end of the bolt. Two 'long' bolts passing though the mylonite zone on the left side of the cavern, because of displacement distribution in this region, which causes contraction of this zone, are loaded under compression.
From Figures 11 and 12 it can be concluded that the bolts set in the crown are not fully loaded, but their role is also to prevent possibly instable rock blocks from falling. In this calculation this problem was not taken into consideration, because the position and direction of fault systems was unknown and it will make calculation too complicated.
The effect of rock bolts on the normal force in the shotcrete lining is shown in Figure 13. Installed bolts reduce the lining deformations and thus normal forces in lining, but the rock bolts also exert internal pressure on the lining.

8 CONCLUSION

The rock bolts used in tunnel excavations are mostly untensioned grouted bolts, and this type of

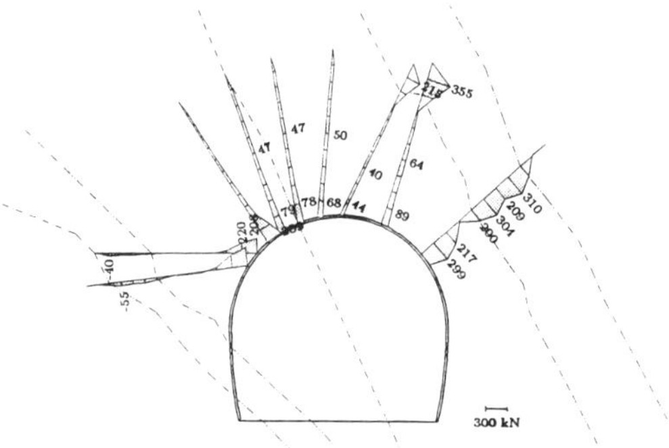

Figure 12: Axial force distribution in 'long' bolts for final loading step

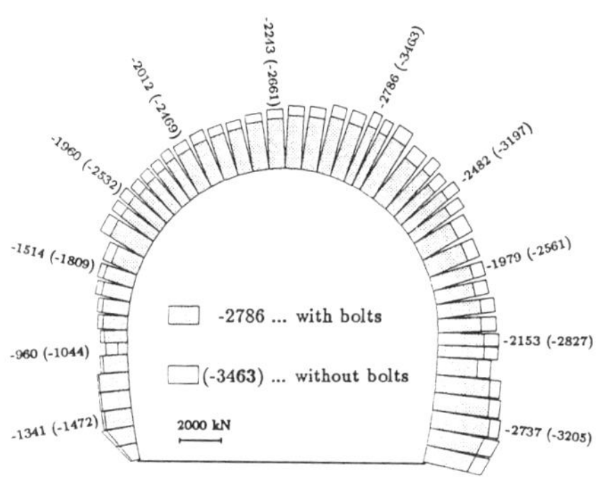

Figure 13: Axial force in shotcrete lining for calculation with and without rock bolts

bolt is the main focus of this work. From the model of the grouted bolt, other types of rock bolts can also be modelled by the theory presented herein. Bolt behavior in intact rock mass is so different from behavior when a bolt intersects a joint that a model with two different elements is suggested for numerical calculation; one element for the bolt in the intact rock mass and one as a bolt intersecting with a joint.

The bolt element is defined as an element consisting of a bar and a grout annulus. The bar can take shear and normal forces and is continuously connected with a surrounding rock through the grout annulus. In underground constructions the bolt is often overloaded and plastic behavior is used for the bar and the grout.

A bolt crossing through the joint increases joint stiffness but also joint strength. Joint strength is enhanced by the increase in the bolt normal force and the 'dowel' effect. Both effects are difficult to describe and depend on many parameters such as the bar, grout and rock material characteristics, geometric parameters of the grout and bar, but also the bolt setting angle. The theory presented includes all of these parameters and both effects can be detected. The bolt in intersection with a joint element is a special element used when a bolt crosses through a joint for description of that complicated behavior. The element connects bar nodes on both sides of the joint, and the element stiffness is defined in a special substructure calculation. The substructure calculation defines the bolt stiffness on one side of the joint for the real shear and normal displacements taking part on the joint. The results agree satisfactorily with experimental data. This confirms that the mechanical behaviour of the grouted rock bolt can be theoretically quantified with satisfactory precision.

When the bolt is not modelled as a special element, the effect of rock bolting can be described as an increase in joint strength, which is effected by increasing joint cohesion without changing the friction angle. The setting angle for maximal mobilised shear resistance is between 50° and 70° and also depends on joint friction angle.

ACKNOWLEDGEMENT

The authors express their gratitude to the 'Jubilaeumsfond der Österreichischen Nationalbank' for promoting this research project (No. 3909) and to the 'Forschungsförderungsfond der gewerblichen Wirtschaft' – Project RD-Anker, which made it possible to conduct essential fundamental investigations on the impact of rock bolts in the New Austrian Tunnelling Method.

REFERENCES

Aydan, Ö., T. Kyoya, Y. Ichikawa, T. Kawamoto, T. Ito T & Y. Shimizu 1988. Three-dimensional simulation of an advancing tunnel supported with forepoles, shotcrete, steel ribs and rockbolts. Numerical Methods in Geomechanics, Swoboda(ed.), Innsbruck: 1481-1486. Rotterdam: Balkema.

Bjurström, S. 1974. Shear strength of hard rock joints reinforced by grouted untensioned bolts. Proc. of 3rd Cong. of the ISRM, Vol II, Part B, Denver: 1194-1199.

Brady, B. & L. Lorig 1988. Analysis of rock reinforcement using finite difference methods. Computers and Geotehnics 5:123-149.

Ghamboussi, J., E.L. Wilson & J. Isenberg 1973. Finite element for rock joints and interfaces. Journal of the Soil Mechanics and Foundations Division, Proc. of ASCE, Vol 99, SM10: 833-848.

Laabmayr, F. & G. Swoboda 1978. The importance of shotcrete as support element of the NATM. Eng. Foundation Conf. Proc. of the 2nd Shotcret Conf.: 65-79. St. Anton 1978.

Marenče, M. 1992. Numerical model for rockbolts under consideration of rock joint movements. Thesis for the Degree of Doctor of Engineering, Universität Innsbruck, Innsbruck, Austria.

Panet, M. 1987. Reinforcement of rock foundations and slopes by active or passive anchors. Proc. 6th Int. Cong. ISRM, Montreal: 1411-1420.

Pells, P.J.N. 1974. The behavior of fully bonded rock bolts. Proc. 3rd Congr. Int. Soc. Rock Mech., Denver: 1212-1217.

Poli, Dei S., S. Ottanà & S. Sesana 1988. Modellazione Non Lineare Dell'azione di Spinotto di Una Barra Immersa in Una Massa Illimitata di Calcestruzzo. Studi e Ricerche, Corso di Perfezionamento per la Costrucioni in Cemento Armato F.lli Pesenti, Vol.10, Politecnico di Milano, Italia.

Schubert, P. 1984. Das Tragvermögen des mörtelversetzten Ankers unter aufgezwungener Kluftverschiebung. Dissertation an der Mountanuniversität Leoben, Österreich.

Siriwardane, H.J. 1989. Numerical analysis of anchors in soil - Application brief. International Journal for Numerical and Analytical Methods in Geomechanics 13: 427-433.

Spang, K. 1988. Beitrag zur rechnerischen Berücksichtigung vollvermörtelter Anker bei der Sicherung von Felsbauwerken in geschichtetem oder geklüftetem Gebirge. Ecole Polytechnique Federale de Lausanne, These No 740.

Swoboda G. 1989. Numerical modelling of tunnels. Numerical Methods and Constitutive Modelling in Geomechanics, CISM Courses and Lectures No. 311, Udine: 277-318.

Swoboda, G. 1990. Programmsystem FINAL - Finite element analysis program for linear and nonlinear structures, Version 6.3. Print University Innsbruck.

Swoboda, G. & M. Marenče 1991. FEM modelling of rock bolts. Proc. of 7th Int. Conf. Computer Methods and Advances in Geomechanics, Cairns: 1515-1520.

Wullschläger, D. & O. Natau 1983. Studies of the composite system of rock mass and non-prestressed grouted rock bolts. Proc. of the Int. Symp. on Rock Bolting: 75-85. Abisko, Sweden.

Yoshinaka, R., T. Shimizu, H. Arai & S. Ariska 1988. Experimental study on the rock bolt reinforcement in jointed rock mass. ISRM Symp. Rock Mechanics and Power Plants, Romana(ed), Madrid: 427-435.

Zenz G. & J. Kummerer 1991. Beitrag zur Dimensionierung tiefliegender Kavernen. Proc 7th Cong. Int. Soc. Rock Mech., Aachen: 1251-1254.

Impact of Computational Mechanics on Engineering Problems, Pulmano & Murti (eds)
© 1993 Balkema, Rotterdam, ISBN 90 5410 324 8

Hardening models: Some implications for elastoplastic analysis by mathematical programming

F. Tin-Loi
University of New South Wales, N.S.W., Australia

ABSTRACT: The implications of using different hardening models in the analysis of discretized elastoplastic structures by mathematical programming methods are reviewed. Both piecewise linearized linear and nonlinear hardening laws are considered. Particular emphasis is given to a class of nonlinear hardening laws for which an extremum approach can be impractical. The alternative approach proposed involves formulation and solution as a nonlinear complementarity problem. This procedure is briefly described within a holonomic context.

1 INTRODUCTION

Elastoplastic analysis is a fundamental problem in structural plasticity. A crucial aspect of this problem is the choice of a model to depict material behaviour, especially when the applied loads are variable in nature. The constitutive model used must be, on the one hand, accurate enough to depict faithfully observed material behaviour and, on the other hand, simple enough to be used in suitable methods of analysis.

The primary aim of this paper is to highlight some of the implications in using certain hardening models within elastoplastic analysis procedures based on mathematical programming (MP). In addition to the general class of piecewise linear associated constitutive laws with translating-interacting yield planes, introduced since the early seventies by Maier (1970), we also propose and consider a simple nonlinear hardening model.

The organisation of this paper is as follows. For simplicity and without undue loss of generality, we assume holonomic (path-independent) behaviour and introduce a general class of plastic constitutive models with nonlinear hardening. It is shown that the nonlinear model proposed is a generalization of Maier's linear hardening representation. Within the convenient matrix-analytical dress provided by conventional finite element description, we then, under the assumption of small displacements, describe the governing relations for the entire structure. Extremum characterization of the analysis is then described and the implications of the particular hardening rule used discussed. A nonlinear complementarity problem (NCP) approach is proposed for the analysis of structures with nonlinear hardening. Finally, an extension of the holonomic constitutive laws to the stepwise holonomic case is outlined.

2 NONLINEAR HARDENING MODEL

Consider a structural system discretized as an aggregate of a finite number of individual parts (bars, frame members, etc.). It is assumed that the material behaviour is directly reflected by the element behaviour (e.g. Corradi, 1978). Following well-known notation and description (e.g. Maier, 1968), we let the vectors $\mathbf{q}^i$ and $\mathbf{Q}^i$ denote, respectively, the so-called natural (Argyris, 1967) generalized strains and stresses pertaining to the generic ith finite element. The behaviour of each individual element is assumed to be holonomic so that its constitutive laws can be defined as follows:

$$\mathbf{Q}^i = \mathbf{E}^i\mathbf{e}^i \tag{1a}$$

$$\mathbf{q}^i = \mathbf{e}^i + \mathbf{p}^i \tag{1b}$$

$$\mathbf{p}^i = \mathbf{N}^i\boldsymbol{\lambda}^i \tag{1c}$$

$$-\phi^i = \mathbf{K}^i - \mathbf{N}^{iT}\mathbf{Q}^i + \mathbf{R}^i(\boldsymbol{\lambda}^i) \geq \mathbf{0} \tag{1d}$$

$$\boldsymbol{\lambda}^i \geq \mathbf{0} \tag{1e}$$

$$\phi^{iT}\boldsymbol{\lambda}^i = 0 \tag{1f}$$

Equation (1a) describes the elastic behaviour through the symmetric, positive definite matrix $\mathbf{E}^i$ of element stiffnesses relating generalized elastic strains $\mathbf{e}^i$ to stresses $\mathbf{Q}^i$. Equation (1b) expresses the additivity of elastic $\mathbf{e}^i$ and plastic $\mathbf{p}^i$ strains; for simplicity, imposed strains are assumed to be zero. The holonomic, path-independent plastic behaviour is described by relations (1c) to (1f). The nonnegative vector $\boldsymbol{\lambda}^i$ plays a key role in the elastoplastic analysis in that it is a measure of the plastic flow due to the activation of the relevant yield hyperplanes. In particular, through (1c) the plastic strain vector $\mathbf{p}^i$ is expressed as a linear function of the plastic multipliers or internal variables contained in $\boldsymbol{\lambda}^i$ by taking $\mathbf{N}^i$ as a constant matrix of unit outward normal vectors. Relation (1c) thus shows that the flow rule is associated. Equation (1d) defines the yield functions for the *i*th element collected in vector ϕ^i. As is apparent, ϕ^i is linear in $\mathbf{Q}^i$ through the constancy of $\mathbf{N}^i$ and, with $\mathbf{K}^i$ representing the vector of positive and constant yield limits, describes in $\mathbf{Q}^i$-space a polyhedral elastic domain $\phi^i \leq \mathbf{0}$ which changes at yielding in accordance with the generally nonlinear hardening rule $\mathbf{R}^i(\boldsymbol{\lambda}^i)$. Finally, condition (1f) ensures the complementarity relation between ϕ^i and $\boldsymbol{\lambda}^i$ and applies componentwise in view of the sign constraints on the vectors involved.

Some comments regarding the hardening rule are appropriate at this stage. In general, this rule expresses a nonlinear dependence of the changes in the yield limits on the plastic multipliers. The hardening functions can be chosen to accurately fit experimentally observed plastic behaviour. In practice, however, the selection of a constitutive model is usually made to ensure a good compromise between realism and simplicity. To this end, a simple, yet powerful, expression was proposed by Tin-Loi and Pang (1992). It is an extension of Maier's remarkable matrix representation of linear hardening (e.g. Maier, 1970) and has the form

$$\mathbf{R}^i(\boldsymbol{\lambda}^i) = \mathbf{H}^i\mathbf{f}(\boldsymbol{\lambda}^i) \tag{2}$$

The hardening matrix $\mathbf{H}^i$ is assumed to be constant; any nonlinearity is expressed through the vector $\mathbf{f}(\boldsymbol{\lambda}^i)$. Expression (2) is clearly a generalization of well-known linear constitutive models. Some important examples of different models which can be represented through (2) are:

(a) $\mathbf{H}^i = \mathbf{0}$ describes the traditional piecewise linear approximation of fixed yield surfaces of perfectly plastic materials.

(b) $\mathbf{f}(\boldsymbol{\lambda}^i) = \boldsymbol{\lambda}^i$ implies Maier's linear hardening representation with the possibility of various hardening rules being defined by giving the constant hardening matrix $\mathbf{H}^i$ different entries. For instance (Maier, 1970), a diagonal matrix gives Koiter's noninteracting hardening, while nondiagonal entries can be used to simulate various models of cyclic hardening, of which isotropic hardening and Prager's kinematic law are the most well-known.

(c) Constant $\mathbf{H}^i$, as in (b), and nonlinear $\mathbf{f}(\boldsymbol{\lambda}^i)$ give rise to the class of nonlinear hardening rules considered by Tin-Loi and Pang (1992). As a simple example, consider a uniaxial stress case as for a truss member. Only one natural generalized stress and one natural generalized strain govern its behaviour. Two yield modes (tension and compression) are possible and the matrix of outward normals is given by $\mathbf{N}^i = [1 \quad -1]$. The well-known Ramberg-Osgood type of hardening with no interaction (Koiter's hardening) can then be simply described by a 2 x 2 diagonal $\mathbf{H}^i$ matrix and $\mathbf{f}(\boldsymbol{\lambda}^i) = [\lambda_1^{im} \quad \lambda_2^{im}]^T$ where *m* is a material constant typically assumed to be in the range 0<*m*<1. Cyclic interactive hardening, which includes isotropic and kinematic types of behaviour as special cases, can be simulated as for (b).

It should finally be noted that significant work by Maier and his co-workers (e.g. Comi and Maier, 1990; Maier and Nappi, 1989; Maier and Novati, 1990) considered more general nonlinear hardening laws than given by (2). However, they impose a restriction on the constitutive law by assuming that the hardening is reciprocal. Mathematically, this condition requires the Jacobian matrix of $\mathbf{R}^i(\boldsymbol{\lambda}^i)$ to be symmetrical.

3 DISCRETE STRUCTURAL MODEL

The relations governing the overall quasistatic behaviour of the discretized structural model consisting of n finite elements can be assembled in a conventional manner from element contributions. For instance, the constitutive laws (1a) to (1f) can be expressed compactly as follows:

$$\mathbf{Q} = \mathbf{E}\mathbf{e} \tag{3a}$$

$$\mathbf{q} = \mathbf{e} + \mathbf{p} \tag{3b}$$

$$\mathbf{p} = \mathbf{N}\boldsymbol{\lambda} \tag{3c}$$

$$-\boldsymbol{\phi} = \mathbf{K} - \mathbf{N}^T\mathbf{Q} + \mathbf{R}(\boldsymbol{\lambda}) \geq \mathbf{0} \tag{3d}$$

$$\boldsymbol{\lambda} \geq \mathbf{0} \tag{3e}$$

$$\boldsymbol{\phi}^T\boldsymbol{\lambda} = 0 \tag{3f}$$

The indexless symbols cover all n elements simultaneously, with the self-evident definitions that vectors contain as nonintersecting subvectors the contributions of each element and matrices have a block diagonal structure, with each block referring to one element; e.g. $\mathbf{Q}^T = [\mathbf{Q}^{1T}\ldots\mathbf{Q}^{nT}]$, $\mathbf{E} = \text{diag}[\mathbf{E}^1\ldots\mathbf{E}^n]$, $\mathbf{N} = \text{diag}[\mathbf{N}^1\ldots\mathbf{N}^n]$, etc.

For infinitesimal displacements, as assumed in the present instance, the equations of geometric compatibility and equilibrium for the whole structure can be written, respectively, as

$$\mathbf{q} = \mathbf{C}\mathbf{u} \tag{4a}$$

$$\mathbf{C}^T\mathbf{Q} = \mathbf{P} \tag{4b}$$

where $\mathbf{u}$ represents the vector of the free nodal displacements and $\mathbf{P}$ is the vector of corresponding equivalent nodal loads. Matrix $\mathbf{C}$ is commonly termed the compatibility matrix and its transpose the equilibrium matrix.

4 EXTREMUM CHARACTERIZATION

The sets of equations (3) and (4) describe fully the finite, holonomic behaviour of the discrete system. The aim of the analysis is to find $\mathbf{u}$ and $\mathbf{Q}$ for a given load history $\mathbf{P}$. A popular approach employed in MP-based schemes is to formulate and solve the analysis as an extremum problem (numerous references are given in Lloyd Smith, 1990; Maier and Lloyd Smith, 1986; Maier and Munro, 1982). We introduce this concept by first recalling some familiar notions in MP theory.

The problem in a more general form consists of finding a finite dimensional vector $\mathbf{x}$ such that

$$\mathbf{x} \geq \mathbf{0};\ \mathbf{F}(\mathbf{x}) \geq \mathbf{0};\ \mathbf{x}^T\mathbf{F}(\mathbf{x}) = 0 \tag{5}$$

where $\mathbf{F}(\mathbf{x})$ is a vector-valued function of the same dimension as $\mathbf{x}$. Obviously, any solution of this NCP is a globally optimal solution of the nonlinear program

$$\text{minimize} \quad \mathbf{x}^T\mathbf{F}(\mathbf{x}) \tag{6a}$$

$$\text{subject to} \quad \mathbf{x} \geq \mathbf{0};\ \mathbf{F}(\mathbf{x}) \geq \mathbf{0} \tag{6b}$$

Conversely, any (globally) optimal solution of the latter program with a zero (optimal) objective value solves the above NCP (5). In the special case where the function $\mathbf{F}(\mathbf{x})$ is continuously differentiable and its Jacobian matrix $\nabla\mathbf{F}(\mathbf{x})$ is symmetric for all $\mathbf{x}$, then the NCP becomes the stationarity conditions of the following nonnegatively constrained nonlinear program:

$$\text{minimize} \quad \theta(\mathbf{x}) \tag{7a}$$

$$\text{subject to} \quad \mathbf{x} \geq \mathbf{0} \tag{7b}$$

where $\theta(\mathbf{x})$ is the potential function corresponding to $\mathbf{F}(\mathbf{x})$; that is, $\nabla\theta(\mathbf{x}) = \mathbf{F}(\mathbf{x})$. If the Jacobian matrix is not symmetric, there cannot exist a real-valued potential function with the latter property; hence the nonlinear program (7) would not be available in this case. There are various analytical and computational difficulties associated with the use of the nonlinear program (6) for solving the NCP; foremost among these is the lack of convexity in (7) for a general nonlinear function $\mathbf{F}(\mathbf{x})$. One implication of this is that we generally would need a global optimization procedure for solving the nonlinear program (6) in order to obtain a solution to the NCP; such a procedure does not exist.

An application of the above to our plasticity problem leads to the important conclusion that it is only practicable to perform the analysis as an extremum problem when the Jacobian of the hardening matrix is symmetrical. Indeed, the quadratic programming (QP) and other more general constrained optimization approaches proposed to date (e.g. Comi and Maier, 1990; Maier, 1970) assume this property for the adopted hardening rules. In our case, the nonlinear hardening rule (2) does not always satisfy such a symmetry property and hence fail to satify the requirement of the extremum formulation (7). The approach proposed by Tin-Loi and Pang (1992) for treating the general system described by (3) is to solve the NCP directly. The next section shows briefly how this can be done. Of course, such properties as existence and uniqueness of solution need to be first established for the particular hardening rule chosen, but will not be discussed in this paper.

5 NCP FORMULATION AND SOLUTION

The formulation follows an analogous path to that used for the finite, holonomic, piecewise linear elastic-plastic problem (e.g. De Donato and Maier, 1973). In particular, the analysis is first expressed in terms of the unknown $\boldsymbol{\lambda}$ variables which, after being evaluated, will supply the values of all quantities of interest. Without going into the simple algebraic manipulations involving relations (3) and (4), the holonomic analysis can be expressed in the form of the following NCP:

$$\mathbf{F} = \mathbf{K} - \mathbf{N}^T\mathbf{Q}^e + [\mathbf{R}(\boldsymbol{\lambda}) - \mathbf{N}^T\mathbf{Z}\mathbf{N}\boldsymbol{\lambda}] \geq \mathbf{0} \tag{8a}$$

$$\boldsymbol{\lambda} \geq \mathbf{0} \tag{8b}$$

$$\mathbf{F}^T\boldsymbol{\lambda} = 0 \tag{8c}$$

where $\mathbf{F} = -\boldsymbol{\phi}$; $\mathbf{Q}^e = \mathbf{ECS}^{-1}\mathbf{P}$ is the elastic response to the given loads; $\mathbf{S} = \mathbf{C}^T\mathbf{EC}$ is the familiar, symmetric, positive definite elastic stiffness matrix of the structure; $\mathbf{Z} = \mathbf{ECS}^{-1}\mathbf{C}^T\mathbf{E} - \mathbf{E}$ is a well-known (Maier, 1968) symmetric operator (negative semidefinite in the absence of geometric effects; $\mathbf{N}^T\mathbf{ZN}$ is therefore also symmetric, negative semidefinite). Solution of the NCP (8) furnishes $\boldsymbol{\lambda}$ from which the quantities $\mathbf{Q} = \mathbf{Q}^e + \mathbf{ZN}\boldsymbol{\lambda}$ and $\mathbf{u} = \mathbf{S}^{-1}\mathbf{P} + \mathbf{S}^{-1}\mathbf{C}^T\mathbf{EN}\boldsymbol{\lambda}$ can be calculated. Note also that relation set (8) becomes a linear complementarity problem (LCP) when no or linear hardening is assumed. The analysis can then be solved as such or through an equivalent QP formulation. It is interesting to note that some QP algorithms operate in quite a similar way to the so-called direct methods of initial strain (Argyris and Scharpf, 1972) or initial stress (Zienkiewicz et al., 1969).

Numerous methods exist for solving the NCP (e.g. Harker and Pang, 1990). One of these is the basic Newton method (Pang and Chan, 1982) which involves solving a sequence of LCP subproblems. More specifically, a converging sequence $\{\boldsymbol{\lambda}^k\}$ is generated such that for known $\boldsymbol{\lambda}^k$, $\boldsymbol{\lambda}^{k+1}$ is defined to be the solution of the LCP

$$\mathbf{w} = \mathbf{F}(\boldsymbol{\lambda}^k) + \nabla\mathbf{F}(\boldsymbol{\lambda}^k)(\boldsymbol{\lambda} - \boldsymbol{\lambda}^k) \geq \mathbf{0} \tag{9a}$$

$$\boldsymbol{\lambda} \geq \mathbf{0} \tag{9b}$$

$$\mathbf{w}^T\boldsymbol{\lambda} = 0 \tag{9c}$$

Subproblem (9) can be solved by any standard LCP routine such as one based on Lemke's or Cottle's well-known principal pivoting algorithms. This method of solving the NCP preserves some of the nice features associated with the classical Newton scheme, such as simplicity of computer implementation and quadratic rate of convergence. Some drawbacks include the necessity of evaluating the Jacobian matrix and solving LCP (9). Alternative schemes for soving an NCP exist but will not be discussed here.

6 STEPWISE HOLONOMIC ANALYSIS

Holonomy is a reasonable assumption only when a monotonic increase in stresses occurs. For situations where this is not so, as in cyclic loading, a holonomic analysis is grossly inappropriate. A nonholonomic analysis involves, in the general case, integrating in space and time a system of nonlinear partial differential equations governing the incremental plasticity laws. Few problems are amenable to such a direct method of solution. Typically, the strategy followed is one of approximate, albeit acceptable, numerical solutions resting on discretizations in space and time. The former discretization is carried out, as shown earlier in this paper, through a conventional finite element model. The time discretization, however, is more difficult and is still the subject of intensive research (e.g. Martin et al., 1987; Ortiz and Popov, 1985; Simo and Taylor, 1986). Fairly recent contributions, mostly by Martin and his co-workers (e.g. Martin et al., 1987), have shown that a backward difference time integration scheme is particularly attractive because of its good convergence and stability properties. Mechanically interpreted, this numerical method is equivalent to replacing the nonholonomic constitutive rate laws by a holonomic counterpart within a finite loading step; any loading-unloading occurrence is allowed for between two consecutive steps. It is also significant to note that, in the context of linear yield functions and a fairly general class of linear hardening models (Maier, 1970), the backward difference analysis was introduced by Maier and his co-workers in the seventies under the labels of stepwise holonomic or multistage analysis (e.g. De Donato and Maier, 1973).

Whilst it appears that the load increments need to be small in a stepwise holonomic analysis so that the rate solution is followed as closely as possible, such a restriction is often not necessary in practice. Even loading histories which are far from proportional can be subdivided such that nearly proportional loading is reasonably expected within each step. In general, adequate accuracy is provided through the assumption of holonomic behaviour within steps of finite amplitude. A stepwise holonomic or multistage formulation of the NCP (8) then becomes an attractive method of solution. The fundamentals are similar to the linear case and will only be briefly outlined in the following.

The constitutive laws are expressed in finite terms with the loading history subdivided into a finite steps. Holonomy is assumed to hold within each step and any loading-unloading is allowed for between steps. The entire evolution from a known state Σ_n to the unknown state Σ is then analysed as a sequence of holonomic problems, each problem concerning a finite configuration change $\Delta\Sigma$ due to a finite load increment. The following inequality, corresponding to (8a), is fulfilled for Σ_n:

$$\mathbf{F}_n = \mathbf{K} - \mathbf{N}^T\mathbf{Q}_n^e + [\mathbf{R}(\boldsymbol{\lambda}_n) - \mathbf{N}^T\mathbf{Z}\mathbf{N}\boldsymbol{\lambda}_n] \geq \mathbf{0} \tag{10}$$

After the finite step $\Delta\Sigma$, in which finite changes $\Delta\mathbf{Q}^e$ and $\Delta\boldsymbol{\lambda}$ are experienced, the final value of $\mathbf{F}$ can be expressed, after some rearrangement, as

$$\mathbf{F} = \mathbf{F}_n - \mathbf{N}^T\Delta\mathbf{Q}^e + [\mathbf{R}(\boldsymbol{\lambda}_n + \Delta\boldsymbol{\lambda}) - \mathbf{R}(\boldsymbol{\lambda}_n)] - \mathbf{N}^T\mathbf{Z}\mathbf{N}\Delta\boldsymbol{\lambda} \tag{11}$$

The following NCP, by a similar path of reasoning as in De Donato and Maier (1973), now enforce the stepwise holonomic behaviour for step $\Delta\Sigma$:

$$\mathbf{F} \geq \mathbf{0};\ \Delta\boldsymbol{\lambda} \geq \mathbf{0};\ \mathbf{F}^T\Delta\boldsymbol{\lambda} = 0 \tag{12}$$

NCP (12) can be solved for $\Delta\boldsymbol{\lambda}$ as described earlier.

7 CONCLUSIONS

The choice of a hardening rule not only influences the predictive accuracy of an elastoplastic analysis but can also dictate the method of solution to be adopted when using MP-based formulations. In particular, it is shown that it is only practical to cast and solve the analysis in the form of an extremum problem when reciprocal hardening, evidenced by the symmetry of the Jacobian matrix of the hardening function, is assumed. This paper considers a simple, not necessarily reciprocal nonlinear hardening model, and proposes that the analysis, cast in holonomic form, be solved directly as an NCP. A simple Newton iterative scheme involving a sequence of LCPs is outlined as a possible method of solution to the NCP. Also described briefly is an extension to stepwise holonomic analysis which is necessary in the presence of local unloadings and cyclic phenomena. Limited numerical experience, not described herein, supports the viability of the NCP approach.

ACKNOWLEDGMENTS

This work was supported by a grant from the Australian Research Council. I would also like to thank Professor J.S. Pang, Department of Mathematical Sciences, The Johns Hopkins University for many enlightening discussions on nonlinear complementarity problems.

REFERENCES

Argyris, J.H. 1967. Continua and discontinua, Proceedings, Conference on Matrix Methods in Structural Mechanics, Wright-Patterson A.F.B., Dayton, Ohio, October 1965, 11-189.

Argyris, J.H. and Scharpf, D.W. 1972. Methods of elastoplastic analysis, ZAMP 13: 517-551.

Comi, C. and Maier, G. 1990. Extremum theorem and convergence criterion for an iterative solution to the finite-step problem in elastoplasticity with mixed nonlinear hardening, European Journal of Mechanics, A/Solids 9: 563-585.

Corradi, L. 1978. On compatible finite element models for elastic plastic analysis, Meccanica 13: 133-150.

De Donato, O. and Maier, G. 1973. Finite element elastoplastic analysis by quadratic programming: the multistage method, Proceedings, 2nd International Conference on Structural Mechanics in Reactor Technology (SMiRT), Berlin, 1973, Vol. V, Part M, Paper M2/8.

Harker, P.T. and Pang, J.S. 1990. Finite-dimensional variational inequality and nonlinear complementarity problems: a survey of theory, algorithms and applications, Mathematical Programming 48: 161-220.

Lloyd Smith, D. (ed.) 1990. Mathematical programming methods in structural plasticity, Springer-Verlag, New York.

Maier, G. 1968. Quadratic programming and theory of elastic-perfectly plastic structures, Meccanica 3: 265-273.

Maier, G. 1970. A matrix structural theory of piecewise linear elastoplasticity with interacting yield planes, Meccanica 5: 54-66.

Maier, G. and Lloyd Smith, D. 1986. Update to "Mathematical programming applications to engineering plastic analysis", Applied Mechanics Update 377-383.

Maier, G. and Munro, J. 1982. Mathematical programming applications to engineering plastic analysis, Applied Mechanics Reviews 35: 1631-1643

Maier, G. and Nappi, A. 1989. Backward difference time integration, nonlinear programming and extremum theorems in elastoplastic analysis, Solid Mechanics Archives 14: 37-64.

Maier, G. and Novati, G. 1990. Extremum theorems for finite-step backward-difference analysis of elastic-plastic nonlinearly hardening solids, International Journal of Plasticity 6: 1-10.

Martin, J.B., Reddy, B.D., Griffin, T.B. and Bird, W.W. (1987). Applications of mathematical programming concepts to incremental elastic-plastic analysis, Engineering Structures 9: 171-176.

Ortiz, M. and Popov, E.P. 1985. Accuracy and stability of integration algorithms for elastoplastic constitutive relations, International Journal for Numerical Methods in Engineering 21: 1561-1576.

Pang, J.S. and Chan, D. 1982. Iterative methods for variational and complementarity problems, Mathematical Programming 24: 284-313.

Simo, J.C. and Taylor, R.L. 1986. A return mapping algorithm for plane stress elastoplasticity, International Journal for Numerical Methods in Engineering 22: 649-670.

Tin-Loi, F. and Pang, J.S. 1992. Elastoplastic analysis with nonlinear hardening: a nonlinear complementarity approach, Computer Methods in Applied Mechanics and Engineering (in print).

Zienkiewicz, O.C., Valliappan, S. and King, I.P. 1969. Elasto-plastic solutions of engineering problems: initial stress, finite element approach, International Journal for Numerical Methods in Engineering 1: 75-100.

Impact of Computational Mechanics on Engineering Problems, Pulmano & Murti (eds)
© 1993 Balkema, Rotterdam, ISBN 90 5410 324 8

Fractal analysis on the fracture damage propagation of rock

Zhou Weiyuan
Tsinghua University, People's Republic of China

[ABSTRACT] The Damage tensors are ralated to joints of rock mass which could be expressed by fractal geometry. Damage propagations in rock mass were found from rock sample tests in terms of fractal element. In this paper, fractal dimension analysis, which are used to describe the fractures in rock mass, is presented. Fractal Geometry as a novel method to dictate the propagation of damage of rock mass is described.

1. INTRODUCTION

In recent years, fractal geometry as a main tool to analyze the rock joints has been developed. In nature, the jionts in rock mass could be devided into basic elements of fractal geometry. An approach of analyzing joint nets, using fractal geometry, is presented in this paper.

The main concept for rock joints is the fractal demension and related to it, self-affinity character in rock. After introducing fractal geometry to rock mechanics, the joint structures in rock could be described by some thing like more exactly, and their mechanical behaviors could be described by fracture toughness and strength of rock.

In this paper, at first, the rock mechanical behavior of rock structures in terms of fractal grometry, are presented. Then, the relationship between fracture toughness and fractal dimension is established. Finally, the damage propagation pattern of rock is related to fracture damage toughness.

2 GEOMETRIC PATTERN CHARACTERISTICS OF ROCK JOINTS USING FRACTAL DIMENSION IN ROCK MASS

In rock mass, fractal dimension D and basic fractal element, could be found as follows. At first, the rock joints must be devided into defferent groups according to their orientation, waivness and termination. They could be specified within limit. Then, the basic structure should be found by try and error method. As shown in fig(1). There are two sets of joints which may be interpreted. Four geometric factors were examined: the number of joints in unit area, joint lengths, orientations and spatial distribution.

The basic fractal geometry is defined as

$$N(r)=Ar^{-D} \qquad (1)$$

where r is the joint distribution distance, N is the number of joint generator. In fig(1), after a statistical calculation, a generator, a basic geometric shape which consists of two lines, could be established. The generator is shown in fig(2a). To constitute the rock joint set, the following rule is established that at succesive levels of development, each line on the existing geometry shape is replaced by the entire original geometric shape at an appropriately reduced scale, the fractal shown in fig(2b) would be generated after four levels recursion. The joint geometric distribution is establbished by superposity of succesive joints with different length r which follows the formulation (1).

At first, the joints is likely to be decomposed into two sets with defferent orientations. A procedure of statistics of joint were completed. The fractal demension D coould be found. In fig(3), the relationship of the rock jionts between N and r is expressed with D(2.68). A generator with which the joint map is constituted , is shown in fig(2a).

By the similar procedure and with the same approach, the fractal dimension for the joint length density and spatial spacing distribution are established. By a superposition process, the joint map is plotted and formed, as shown in fig (1).

3 DAMAGE PROPAGATION AND FRACTURE TOUGHNESS IN TERMS OF FRACTAL DIMENSION

Since the mechanical effect of crack damage is reflected in stiffness degradation of damaged bodies and usually such degradation is anisotropic, a damage tensor through the elasto-flexibility $\underline{C}^0$ of intact rock as undamaged material whereas $\underline{C}^{0-d}$ of jointed rock as damaged material is defined as

$$\underline{DD}=\underline{I}-(\underline{C}^{0-d}):\underline{C}^0 \qquad (2)$$

where $\underline{DD}$ is the asymmetric forth order tensor characterizing the 3-D anissotropic damage. For jointed rock mass, C^0 could be obtained by tests of mechanical properties of intact rock whereas $\underline{C}^{0-d}$ could be

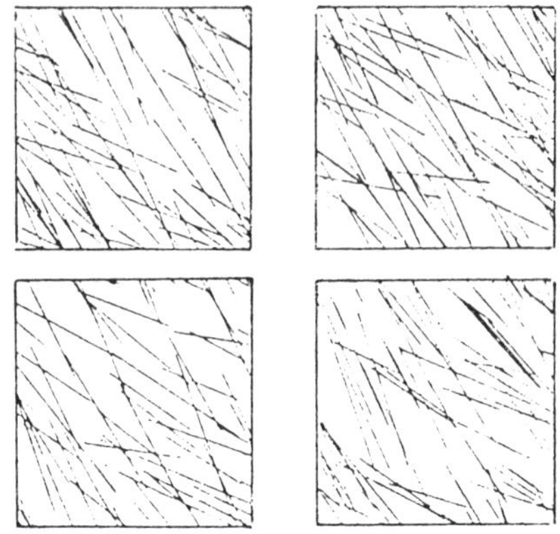

Fig.1 The joint distribution in rock

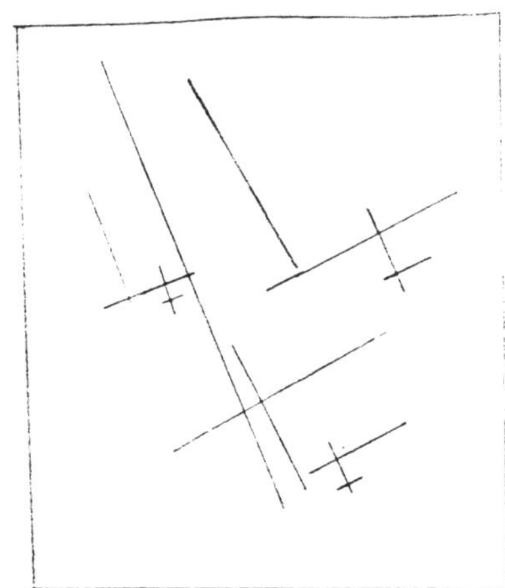

Fig.2a The Fractal Geometry generator of joints

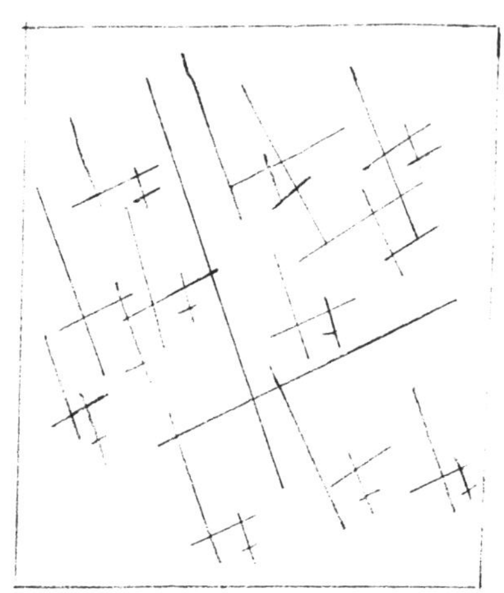

Fig.2b Successive superposition of joint generator

determined by accounting the interaction between joints cracks and intact rock. $\underline{C}^{0-d}$ is defined as the equivalent elastic-damage flexibility of rock masses. By introduction of strain energy associated with cracks on the basis of self-consistent method, $\underline{C}^{0-d}$ could be calculated. Generally, it may be expressed as below [5]

$$C^{0-d}_{ijkl}=C^{0}_{ijkl}+\frac{1}{2E}\sum_{k=1}^{m}\rho^{(k)}a^{(k)}B^{(k)}C_s^{(k)}$$

$$(\delta_{jk}n_i^{(k)}n_l^{(k)}+\delta_{jl}n_i^{(k)}n_k^{(k)}+\delta_{il}n_j^{(k)}n_k^{(k)}+$$

$$\delta_{jk}n_j^{(k)}n_l^{(k)}-4n_i^{(k)}n_j^{(k)}n_k^{(k)}n_l^{(k)}) \quad (3)$$

The propagation of cracks will affect the damage tensors of rock mass, which could be expressed in terms of fractal dimension of fracture nets. For example, the cracks in rock is similar to a tree shaped mesh, as shown in fig (4). The fractal dimension of the mesh is given:

$$D\ \frac{\log Mi}{\log Li}\approx 1.6$$

where M_1, L_i are refered to the number of failure elements and the length of various elements respectively.

According to LEFM, the surface fracture energy is

$$G_s=2r_s$$

where rs is the unit surface energy. Its fractal expression may be shown as:

$$G_s=2A(\varepsilon)/A_0r_s$$

where $A(\varepsilon)$ is the true fracture area, A0 the fracture macroscopic area and r_s the measuring scale.

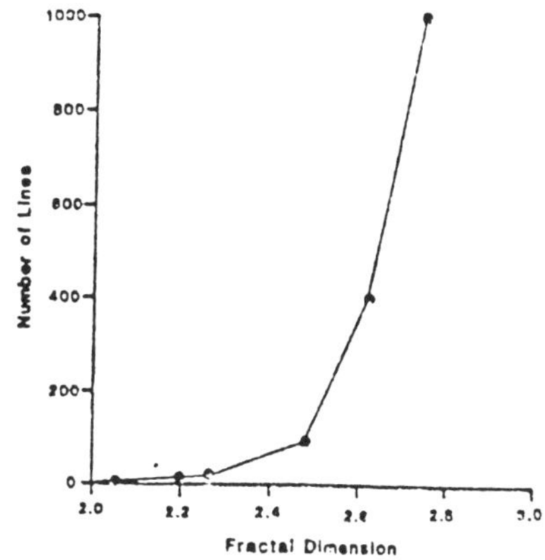

Fig.3 Joint Density versus fractal dimension

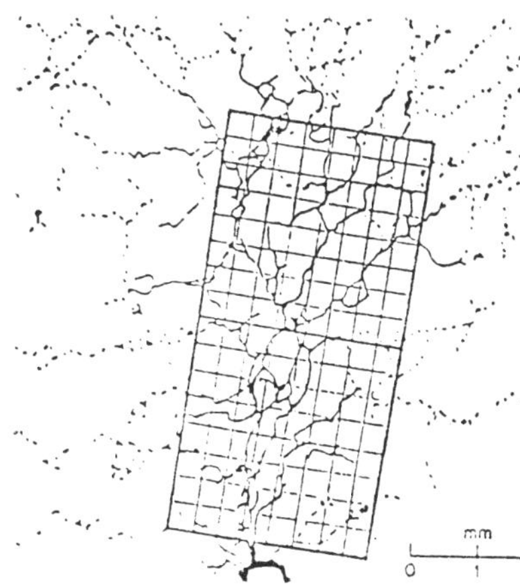

Fig.4 The microcracks in rock under tension

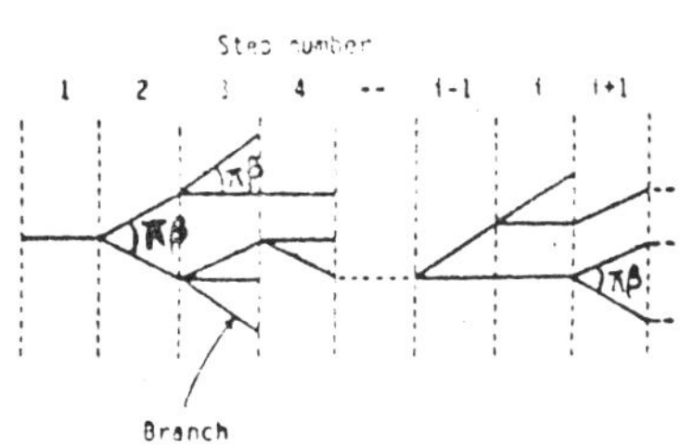

Fig.5 The model for joint bifurcation

On assuming $A(\xi)=A_0^{1/2}$ (2-D), and D the fractal dimension of mesh, then

$$G_s=2\left(\frac{\varepsilon}{A_0^{1/2}}\right)^{2-D}r_s \qquad (4)$$

and then

$$G_s=2\varepsilon_0^{2-D}r_s \quad (5)$$

From self-consistent energy principle, the dissipation energy is given:

$$U_d=\int_\Omega G d\Omega = \frac{(1-M^2)}{E}\int_\Omega \left(K_I^2+K_{II}^2+\frac{2K_{III}^2}{1-M}\right) dV \quad (6)$$

where K_I K_{II} , are the crack tip stress intensity factors.

Generally, the fracture energy, is related to Kc, fracture toughness, as

$$K_c=AG_s^{1/2} \quad (7)$$

then

$$K_c=K_0\ \varepsilon_0^{(2-D)/2} \quad (8)$$

where A is a coefficient, and $K_0=A\cdot(2\varepsilon_0)^{\frac{1}{2}}$ Dissipation energy of rock is generally known as:

$$W_d=\frac{1}{2}\ \underline{\sigma}:C^{0-d}:\underline{\sigma} \quad (9)$$

W_d is actually caused by the crack propagation. It can be calculated as below when the i crack length become longer, from 1 to 1+i:

$$W_d=\frac{1}{2}\ G_s\ \pi\ l^{(i)}\ i^{(i)} \quad (10)$$

Substitute formula (5) into (10), then

$$W_d=\pi\ l^{(i)}\ \varepsilon_0^{(1-D^{(i)})}\ r_i\ l^{(i)} \quad (11)$$

For each set of joints, in the rock mass, before the crack tip, the crack propagation behaves a fissure net with the same fractal dimension D under corresponding stress condition, then

$$W_d=\sum_{i=1}^{M}\ \rho^{(i)}\ \pi\ \varepsilon_0^{2-D^i}\int_0^{l(i)} f^{(i)}(l)\ l(i)\ l'(i)\ dl \quad (12)$$

under different stress condition, from fracture mechanics theory, l(i) may be obtained in terms of $\underline{\sigma}$

$$l(i)=F(l,\underline{\sigma}):\underline{\sigma} \quad (13)$$

substitute (13) into (12), we have

$$W_d=\underline{Y}(l,\underline{\sigma}):\underline{\sigma} \quad (14)$$

where

$$\underline{Y}(l,\underline{\sigma})=\sum_{i=1}^{M}\rho^{(i)}\ \pi\ \varepsilon_0^{2-D^i}\ r_i\int_0^{l(i)} f^{(i)}(l)\ \ l\ \underline{F}(l,\underline{\sigma})\ dl \quad (15)$$

The term Y gives the relation of dissipation energy between stress state, and crack distribution.

From (14), then

$$C^{0-d}=\frac{2\partial W_d}{\partial(\underline{\sigma}\ \ \underline{\sigma})}$$

and finally,

$$C^{0-d}=\frac{2\partial\ (Y(l,\sigma):\dot{\sigma})}{\partial(\underline{\sigma}\ \ \underline{\sigma})} \quad (16)$$

The above formula(16), express the variation C^{0-d} with D

From dessipation potential for jointed rock mass, the yielding function for rock mass is established

$$Fy=Fy(\sigma_{ij}\ \ \bar{H}\ \ D_{ijkl})$$

where plastic internal variable

$$\bar{H}=\sigma_{ij}\ \varepsilon'^{p}_{ij}$$

The constitutive relation is written as

$$\sigma_{ij}=K_{ijkl}\ \varepsilon'_{kl}$$

where

$$K_{ijkl}=K^{0-d}_{ijkl}-K^{0-d}_{ijpq}\left(\frac{\partial Fy}{\partial\sigma_{st}}+\frac{\partial F}{\partial D_{mnrw}}H_{mnrwst}-\frac{\partial Fy}{\partial\bar{H}}\ \sigma_{uv}\ \frac{\partial C^{0-d}_{uv\alpha\beta}}{\partial D_{mnrw}}\ \sigma_{\alpha\beta}H_{mnrwst}\right)K^{0-d}_{stkl}\ \frac{\partial Fy}{\partial\sigma_{pq}} \quad (17)$$

The constitutive relation (17) is characterized by taking account of the brittle coupled strain rate and therefore reflects the elasto-plastic coupling and strain softening properties possesed by jointed rock masses.

As shown in fig (4), the microcracks before the fracture tip may be modelled with a tree shaped distribution in plane. As pointed out by Watanabe, the tree shaped microcracks are formed with the propagation of fracture flow, ie. for each crack, the bifurcation of cracks starts with the propagation of them, and the angles are formed by the crack branches. This is the so-called "self-affinity" of fractal geometry of joints. From statistical calculation, the fractal dimension may be expressed as below.

$$N = 3 \qquad r = \frac{1}{2\cos(\frac{\pi\beta}{2})}$$

$$D_2 = \frac{\ln N}{\ln(1/r)} = \frac{\ln 3}{\ln((2\cos\frac{\pi\beta}{2}))}$$

where D_2 means the fractal dimension in 2D plane. It varies proportionally with π/β. From Tab.1 (from Dr. Shy), K_c/K_0 with various values of $\pi\beta$ can be seen.

CONCLUSION

The jointed rock mass has fractal mathematics, which have been found by tests. In this paper, a model of damage mechanics for rock with fractal geometry is proposed. Using fractal geometry, the rock structures distribution could be precisely narrated and the most favorite point implies that the scale effect, for rock mass could be prevented.

ACKNOWLEDGEMENTS

The author is grateful to the funds from China Natural Scientific Research Commission and Acadmia Sinica.

References

[1] Fractal in Physics, Proceedings of the 6th Triest Int. Symposium on Fractals in Physics, Italy 1985

[2] Mandelbrot B.B., The Fractal Geometry of Nature, 1982, New York

[3] E. Louis, F.Guinea and F. Flores, The Fractal of Fracture, Proc. of the 6th Trieste Int. Symposium on Fractal in Physics

[4] M.M. Poulton et al, Scale Invariant behavior of Massive and Fragmented Rock, Int. J. Rock Mech. Min. Sci.& Geomech. Abstr., Vol 27, 1990

[5] Zhou Weiyuan et al, The Fractal in Rock Mechanics------Theory and Practice of Geotechnical Enginering, Nov. 1992, Zhejiang Univ. Press.

Tab.1 (from Dr. Shy)

K/K_0 \ $\pi\beta$ / a_0	10°	30°	50°	60°
10^{-2} cm	3.93	4.66	7.03	10.0
10^{-3} cm	15.42	21.68	49.43	100.0
10^{-4} cm	60.53	100.9	347.5	1000.0

Impact of Computational Mechanics on Engineering Problems, Pulmano & Murti (eds)
© 1993 Balkema, Rotterdam, ISBN 90 5410 324 8

Fatigue damage analysis of hammer foundation

Zhang Wohua
School of Civil Engineering, University of New South Wales, N.S.W., Australia

Abstract: The dynamic response of hammer-foundation system has been studied based on the damage mechanics concepts. The fatigue damage growth and damage fatigue life problems of the hammer-foundation system have been investigated in this paper. The influences of damage due to a hammer blow on the ground near the foundation has been numerically analysed.

1. Introduction

When a structural component is subjected to impact or shock loading, the transient stress waves are generated and propagated The response due to dynamic loading can cause elevation of stress level especially in a damaged zone or in local regions surrounding the cracks and defects. In particular, within the damaged zone, the micro structure of damaged material is significantly changed from the undamaged zone due to the onset and growth of damage. The dynamic response of a damaged structural component is considerably different from the corresponding undamaged one due to the changing of micro structure [1].

Hammers are most typical of impact machines and are widely used in mechanical production industry. The fatigue life problem of a hammer-foundation system is one of the important design objects for the impact machines. In order to apply the damage mechanics to the problems of machine foundation design, a damage fatigue problem of a hammer-foundation system considering with fatigue damage growth is investigated in this paper. It is important to predict the effects of the damage growth on the fatigue process of the hammer foundation especially when subjected to long-term repeated impact loading. The damage response of a damaged structural component and the dynamic behaviour of the damaged hammer foundation will be discussed in this paper using the concept of damage mechanics.

2. Model of Hammer Foundation

The hammer-foundation system consists of a frame, a falling weight known as "tup", the anvil and the foundation block [2]. To reduce the transmission of impact stresses in the concrete block and the frame, an elastic pad consisting of rubber, felt, cork or timber is generally provided between the anvil and the foundation block. The foundation block is mostly designed to rest directly on the soil. In the design procedure of hammer-foundation, the dynamic response is computed by modelling them as lumped-mass-dashpot system based on assumptions given in [3]. This simplified system is the two degree of freedom vibration model where the anvil rests on an elastic pad and the foundation block rests directly on the soil, as shown in Fig.-1.

The equations of motion of the hammer-foundation model in the case of free vibration may be written as

$$\begin{bmatrix} m_1 & 0 \\ 0 & m_2 \end{bmatrix} \begin{Bmatrix} \ddot{\xi}_1 \\ \ddot{\xi}_2 \end{Bmatrix} + \begin{bmatrix} C_1^* + C_2^* & -C_2^* \\ -C_2^* & C_2^* \end{bmatrix} \begin{Bmatrix} \dot{\xi}_1 \\ \dot{\xi}_2 \end{Bmatrix} + \begin{bmatrix} K_1^* + K_2^* & -K_2^* \\ -K_2^* & K_2^* \end{bmatrix} \begin{Bmatrix} \xi_1 \\ \xi_2 \end{Bmatrix} = \{0\} \tag{1}$$

where m_1 , m_2 are the masses of the foundation and anvil. K_1^* is the stiffness of the equivalent soil spring below the foundation block for vertical vibration with consideration of damage Ω_s in soil and K_2^* is the stiffness of the equivalent spring of the pad below the anvil with consideration of damage Ω_p in the pad. C_1^*, C_2^* are the equivalent viscous damping of soil ground. and the pad absorber. ξ_1, ξ_2 are the displacement of the foundation block and the anvil.

The value of the equivalent soil spring stiffness K_1^*, K_2^* can be obtained from the elastic half-space method as

$$K_1^* = \frac{4G^* r_o}{1-v^*} \qquad K_2^* = \frac{E^*}{b} A_2 \tag{2}$$

where $G^* = (1-\Omega_s)^2 G$ and v^* are damaged shear modulus and Poisson's ratio of the soil. $E^* = (1-\Omega_p)^2 E$ is the damaged Young's modulus of the pad absorber. b is the thickness of the pad. A_2 is the area of the anvil in contact with the pad.

The two natural frequencies ω_1^* and ω_2^* of free vibration of the hammer-foundation model can be determined from the frequency equation given below,

$$\omega_{2,1}^{*2} = \frac{q_1+q_3}{2} \pm \left[\left(\frac{q_1-q_3}{2} \right)^2 + q_2 q_3 \right]^{\frac{1}{2}} \tag{3}$$

where

$$q_1 = \frac{K_1^* + K_2^*}{m_1} \qquad q_2 = \frac{K_2^*}{m_1} \qquad q_3 = \frac{K_2^*}{m_2} \tag{4}$$

The equivalent damping ratio of the system can be evaluated by

$$\zeta_1^* = \frac{C_1^* + C_2^*(1-a_1)^2}{2\omega_1^*(m_1 + m_2 a_1^2)} \qquad \zeta_2^* = \frac{C_1^* + C_2^*(1-a_2)^2}{2\omega_2^*(m_1 + m_2 a_2^2)} \tag{5}$$

The response of the anvil and the foundation block may be computed by considering the free vibration of the anvil-foundation soil system as being triggered by the initial velocity V_a imparted to the anvil by impact of ram on the metal piece being forged. Thus, the displacement response of the anvil and foundation block has form as

$$\xi_1 = \frac{V_a}{a_1-a_2}\left[e^{-\zeta_1^*\omega_1^* t}\frac{\sin(\bar{\omega}_1 t)}{\bar{\omega}_1} - e^{-\zeta_2^*\omega_2^* t}\frac{\sin(\bar{\omega}_2 t)}{\bar{\omega}_2} \right] \tag{6a}$$

$$\xi_2 = \frac{V_a}{a_1-a_2}\left[e^{-\zeta_1^*\omega_1^* t}\frac{a_1\sin(\bar{\omega}_1 t)}{\bar{\omega}_1} - e^{-\zeta_2^*\omega_2^* t}\frac{a_2\sin(\bar{\omega}_2 t)}{\bar{\omega}_2} \right] \tag{6b}$$

where

$$\bar{\omega}_i = \sqrt{1-\zeta_i^{*2}}\,\omega_i^* \qquad (i=1,2) \tag{7}$$

$$a_{1,2} = \frac{1}{q_2}\left\{ \frac{q_1-q_3}{2} \pm \left[\left(\frac{q_1-q_3}{2} \right)^2 + q_2 q_3 \right]^{\frac{1}{2}} \right\} \tag{8}$$

where V_a is the initial velocity of the anvil motion. It can be determined from Ref. [3].

A simple relationship of damaged and undamaged damping ratio $\eta_\zeta = \zeta^* / \zeta$ has been found as a function of the damage variable Ω and the ratio of the undamaged natural frequency ω_1 / ω_2 [1]

$$\eta_\zeta = \frac{\zeta^*}{\zeta} = \frac{\dfrac{1}{(1-\Omega)} + (1-\Omega)\dfrac{\omega_1}{\omega_2}}{1 + \dfrac{\omega_1}{\omega_2}} \tag{9}$$

The compressive stress in the pad is expressed by

$$\sigma_P = K_2^*(1-\Omega_p)\frac{\xi_1+\xi_2}{A_2} + C_2^*\frac{\dot{\xi}_1+\dot{\xi}_2}{A_2} \tag{10}$$

and the pressure transmitted to the soil through the foundation block due to one blow is expressed as

$$p_s = \frac{K_1^*(1-\Omega_s)\xi_1 + C_1^*\dot{\xi}_1}{A_1} \tag{11}$$

It is necessary to note that the contribution of the higher of the two natural frequencies ω_1^* and ω_2^*, towards the amplitude motion is small and may be neglected for all practical purposes.

3. Fatigue Damage of Hammer foundation

It should be noted that in the fatigue failure process only the excess stress beyond the fatigue strength in a material can produce the fatigue damage. Generally, the fatigue strength (σ_N) is considered to be a fraction of fatigue limit (σ_{-1}), that is, $\sigma_N = k\sigma_{-1} = (0.5 \sim 0.7)\sigma_{-1}$. Therefore, in this particular case of fatigue damage, damage growth equation can be written as

$$\frac{d\Omega}{dt} = \begin{cases} A\left(\dfrac{|\sigma(t)|}{1-\Omega} - k\sigma_{-1} \right)^n & \text{if } |\sigma(t)| > k(1-\Omega)\sigma_{-1} \\ 0 & \text{Otherwise} \end{cases} \tag{12}$$

where k and n are material constants which can be obtained from fatigue tests. Because of the damping effect of the system, the stress amplitude produced by one blow exceeds the fatigue strength in the beginning and then decreases below the value. Only the shaded area of the stress distribution may provide the contribution to damage growth (as shown in Fig.-2). In order to evaluate an average of damage increment during the i-th blow, Eqn.(12) can be rewritten as

$$\Delta\Omega_i = \frac{1}{T}\int_0^T A\left(\frac{|\sigma_i(t)|}{1-\Omega_i} - k\sigma_{-1} \right)^n H\left[|\sigma_i(t)| - k(1-\Omega_i)\sigma_{-1} \right] dt \tag{13}$$

where $H[x]$ is the unit step function. From Fig.-2, it

can be seen that that there is no shaded area after $t > T_i$, where T_i can be determined from the the last crossing point of the curve $\sigma_i(t)$ and the lines $\pm \mid k(1-\Omega_i)\sigma_{-1} \mid$ as

$$T_i = -\frac{\sigma_{max}}{\zeta_{i1}^* \omega_{i1}^*} \ln(k(1-\Omega_i)\sigma_{-1}) \quad (14)$$

During one blow, the variation of damage Ω_i is very small, the contribution of Ω_i to the integration in Eqn.(13) can be considered as a constant. Then $(1-\Omega_i)^n$ can be approximately taken out from the integration.

Defining a function $J(\Omega_i)$ as

$$J(\Omega_i) = \frac{1}{T_i}\int_0^{T_i} (\mid\sigma_i(t)\mid - k(1-\Omega_i)\sigma_{-1})^n \, H[\mid\sigma_i(t)\mid - k(1-\Omega_i)\sigma_{-1}]dt \quad (15)$$

and introducing a dimensionless factor

$$j(\Omega_i) = \frac{J(\Omega_i)}{J_0} \quad (16)$$

in which J_0 is the initial value of Eqn.(15) corresponding to the case without damage Eqn.(13) can be modified as

$$\Delta\Omega_i = \frac{A J_0 j(\Omega_i)}{(1-\Omega_i)^n} \quad (17)$$

Since $\Omega_{i+1} = \Omega_i + \Delta\Omega_i$, the recurrence formula of damage state can be obtained as

$$\Omega_{i+1} = \Omega_i + \frac{A J_o j(\Omega_i)}{(1-\Omega_i)^n} \quad (18)$$

Eqn.(18) in fact gives the accumulated damage in a material. Thus, rewriting this equation as a summation of damage for a number of cycles, say, N, will give

$$A J_o \sum_{i=1}^{N} \frac{j(\Omega_i)}{(1-\Omega_i)^n}$$

From the above equation, we can determine the fatigue life, that is, the number of blows up to fatigue damage which will be the difference between the rupture damage Ω_c and initial damage, Ω_o. Thus, we have

$$\sum_{i=1}^{N} \frac{j(\Omega_i)}{(1-\Omega_i)^n} \geq \frac{(\Omega_c - \Omega_o)}{A J_o} \quad (19)$$

It should be noted that there are minimum and maximum values in function $j(\Omega_i)$. Therefore, it can be said, that the fatigue life of the pad and the soil foundation would also have the minimum and maximum values based on the level of damage. These minimum and maximum values can be determined by considering the relation $(1-\Omega_i)^n \geq (1-\Omega_c)^n$, as follows

$$j_{smin} N_s \leq \sum_{i=1}^{N_s} \frac{j_s(\Omega_{si})}{(1-\Omega_{si})^{n_s}} \leq N_s \frac{j_{smax}}{(1-\Omega_{sc})^{n_s}} \quad (20a)$$

$$j_{pmin} N_p \leq \sum_{i=1}^{N_p} \frac{j_p(\Omega_{pi})}{(1-\Omega_{pi})^{n_p}} \leq N_p \frac{j_{pmax}}{(1-\Omega_{pc})^{n_p}} \quad (20b)$$

where j_{smin}, j_{smax} and j_{pmin}, j_{pmax} are the minimum and maximum values of $j(\Omega)$ for soil and pad respectively.

Substituting Eqn.(18) into (20), the fatigue life of the foundation block and the pad, (N_s and N_p), can be estimated as

$$N_{smin} \leq N_s \leq N_{smax} \quad (21a)$$

$$N_{pmin} \leq N_p \leq N_{pmax} \quad (21b)$$

where the expressions of the upper and lower limit of fatigue life of the foundation block and the pad are defined as

$$N_{smin} = INT\left[\frac{(\Omega_{sc} - \Omega_{so})(1-\Omega_{sc})^{n_s}}{A_s J_{so} j_{smax}}\right] \quad (22a)$$

$$N_{smax} = INT\left[\frac{(\Omega_{sc} - \Omega_{so})}{A_s J_{so} j_{smin}}\right] \quad (22b)$$

$$N_{pmin} = INT\left[\frac{(\Omega_{pc} - \Omega_{po})(1-\Omega_{pc})^{n_p}}{A_p J_{po} j_{pmax}}\right] \quad (23a)$$

$$N_{pmax} = INT\left[\frac{(\Omega_{pc} - \Omega_{po})}{A_p J_{po} j_{pmin}}\right] \quad (23b)$$

4. Damage Response of Ground Near the Foundation

The damage response of ground near the hammer-foundation system can be investigated by combining the above simplified model and finite element method. The equation of an anisotropic damaged ground under dynamic loading can be written as

$$[\mathbf{M}]\{\ddot{U}\} + [\mathbf{C}^*(\Omega(t))]\{\dot{U}\} + [\mathbf{K}^*(\Omega(t))]\{U\} = \{\mathbf{P}(t)\} \quad (24)$$

where $\{\mathbf{P}(t)\}$ is known as the general nodal force vector due to the pressure of foundation block on the soil ground. $[\mathbf{M}]$ is the mass matrix for a damaged element in the ground. $[\mathbf{C}^*(\Omega(t))]$ is the time-dependent damping matrix for a damaged element in the ground.

$$[\mathbf{K}^*(\Omega(t))] = \int_{V_e} [\mathbf{B}]^T [\mathbf{T}]^T [\hat{\mathbf{D}}^*] [\mathbf{T}] [\mathbf{B}] dv \quad (25)$$

is the time-dependent stiffness matrix for a damaged element in the ground. $[\mathbf{T}]$ is the coordinate transformation matrix. $[\hat{\mathbf{D}}^*]$ is the anisotropic damaged material matrix [4]. The components of matrix $[\hat{\mathbf{D}}^*]$ can be determined by the effective elastic properties of anisotropic damaged material in terms of undamaged anisotropic elastic properties and the principal anisotropic damage vector $\{\Omega\}$ [5]. The initial principal anisotropic damage vector can be obtained from the anisotropic damage tensor $[\ \Omega\ ]$ through the eigenvalue analysis [6].

So far, the influence of damage on material damping has not been discussed in the context of either experimental or analytical investigation in any published work. In order to discuss this problem from the point of view of numerical analysis, it is convenient to assume the Rayleigh damping and equivalent viscous damping as follows.

For Rayleigh damping matrix, one can adopt the usual assumption

$$[\ \mathbf{C}^*(\ \Omega(t)\ )\ ] = \alpha^*\ [\ \mathbf{M}\ ] + \beta^*\ [\ \mathbf{K}^*(\ \Omega(t)\ )\ ] \tag{26}$$

For equivalent viscous damping matrix, we have

$$[\ \mathbf{C}^*(\ \Omega(t)\ )\ ] = \int_{V_e} \gamma^*(t)\ [\ \mathbf{N}\ ]^T\ [\ \mathbf{N}\ ]\ dv \tag{27}$$

where α^* , β^* are Rayleigh damping parameters of damaged material. $\gamma^*(t)$ is the time-dependent equivalent viscous damping coefficient of damaged material. These damaged damping parameters can be determined from the undamaged parameter and using Eqn.(9)

In general, for the finite element analysis of dynamic problems, the damping matrix should include both material damping (due to energy dissipation) and geometric damping (due to absorption of energy at the boundary) [7]. Hence, in the present analysis the model of a footing on damaged soil ground with material damping (Rayleigh damping) and a perfect energy absorbing boundary [7] has been considered. Thus, the total damping matrix can be considered as

$$[\ \mathbf{C}^*\ ] = [\ \mathbf{C}^*(\Omega)\ ] + [\ \mathbf{C}_b^*\ ] \tag{28}$$

where

$$[\ \mathbf{C}_b^*\ ] = \int_{S_2} [\ \mathbf{N}\ ]^T\ [\ \mathbf{D}_b^*\ ]\ [\ \mathbf{N}\ ]\ ds \tag{29}$$

and the matrix $[\ \mathbf{D}_b^*\ ]$ is formulated on the base of shear modulus (similar in [7]) for the damaged material.

5. Numerical Results

Fig.-3 shows the dynamic response records of displacement for the anvil and foundation block of a hammer-foundation [2]. Record 1 in Fig.-2 is for the top of the foundation block, and record 2 is for the middle point of the anvil. Fig.-4 and Fig.-5 show the values obtained from Eqn.(6).

The response of cyclic stress in the pad and the soil due to one blow of the hammer is shown in Fig.-6. It can be seen that the stress decreases over the period of time. Even though there may be some damage (fatigue) during early stages, this can probably be omitted for the case of one blow. However, if the blows are repeated, then the damage will be accumulated and the damage growth occurs. This will be a severe case in fatigue loading.

Fig.-7 (a) to (c) show the estimation of the damage fatigue life for the hammer-foundation system. Plots in Fig.-7 (a) and (b) present ratios $A_s\ N_{smin}\ /\ J_{so}$ and $A_p\ N_{pmin}\ /\ J_{po}$ versus the values of failure damage Ω_{sc} and Ω_{pc} for different values of parameter n_s and n_p respectively. The lower limit of fatigue life N_{smin} (for the foundation block) and N_{pmin} (for the pad) can be estimated from the given material constants of Ω_{sc} , A_s , n_s and Ω_{pc} , A_p , n_p. Fig.-7(c) presents the ratios $A_s\ N_{smax}\ /\ J_{so}$ and $A_p\ N_{pmax}\ /\ J_{po}$ versus the value of damage threshold Ω_{sc} and Ω_{pc}. This figure can be used to estimate the upper limit of fatigue life N_{smax} (for the foundation block) and N_{pmax} (for the pad) for the given material constants of Ω_{sc} , A_s and Ω_{pc} , A_p). The values J_{so} and J_{po} can be calculated from Eqn.(15).

The influence of damage due to one hammer blow on the ground near the foundation can be analysed by F. E. method. The finite element representation is shown in Fig.-8. The loading on the footing (representing the hammer blow) is applied as the pressure P_s from the foundation block on the soil as presented in Eqn.(11) and Fig.-6. Comparisons of experimental results from Ref. [2] and numerical results from the finite element analysis are presented in Fig.-9. Plots in Fig.-9(a) are ratios of maximum amplitudes along ground surface and maximum amplitude at the center of footing versus distance from the center of footing on the surface. Plots in Fig.-9(b) are ratios of maximum amplitudes along depth and maximum amplitude at the center of footing versus depth from ground surface. The computed results in Fig.-9 are calculated in the case of undamaged state $\Omega = 0$ and damaged state $\Omega = 0.2$. The results in Fig.-9 show that the numerical results obtained by finite element analysis agree well with the experimental results obtained from [2]. Fig.-10 presents the influence of damage growth in the soil on the maximum amplitude in the ground. The figure shows the ratio of damaged maximum amplitudes along ground surface (or along depth) and the undamaged maximum amplitude at the center of footing versus damage in soil ground.

References

[1] Zhang Wohua, *Numerical Analysis of Continuum Damage Mechanics*, Ph.D. thesis, University of New South Wales, Australia, (1992).

[2] Junping Jang, *Computing Method of Vibration and Isolation Design*, Chinese Building Industry Press, pp 164-175, (in Chinese) (1985).

[3] Parkash S. and Puri V., *Foundation for Machines: Analysis and Design* , A Wily-Inter science Publication, (1988).

[4] Valliappan S., Zhang Wohua and Murti V., Finite Element Analysis of Anisotropic Damage Mechanics Problems, J. of Engg. Frac. Mech., Vol. 35, pp 1061-1076, (1990).

[5] Zhang Wohua, Valliappan S., Analysis of Random Anisotropic Damage Mechanics Problems of Rock Mass, Part II -- Statistical Estimation, Int. J. Rock Mech. and Rock Engg. Vol. 23 , pp 241-259, (1990).

[6] Zhang Wohua, Valliappan S., Analysis of Random Anisotropic Damage Mechanics Problems of Rock Mass, Part I -- Probabilistic simulation, Int. J. Rock Mech. and Rock Engg., Vol. 23, pp 91-112, (1990).

[7] White W., Valliappan S. and Lee I., Unified Boundary for Finite Dynamic Models, ASCE, J. of the Engg. Mech. Division, Vol. 103, pp 949-964, (1977).

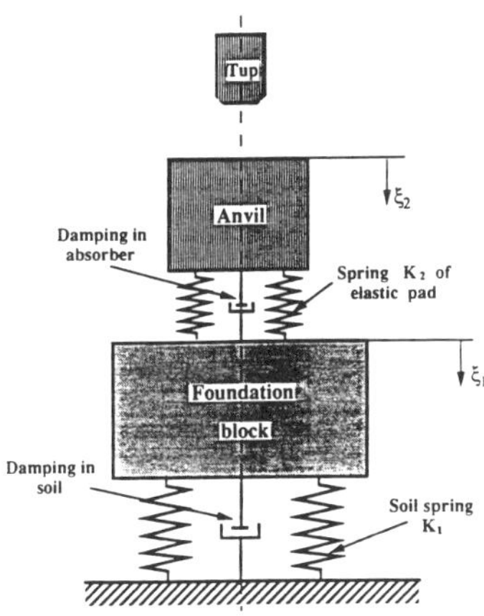

Fig.-1: The simplified model of hammer-foundation.

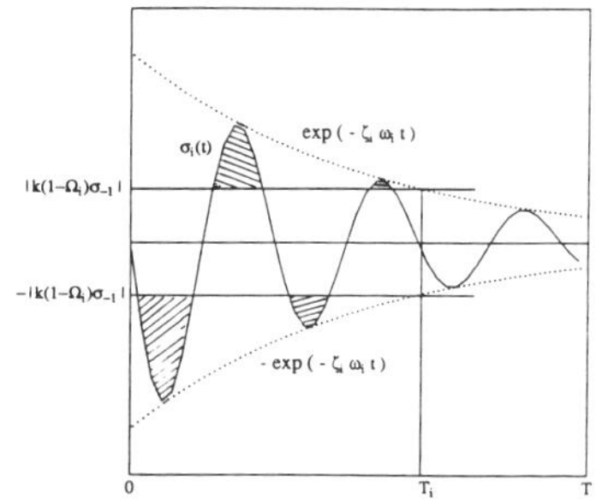

Fig.-2: Illustration of integral region

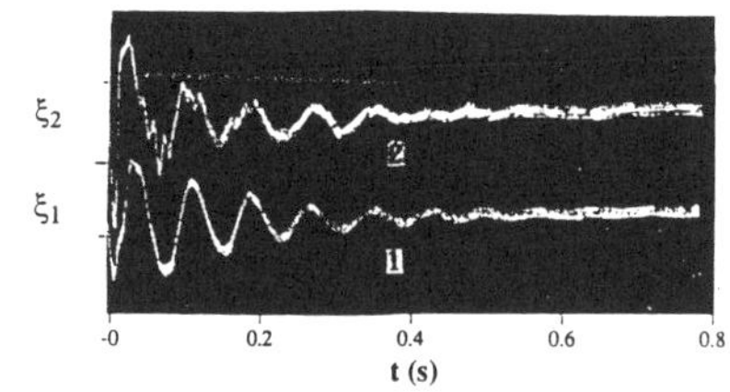

Fig.-3: Records of response of the hammer-foundation [2].

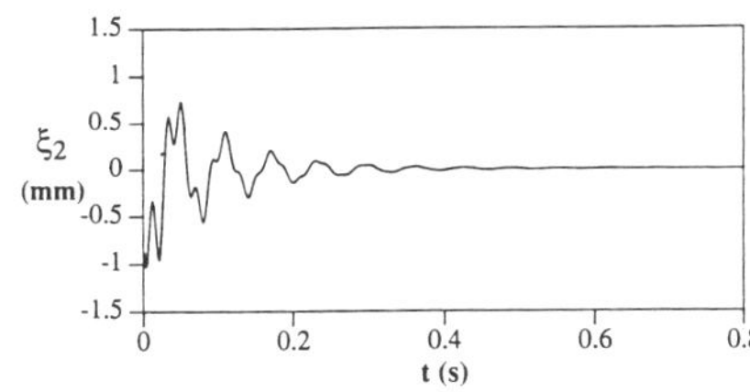

Fig.-4: Computed response of the anvil on the foundation.

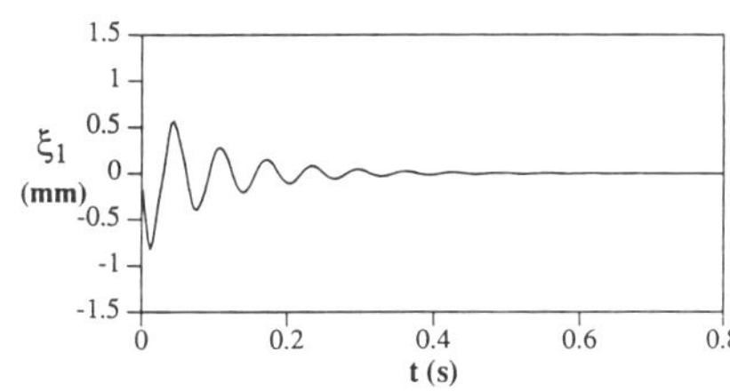

Fig.-5: Computed response of the foundation block.

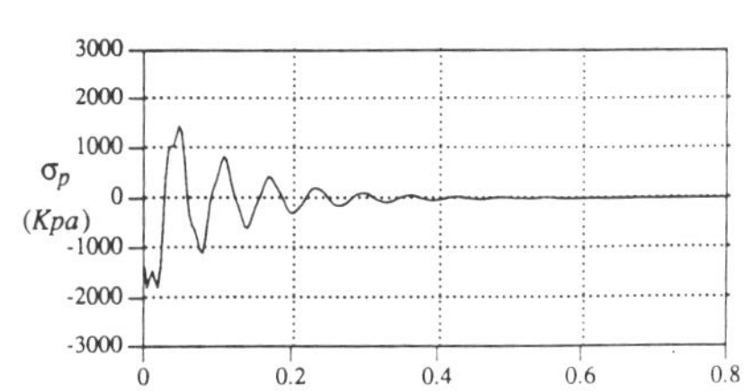

Fig.-6(a): Stress response in the pad due to one blow.

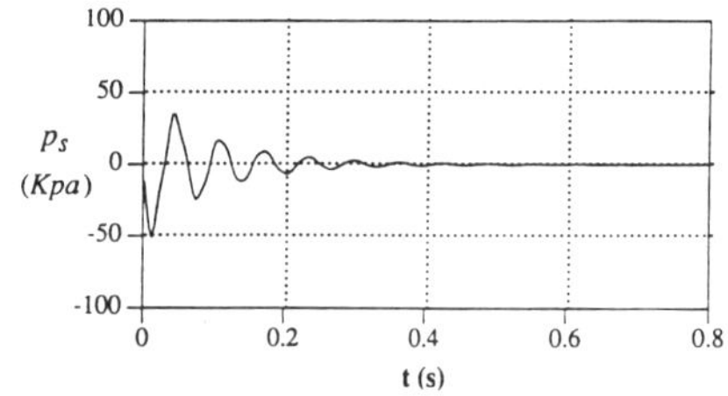

Fig.-6(b): Pressure on soil due to one blow.

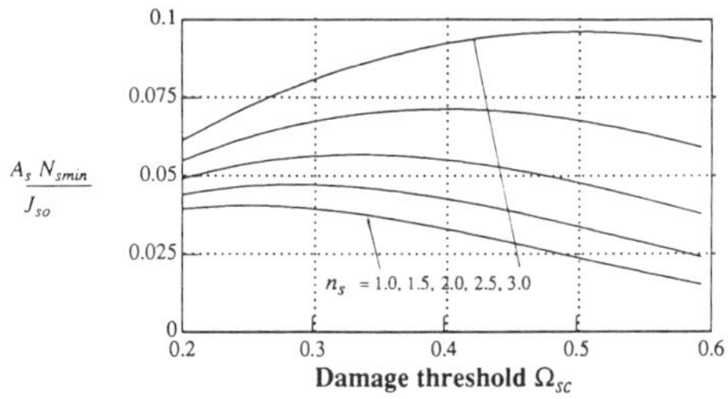

Fig.-7(a): The lower limit fatigue life of foundation block

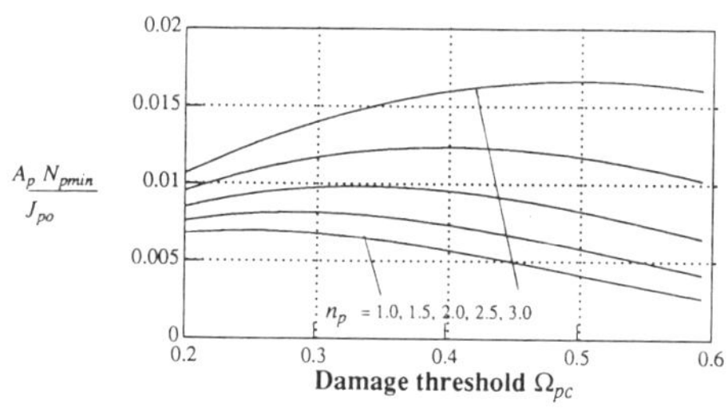

Fig.-7(b): The lower limit fatigue life of pad.

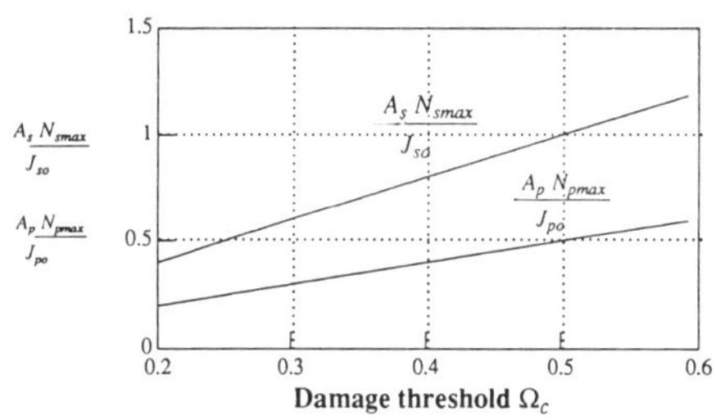

Fig.-7(c): The upper limit fatigue life of hammer-foundation.

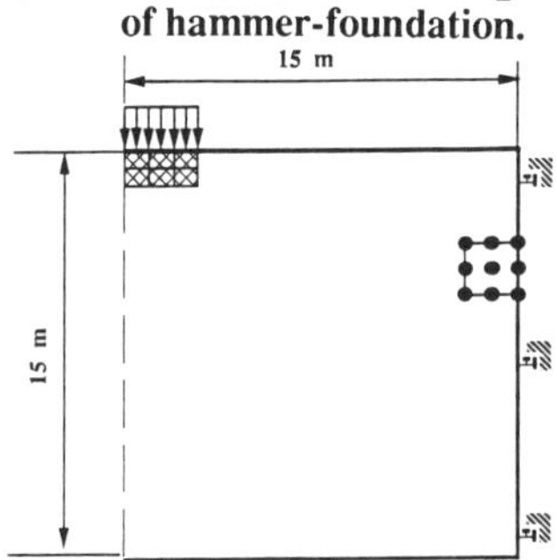

Fig.-8: F. E. mesh of the hammer-foundation.

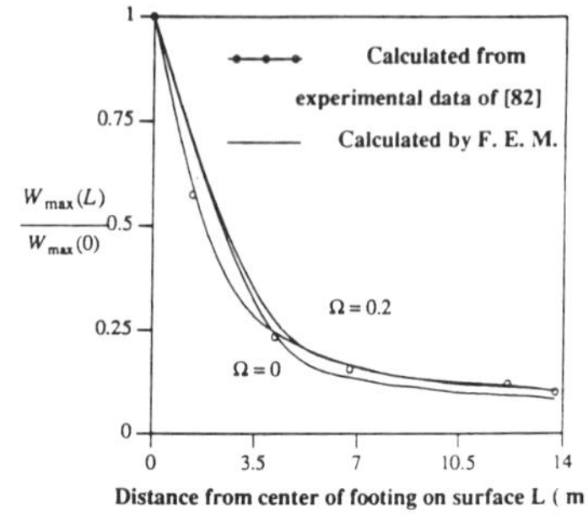

Fig.-9(a): Ratio of the maximum amplitude along ground surface.

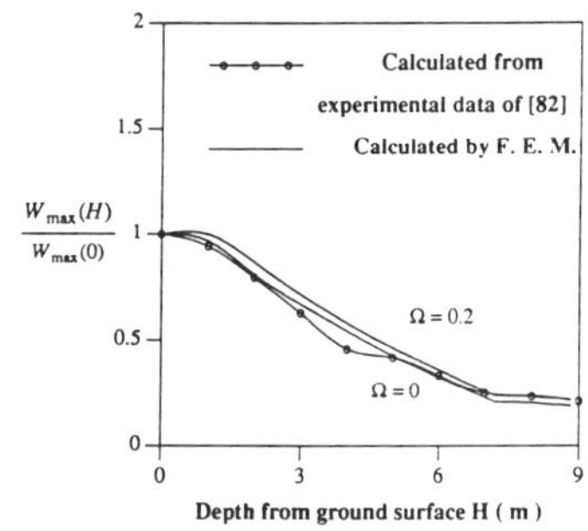

Fig.-9(b) : Ratio of the maximum amplitude along the depth.

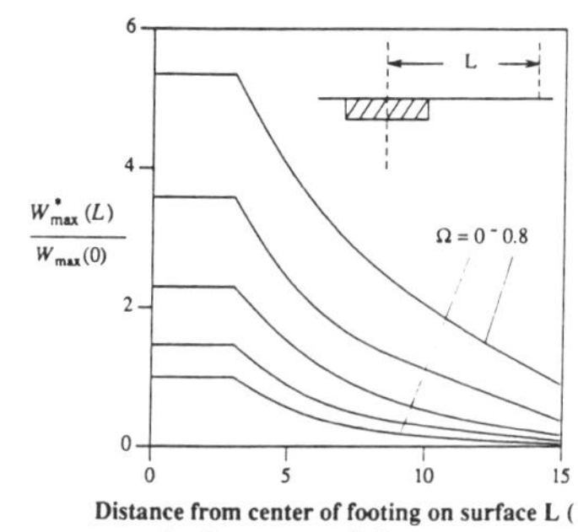

Fig.-10(a): Influence of damage on ground response along surface.

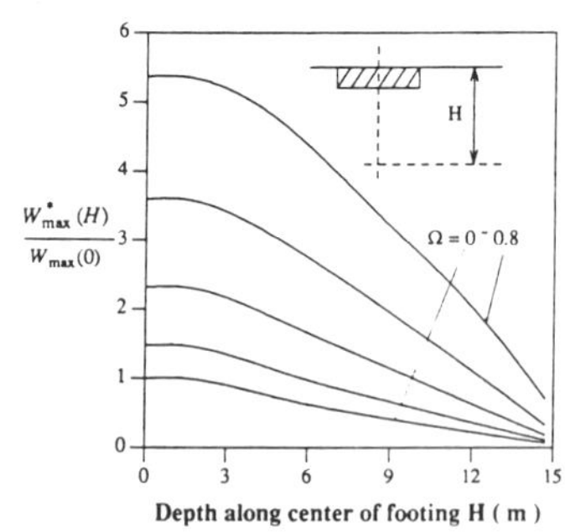

Fig.-10(b): Influence of damage on ground response along depth.

Impact of Computational Mechanics on Engineering Problems, Pulmano & Murti (eds)
© 1993 Balkema, Rotterdam, ISBN 90 5410 324 8

Numerical evaluation of bridge approach settlement

M. Zaman, D. Bhat & J. Laguros
The University of Oklahoma, Okla., USA

ABSTRACT: Differential movements between bridge abutment and approach pavement lead to excessive settlement. The resulting unsafe riding surface requires frequent and costly maintenance. Of the many contributing factors to bridge approach settlement, consolidation settlement of layered foundation soils and the embankment settlement caused by its self weight and vehicular traffic load are two major factors. The former accounts for about 40 to 85% of the total settlement, while the latter accounts for 10 to 50%. A nonlinear finite element algorithm, which includes the infinite element concept, is developed to accurately predict consolidation settlement of the approach foundations and embankments. To overcome the difficulty in preparation of finite element data for the analysis, an interactive, user-friendly preprocessor is developed. Also, a postprocessor is developed to graphically facilitate interpretation of finite element results.

1 INTRODUCTION

Settlement at bridge approaches has been a major problem facing the transportation agencies around the globe. Bridge abutments are generally founded on hard bed-rock or pile-groups driven to hard bed-rock and thus the abutments have a lesser potential for settlement. Highway pavements, on the other hand, resting on embankments and in turn on foundation soil, have a greater potential for settlement. The movements between the bridge deck and the approach lead to an excessive differential settlement. Particularly on highways and freeways where the traffic volume and speeds are very high, the surface faults occuring on pavements can be hazardous to traffic. This not only creates an uncomfortable riding surface, but can also damage the bridge deck and/or the bridge structure itself. Approach settlements usually require frequent expensive maintenance which also hinder the regular flow of traffic.

Many factors are believed to affect the bridge approach settlements including age of the approach, height of embankment, thickness of foundation soil, type of embankment and foundation material, construction technique employed, traffic count, etc. Results of some previous studies (6) indicated that the primary and secondary consolidation of the foundation soils contribute significantly to bridge approach settlement. In addition, improper compaction of the embankment soil and undesirable level of embankment material properties add to the settlement of embankment itself and hence the approach. A comprehensive research project was undertaken at the University of Oklahoma to investigate the various causes of bridge approach settlements and to find remedial measures. As a part of this project, the authors have developed a user-friendly finite element computer program for analyzing consolidation settlement of foundation soil including embankment-foundation interaction effects and treating consolidation as a coupled (two phase) problem. This paper is comprised of four basic elements:

1. A user-friendly preprocessor to facilitate data input in an interactive manner. The preprocessor also creates the finite element mesh data and the boundary condition data.
2. A data converter which converts the preprocessor generated data into a suitable form which is compatible for the main finite element (FE) program, including embankment and foundation data.
3. Finite element main program for the analysis of foundation and embankment. The analysis results include time-settlement history and pore pressure dissipation characteristics of foundation, and the settlement profiles of embankment.
4. A postprocessor for graphical representation of finite element analysis results.

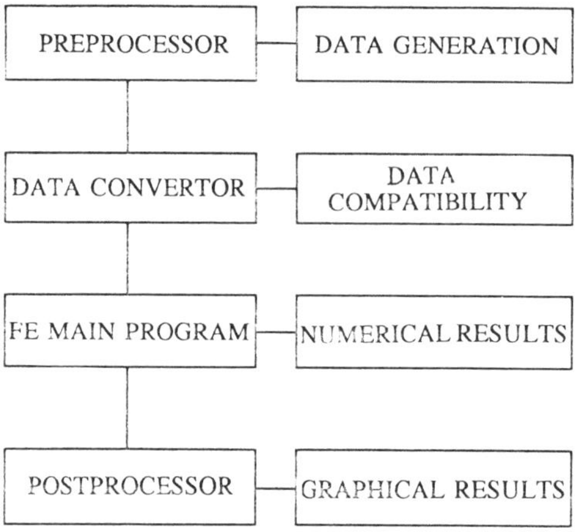

Fig. 1. Flow-Chart Showing Sequence of Analysis

2 FINITE ELEMENT FORMULATION

2.1 *Scope and Limitations*

To effectively conduct the analysis depicted in flow-chart form in Fig. 1., the following assumptions were made:

1. Consolidation of approach foundation is assumed to occur only due to construction of embankment. The in-situ stresses are evaluated from the field test data (e.g., standard penetration test (SPT) values, cone penetration test (CPT) values) in terms of field density, layer thickness, earth pressure coefficient at rest (K_o) and water table depth.

2. Embankment loading is assumed to be a step-wise loading process in which each step of embankment construction is assumed to be completed within a specified time period. The user can select the number of steps so as to simulate the actual construction scenario or may wish to consider various construction scenarios to identify the worst possible cases of approach settlement problem.

3. Nonlinear and inelastic behavior of approach foundation soil is represented by the Cam-Clay model (9, 10, and 12). According to this model, the soil skeleton behaves as an elasto-plastic material. One of the strengths of this model is that it needs relatively fewer material constants compared to other models and that each material constant used in this model has physical meaning.

4. Stress-strain response of the embankment materials is currently represented by the elastic model. However, if the embankment material is clay, then it can also be represented by the Cam-Clay model.

5. The FE formulation is developed within the context of small strain theory. Geometric non-linearity is not taken into account.

6. In the present version of the FE program, the embankment soil is assumed to be isotropic but the foundation soil can be treated as anisotropic.

7. Soil below the water table is assumed to be completely saturated and above the water table unsaturated. In reality, however, the soil above the water table may be partially saturated due to capillary action.

2.2 *Nomenclature*

C^{ep}_{ijkl}	Elasto-plastic constitutive relation tensor
ε_{kl}	Components of strain tensor
$\overline{E_c}$	Displacement function after Laplace transformation of infinite element governing differential equations
G	Shear modulus
i & j	Indices in tensor notation
i,j	Differentiation with respect to j
[J]	Jacobian transformation matrix
k	Coefficient of permeability
k_{ij}	Permeability tensor
$[K_1]$	Stiffness matrix corresponding to soil skeleton
$[K_2]$	Matrix corresponding to the pore pressure
$[K_3]$	Coupled stiffness matrix between soil skeleton and pore pressure
$L_j(\eta)$	Lagrangian polynomial for node j in an infinite element
M	Slope of the failure line in q-p plane
N_i	Shape functions
$\{p_m\}$	Nodal pore pressure vector
q_i	Relative velocity of fluid
$\{q_m\}$	Nodal displacement vector
$\overline{Q}$	Prescribed fluid flow normal to vertical surface
$\{\overline{Q_m}\}$	Prescribed nodal fluid flux vector
{r}	Vectors of displacement and pore pressure
$\{r_p\}$	Nodal pore pressure vector
$\{r_q\}$	Nodal displacement vector
r_*	Real number greater than highest real part of singularities of the Laplace transform of infinite element governing differential equation
{R}	Global load vector

S_{ij} Deviatoric stress tensor
S^* Displacement function
S_c Displacement function after Fourier transformation of infinite element governing differential equation

σ_{ij} Normal stress tensor

$\overline{S_c}$ Displacement function after Laplace transformation of infinite element governing differential equation

$\{\overline{T_i}\}$ Prescribed traction on horizontal surface

$\{\overline{T_m}\}$ Prescribed nodal traction vector

u_i Displacement tensor
x & y Global coordinates

ξ & η Local coordinates

ε_{ij} Strain tensor

ρ Mass density of soil

ρ_w Mass density of pore fluid

* Represents convolution

2.3 *Formulation*

An algorithm is developed based on the nonlinear finite element method (FEM). The infinite element concept is used to idealize the infinite soil domain in lateral direction. The foundation soil domain is divided into near-field and far-field. The near-field is discretized by eight-noded quadrilateral isoparametric elements, while the far-field is discretized by using three-noded infinite elements. The nonlinear foundation soil behavior is represented by modified Cam-Clay constitutive model shown in Fig. 2. The yield surface of this model is an ellipse in the p-q plane and is given by:

$$F = q^2 - M^2 p_0 p + M^2 p^2 \tag{2.1}$$

where,

$$q = \sqrt{3J_{2D}} \tag{2.1a}$$

$$p = (\sigma'_{11} + \sigma'_{22} + \sigma'_{33})/3 \quad \text{and} \tag{2.1b}$$

$$J_{2D} = 0.5\, S_{ij}\, S_{ij} \tag{2.1c}$$

p_0 is the current diameter of the ellipse. An associative flow rule is assumed. The constitutive

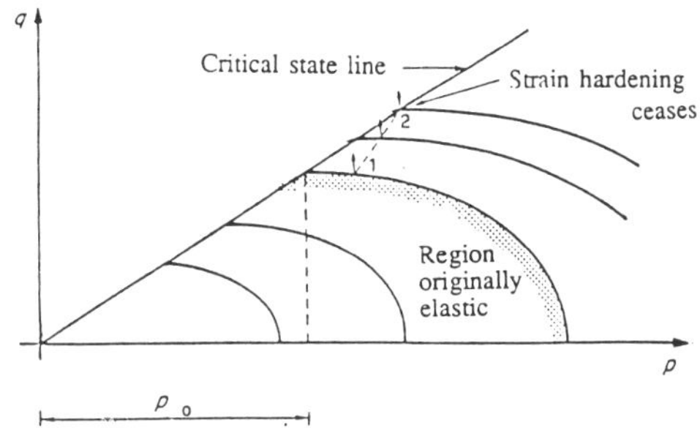

Fig. 2. Modified Cam-Clay Model

equation for the soil skeleton in an incremental form can be written as

$$\Delta\sigma = C^{ep}_{ijkl}\, \Delta\varepsilon_{kl} \tag{2.2}$$

Biot's theory of 3-D consolidation as proposed by Sandhu and Wilson (11) is used in the algorithm. Coupled action of soil skeleton deformations and pore water effect is considered in the formulation and depicted in the governing force equilibrium equation as

$$\sigma'_{ij,j} + p_{,j}\delta_{ij} + \rho F_i = 0 \tag{2.3}$$

The flow equilibrium equation is of the form

$$[k_{ij}(p_{,i} + \rho_w F_i)]_{,j} + \dot{u}_{i,i} = 0 \tag{2.4}$$

Overdot represents the partial derivative with respect to time. The generalized Darcy's law is used to describe the flow behavior. The functional used for the analysis of linearly elastic soil skeleton and incompressible pore fluid is given by

$$\Omega_{(u,p)} = \int_v [\frac{1}{2}\overline{\sigma_{ij}} * \varepsilon_{ij} - \rho F_i^* u_i + p^* u_{i,i} - \frac{1}{2} g^* q_i^* (p_{,i} + \rho_w F_i)] dV - \int_{s1} (\overline{T_i^*} u_{i,i}) dS + \int_{s2} (g^* \overline{Q^*} p) dS \tag{2.5}$$

Repetitive indices are summed from 1 to 3. Comma (,) indicates differentiation, i.e.,

$$\partial u_{i,i} = \frac{\partial u_1}{\partial x_1} + \frac{\partial u_2}{\partial x_2} + \frac{\partial u_3}{\partial x_3}, \tag{2.5a}$$

$$\varepsilon_{ij} = \frac{1}{2}(u_{i,j} + u_{j,i}), \tag{2.5b}$$

The relative velocity of fluid,

$$q_i = k_{ij}(p_j + \rho_w F_i) \tag{2.5c}$$

The asterisk (*) indicates convolution and is given by:

$$A^*B + \int_0^t A(...,\tau)B(...,t-\tau)d\tau \quad (2.5d)$$

Laplace transform is used on differential equations with time as independent variable and the results obtained are divided by the transform variable, S throughout. The Galarkin method of weighted residuals can also be used to obtain the functional. Substituting for all the terms of the functional in Eq. (2.5) and then performing the time integration, the entire system can be written in the matrix form as follows:

$$\begin{bmatrix} [K_1] & [K_3] \\ [K_3]^T & -\alpha\Delta t[K_2] \end{bmatrix} \begin{Bmatrix} r_q(t_n) \\ r_p(t_n) \end{Bmatrix} = \begin{Bmatrix} R_Q(t_n) \\ R_P(t_n) \end{Bmatrix} \quad (2.6)$$

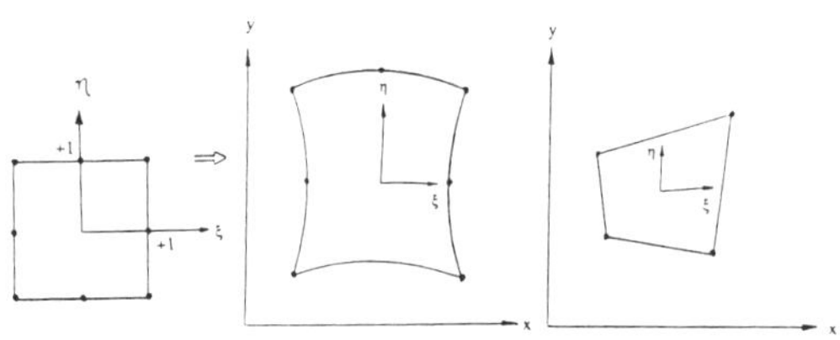

Fig. 3. Mapped Elements for Deformation and Pore Pressure

Fig. 3 represents distorted isoparametric element for displacements and pore pressure mapped from the parent element. The polynomial displacement functions selected for the eight-noded element are of the form

$$u(\xi,\eta) = C_1 + C_2\xi + C_3\eta + C_4\xi\eta + C_5\xi^2$$

$$+ C_6\eta^2 + C_7\xi^2\eta + C_8\xi\eta^2 \quad (2.7)$$

$$v(\xi,\eta) = C_1 + C_2\xi + C_3\eta + C_4\xi\eta + C_5\xi^2$$

$$+ C_6\eta^2 + C_7\xi^2\eta + C_8\xi\eta^2 \quad (2.8)$$

The local coordinates ξ and η take values ± 1. The coefficients of the polynomial Eqs. (2.7 and 2.8) can be obtained by substituting nodal displacements, (u_i), in terms of the shape functions, (N1 through N8) (see ref. 5 for details) as:

$$u = \sum_{i=1}^{8} N_i(\xi,\eta)u_i \quad (2.9)$$

$$v = \sum_{i=1}^{8} N_i(\xi,\eta)v_i \quad (2.10)$$

Similarly the pore pressure functional selected for the four-noded rectangular element is:

$$p(\xi,\eta) = C_1 + C_2\xi + C_3\eta + C_4\xi\eta \quad (2.11)$$

In terms of shape functions,

$$p = \sum_{i=1}^{4} N_i(\xi,\eta)p_i \quad (2.12)$$

Jacobian matrix [J] is used to transform the derivatives from local coordinates to global coordinates and vice versa.

2.4 *Nonlinear Behavior*

The stress-strain relation can be expressed as:

$$[\sigma] = [C]\,[\varepsilon^e] = [C]\,([\varepsilon] - [\varepsilon_p]) \quad (2.13)$$

As the stress-strain contribute to the strain energy portion only, the strain energy part in the functional (Eq. 2.5) shall be changed to account for the nonlinearity (5). The strain energy portion can be rewritten as:

$$U = \frac{1}{2}\iiint_v [\varepsilon]^T[C][\varepsilon]dV$$

$$- \iiint_v [\varepsilon]^T[C][\varepsilon^p]dV$$

$$+ \frac{1}{2}\iiint_v [\varepsilon^p]^T[C][\varepsilon^p]dV \quad (2.14)$$

The total effect of plastic strain is the addition of the load term [Q_0], which represents the contribution of initial plastic strain. The residual load terms are calculated for the contribution of plastic strains at different time steps after the application of external load increment. The modified governing equation including material nonlinearity becomes

$$\begin{bmatrix} [K_1][K_3] \\ [K_3]^T - \alpha\Delta t[K_2] \end{bmatrix} \begin{Bmatrix} r_q(t_n) \\ r_p(t_n) \end{Bmatrix} = \begin{Bmatrix} R_Q(t_n) + Q^p_{0n} \\ R_P(t_n) \end{Bmatrix} \quad (2.15)$$

The interpolation functional selected for infinite element must decay monotonically with distance and must approach zero at infinity. The infinite element algorithm presented by Rajapakse et. al (8) is used in the study. This algorithm, called 'Finite Elements by Singular Contraction', is employed in the present study to model the elasto-static far-field behavior of an infinite soil domain. Fig. 4

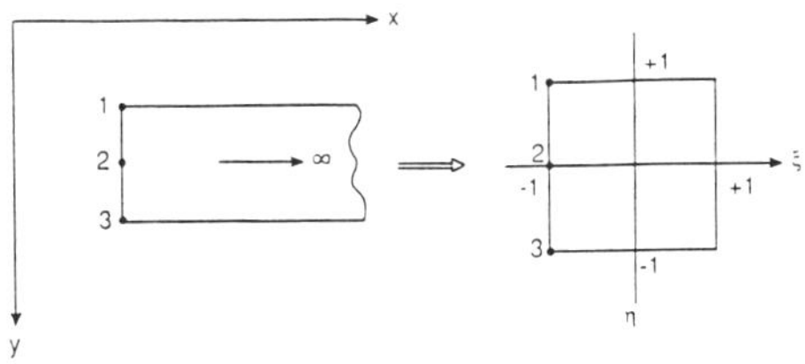

Fig. 4. Singular Contraction of Infinite Element

represents the singular contraction of an infinite element. The following equations are used in the singular contraction algorithm:

$$x = \sum_{j=1}^{3} \frac{2}{(1-\xi)} L_j(\eta) x_j \qquad (2.16)$$

$$y = \sum_{j=1}^{3} L_j(\eta) y_j \qquad (2.17)$$

The same expressions (Eqs. 2.16 and 2.17) can be elaborated in terms of shape functions as:

$$x = \sum_{j=1}^{3} N_j^x x_j \qquad (2.18)$$

$$y = \sum_{j=1}^{3} N_j^y y_j \qquad (2.19)$$

Using appropriate shape functions, the required stiffness matrices are formed in a manner similar to finite element stiffness equation formulation.

The same algorithm is used for embankment analysis but with a slight modification. The pore water pressure terms are not considered. Therefore, the effect of pore fluid and coupling effects between soil skeleton and pore fluid terms are considered to be zero. Also, since the embankment is above the ground, the embankment soil is treated as unsaturated.

3 PREPROCESSOR AND DATA CONVERTER

Manual preparation of data for the FE main program is usually a very tedious process. In addition, the data preparation needs careful checking to make sure that there is no error and that the prepared data is in proper format. In order to overcome these difficulties, a user-friendly preprocessor is developed using ACTOR IV software in a WINDOW environment. To run the preprocessor, an indepth knowledge of the finite element formulation is not essential. In addition, the preprocessor has the capability of generating the finite element mesh by default and the mesh is displayed on-screen. The range of values a parameter can take are stored in the program and whenever the user-specified inputs are out of bounds, the program warns the user and displays the proper range. For a parametric study of any bridge approach site, the preprocessor developed in this study is expected to be a very useful tool.

The following details are needed from the field and the laboratory investigations to run the preprocessor. Sources such as construction records maintenance records have been found to be very helpful in preparation of the data for FE analysis of a given site.

Geometric Configuration: The information about the depth of foundation, encountering different soil layers can be obtained from the construction records or from the boring-log. Findings from the laboratory testing will also be helpful in selecting the thickness of each layer. The previous studies (5,6) performed with different mesh types revealed that a FE mesh with twice the base width of the embankment provides very good results in terms of the settlements and excess pore pressure distributions. In the computer program developed, the width of the soil domain equal to 2.5 times the base width of embankment loading is considered for the purpose of analysis.

Material Properties: The soil properties of each layer are needed for the analysis. Young's modulus (E), coefficient of permeability (k), unit weight (γ), slope of the critical state line (M), compression index (λ), swelling index (κ), and initial void ratio (e_o) are required for defining the nonlinear soil behavior and must be obtained for each layer from appropriate laboratory tests.

Loading: The loading on foundation due to embankment construction can be determined from the embankment height (H_e) and the average unit weight of the embankment material. Since consolidation is a time-dependent process, the time taken for the construction of a particular height of embankment must be defined for the calculation of overburden load on foundation. The construction records, if available, may also be used for computing the embankment load on foundation at a particular time. In the absence of the construction records, empirical formulae relating the embankment height (H_e) to the construction time (T_c) can be used.

The following relationship, given by Hopkins (4) is used in the present study to estimate the loading rate:

$$T_c = 10^{(1.2376 \log_{10} H_e + 0.1122)} \qquad (3.1)$$

For embankment analysis, the major loadings will involve the self weight of the embankment and the vehicular traffic load. Self weight can be computed from the dimensions of the embankment and the unit weight of embankment materials. Maximum vehicular traffic load that is anticipated in the field can be obtained from Highway Standards.

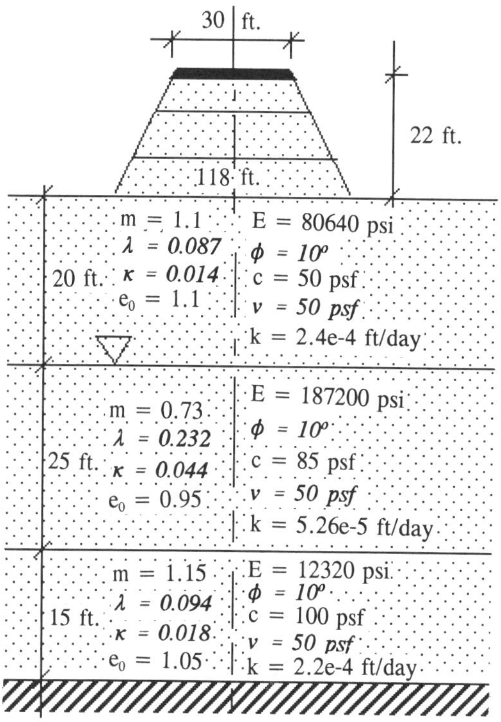

Fig. 5. Foundation Input Data

Fig. 5 gives the sample data prepared for the analysis of bridge approach settlement at East Cordell bridge site, Washita County, Oklahoma, USA.

The data generated by the preprocessor cannot be directly used for the analysis of approach settlements for the following reasons:

* The preprocessor output contains information regarding both foundation and embankment, but the FE main program is capable of analyzing either foundation or embankment at a time.

* The FE main program requires data to be in a specified format. Thus, the output from the preprocessor is compatible for the FE main program.

To convert the preprocessor data, a data converter program is developed in FORTRAN 77. This program takes the data from the preprocessor and separates it into embankment and foundation data in a compatible form for the FE main program.

4 THE POSTPROCESSOR

The finite element (FE) main program gives detailed information on the analysis of both foundation and embankment. The FE outputs are generally in the form of tables which are not very easy to comprehend and to draw inferences from. In order to overcome these difficulties, a 'postprocessor' is developed. The postprocessor helps in presenting the results from the FE main program in the form of easy to visualize graphical plots which are much more convenient to interpret.

The postprocessor makes use of a graphic software package, Graphic 6.0, developed by Scientific Endeavors Corporation, TN, USA. This software contains a number of subroutines which can be invoked for plotting different results. The programs are written in C programming language.

Following are the different plots:

a. Foundation Analysis:
   i. surface settlement versus horizontal distance.
   ii. excess pore pressure versus depth.
b. Embankment Analysis:
   i. surface settlement versus horizontal distance.
   ii. settlement profile along the central section.

5 NUMERICAL/GRAPHICAL RESULTS

The East Cordell bridge site in Washita County, Oklahoma, USA, was selected to demonstrate an application of the aforementioned FE code. The bridge construction was completed in the year 1976. Since the end of construction, it has been reported that, the approach has experienced a total settlement of 18 inches. A cross section of the approach very close to the bridge abutment is shown in Fig. 5. To illustrate the effects of element size on the results, the program is run for two different embankment meshes and the results from these two runs are compared.

The foundation soil domain is discretized by 280 nodes and 90 elements. Due to symmetry, only one-half of the cross-section is considered in the analysis. Beyond 147.5 feet from the center line of the cross-section, the infinite elements are considered. Reports indicate that the 22 feet high embankment was completed in about 50 days. Fig. 6 indicates the variation of settlement along the foundation surface due to embankment construction at various times after the completion of

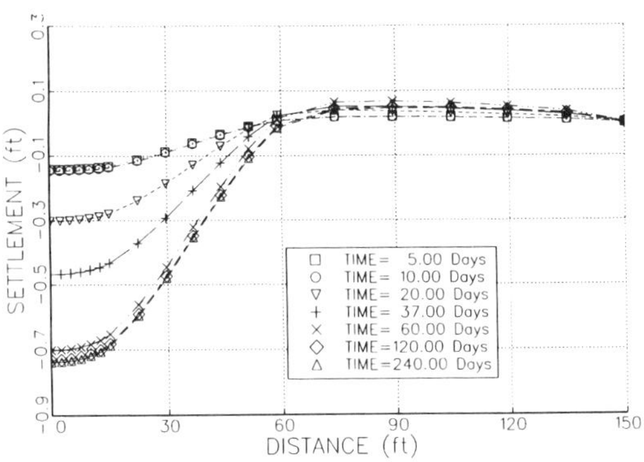

Fig. 6. Foundation Surface Settlement Plots

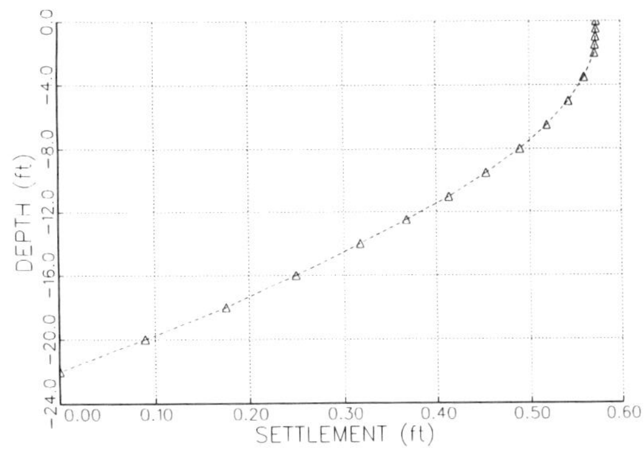

Fig. 9. Embankment Settlement Profile Along Depth

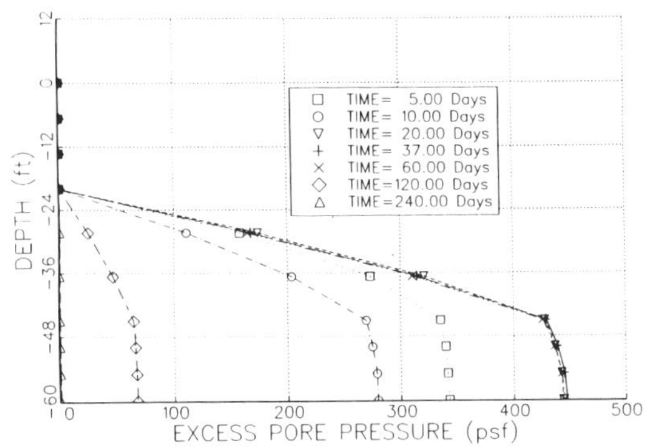

Fig. 7. Foundation Pore Pressure Distribution Plots

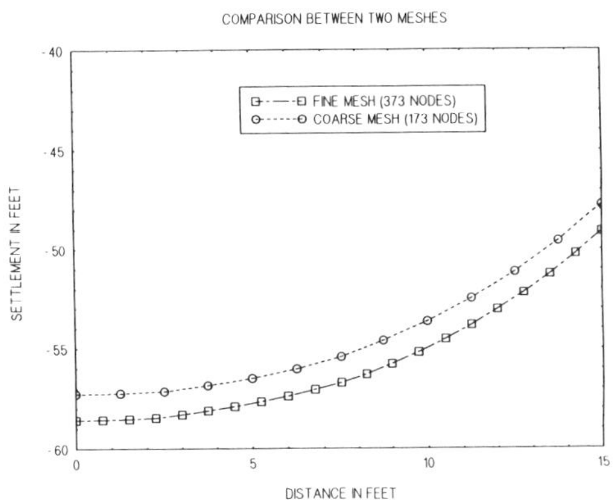

Fig. 10. Comparison Of Coarse and Fine Embankment Mesh Results

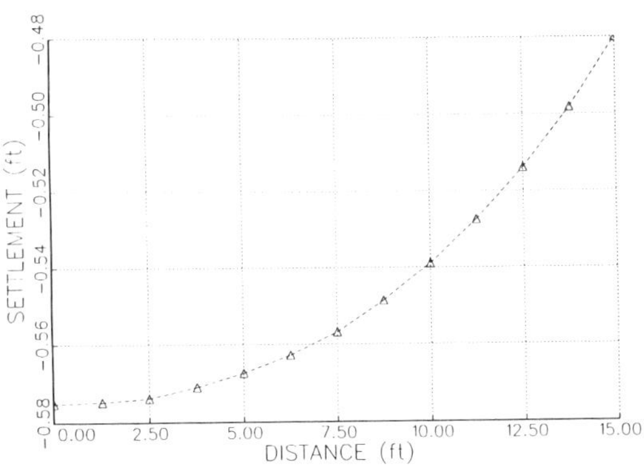

Fig. 8. Embankment Surface Settlement Profile

embankment construction. A maximum settlement of 0.75 feet (9 inches) was observed at the center of the cross-section. This constitutes about 50% of the total observed settlement at the site. Fig. 7 shows the excess pore pressure distribution in the foundation cross-section. Water table was taken at 20 feet depth and since the algorithm assumes that soil is unsaturated above the water table, no excess pore pressure was developed above this level. It took 240 days for complete dissipation of excess pore pressure developed due to consolidation process.

The embankment was discretized by 173 nodes and 48 elements. Maximum axle load is taken as 32 kips. Fig. 8 represents the surface settlement profile and Fig. 9 shows the variation of settlement along the depth of the embankment. A maximum settlement of 0.575 feet (6.9 inches) is observed at the center of the cross-section. This constitutes about 38% of the total observed settlement at the site.

In order to study the effect of element size on the

analysis, the embankment is discretized using a mesh with 373 nodes and 110 elements. Settlements obtained by both meshes are compared in Fig. 10. It is observed that the maximum difference in settlement is only 1.7%. Hence, the first discretization with only 173 nodes and 48 element appears to be adequate for the parametric study (not presented here), as it is economical in terms of computing time.

6 CONCLUSIONS

The analysis of the data obtained in this study leads to the following conclusions:

1. The nonlinear finite element technique is found to be a powerful tool for analyzing consolidation settlements of bridge approach pavements.
2. The infinite element concept better represent the actual situation since the soil domain is considered to be infinite in lateral directions.
3. Addition of preprocessor and postprocessor make the code user-friendly and it makes possible the analysis of a given bridge approach without indepth knowledge of finite element method (FEM).
4. The code can be used in analysis of settlement potential of existing pavements, as well as in design of new pavements.

7 ACKNOWLEDGEMENTS

Financial support for this study was provided in part by the Oklahoma Department of Transportation (ODOT), Oklahoma, USA. Prof. C.S. Desai of University of Arizona provided the original FE code. Technical assistance of Dr. A. Alvappillai and Mr. A. Gopalasingam is appreciated.

REFERENCES

1. Biot, M.A., General Theory of Three Dimensional Consolidation, *J. Applied Physics,* vol. 12, 1941, pp. 155-164
2. Booker, J.R., and Small, J.C., An Investigation of Stability of Numerical Solutions of Biot's Equations of Consolidation, *Int. J. Solids and Structures,* 1975, pp. 151-172.
3. Desai, C.S., and Abel, J.F., *Introduction to Finite Element Method,* Van Nostrand Reinhold Company, New York, 1972.
4. Hopkins, T.C., Long Term Movements of Highway Bridge Approach Embankments and Pavements, *Report UKTRP-85-12,* Kentucky Transportation Research Program, Lexington, Kentucky, 1985.
5. Gopalasingam, A., *M. S. Thesis,* The University of Oklahoma, OK, 1989
6. Laguros, J.G., Zaman, M.M., Alvappillai, A., Kyriakos, E.V., Evaluation of Causes of Excessive Settlements of Pavements Behind Bridge Abutments and Their Remedies - Phase III, *Progress Report,* The University of Oklahoma, OK, 1991
7. McNamee, J., and Gibson, R.E., Displacement Functions and Linear Transforms Applied to Diffusion Through Porous Elastic Media, Q. *J. Mech. Appl. Math.,* Vol. 13, 1960, pp. 98-111.
8. Rajapakse, R.K.N.D., and Karasudhi, P., Elastostatic Infinite Elements for Layered Half Spaces, *J. of Eng. Mech.,* ASCE, Vol 111, 1985, pp. 1144-1158
9. Roscoe, K.H., and Burland, J.B., On the *Generalized Stress Strain Behavior of Wet Clay, Engineering Plasticity,* Cambridge University Press, pp. 535-609.
10. Roscoe, K.H., Schofield, A.N., and Thurairaja, A., Yielding of Clays in States Wetter than Critical, *Geotechnique,* Vol. 13, No. 3, 1963, pp. 211-240.
11. Sandhu, R.S., and Wilson, E.L., Finite Element Analysis of Flow of Saturated Porous Elastic Media, *J. Eng. Mech. Div., ASCE,* Vol. 95, No. EM3, 1969, pp. 641-652.
12. Schofield, A.N., and Wroth, C.P., *Critical State Soil Mechanics,* McGraw-Hill Book Co., London, 1968.
13. Zaman, M.M., Gopalasingam, A., and Laguros, J.G., Consolidation Settlement of Bridge Approach Foundation, *J. Geotechnical Engr.,* ASCE, Geotechnical Engineering Div., Vol. 117, No. 2, Feb. 1991, pp. 219-240.

Impact of Computational Mechanics on Engineering Problems, Pulmano & Murti (eds)
© 1993 Balkema, Rotterdam, ISBN 90 5410 324 8

Seismic response of pile reinforced sections of Sardis dam

W.D.Liam Finn & G.Wu
University of British Columbia, Vancouver, B.C., Canada

R.H.Ledbetter
US Army Corps of Engineers, Waterways Experiment Station, Vicksburg, Miss., USA

ABSTRACT: The Sardis dam has the potential for large upstream deformations because of the hazard of liquefaction in a thin foundation layer during the design earthquake. One option for remediation is to nail the upstream slope using piles. The procedures for estimating the moments and shears resulting from liquefaction and seismic accelerations are described.

1 INTRODUCTION

Previous studies have shown that the upstream section of Sardis Dam, Mississippi, has the potential to undergo large deformations upstream along a layer of weak clayey silt in the upper stratum clay (Fig. 1) should the design earthquake occur (Finn et al., 1991). Due to liquefaction, the shear strength in this layer drops from an initial value in the region of 2000 psf to a residual value of $S_{ur} = 0.075\ \sigma'_{vo}$ where σ'_{vo} is the effective overburden pressure. One option being considered for restraining this deformation is to drive prestressed concrete piles through the upstream slope and the weak layer to an adequate penetration in the stronger layers below.

These piles must be designed to resist the combined effects of the pressures resulting from the restraint of the potentially large flow deformations and the moments and shears developed by seismic shaking during the design earthquake.

The key factors controlling the feasibility and cost of pile installations to restrain flow deformations are pile length and spacing, stiffness and strength of unliquefied soils surrounding the piles, residual strength of liquefied soils, the geometry of the structure, and the intensity of shaking after liquefaction has occurred. The ability to analyze such a complex problem while taking into account nonlinear behaviour of soil, potentially large strain deformations in unremediated

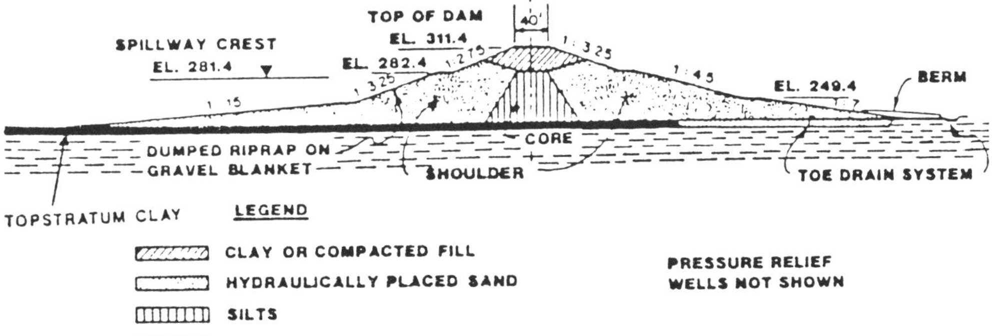

Figure 1. Typical section of Sardis dam.

parts of the structure and a realistic interaction between piles and soil during both static and seismic loading is the essential requirement for determining the best location for the piles, an appropriate length and size and for categorizing the effects of soil properties.

Very little is known about the behaviour of piles under these complex conditions. Most of the evidence is from Japan where pile foundations have been severely damaged in liquefied ground as a result of ground displacements. However these piles were designed for vertical static loading only and both piles and the connections to the pile caps were inadequately designed to resist horizontal loading. Piles in Oakland Harbor were similarly damaged during the Loma Prieta Earthquake of 1989.

Detailed analytical studies were conducted to obtain data on the potential performance of piles resisting flow deformations in Sardis Dam and hence to provide a framework of understanding for the design of cost effective remediation using piles. These studies were conducted using the computer programs TARA-3 (Finn et al., 1986) and TARA-3FL (Finn and Yogendrakumar, 1989) which have proved useful in the analysis of flow deformations, settlements and seismic response of earth structures and soil-structure interaction systems (Finn, 1988, 1990). The programs are not described here but detailed information is found in Finn (1988 and 1990).

The use of these analyses has allowed an acceptable displacement criterion to be used in defining satisfactory performance in addition to the common global factor of safety approach. A problem with remediation measures designed solely on the basis of global factors of safety is that there is little understanding of what some of these factors mean in terms of displacements of the structure. The use of the TARA programs allows a state of the art implementation of Newmark's (1965) recommendation that the performance of earth structures be assessed using deformation criteria.

2 ANALYSIS OF DEFORMATION RESTRAINING PILES

A preliminary design option for restraining the flow deformations in Sardis Dam is to drive prestressed concrete piles, 2 ft by 2 ft in cross-section, through the upstream slope and the weak layer to a penetration of 15 ft below the weak layer. A feasible location for the piles is shown in Fig. 2.

The piles are assumed to be at 12 ft centres giving the plan layout shown in Fig. 3. Each row of piles normal to the axis of the dam is assumed to resist the potential movement of a plane strain section 12 ft wide. Analyses were conducted using a 1 ft thick section.

Analysis of pile response is conducted in two stages. First the effects of triggering residual strength in the layer of clayey silt are analyzed using the program TARA-3FL (Finn and Yogendrakumar, 1989). This analysis gives the moments and deflections in the piles arising from restraining the potential upstream movement of the embankment. Then the dam and piles, in the deformed condition, are subjected to earthquake shaking representative of design conditions by using the S00E component of the 1940 El Centro acceleration record scaled to 0.2g as input motions for a dynamic analysis using the program TARA-3 (Finn et al., 1986). This procedure for analysis

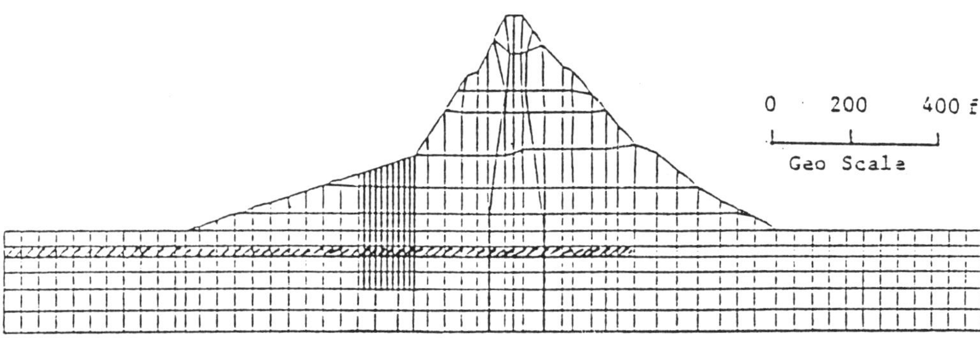

Figure 2. Section of Sardis dam showing weak layer and restraining piles.

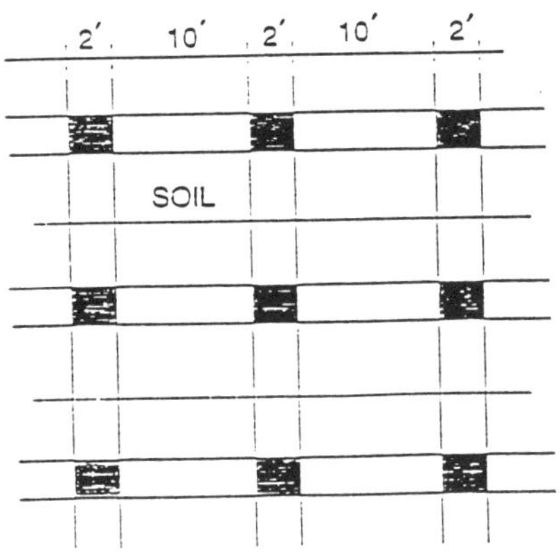

Figure 3. Layout of restraining piles.

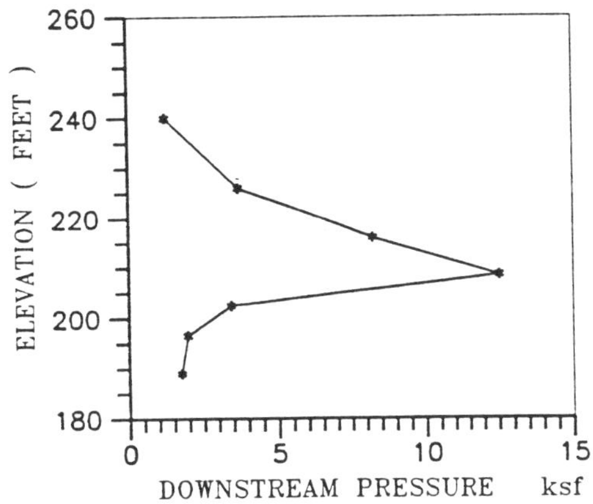

Figure 4. Lateral pressures on lead piles after liquefaction.

simulates the condition where liquefaction and strength loss occur early in the earthquake and is the most conservative case.

3 ANALYSIS OF SARDIS DAM

The first step in the analysis is to construct the dam in layers to establish the initial stress conditions in the dam before remediation. Then the piles are inserted at the appropriate locations.

The liquefaction process is initiated and the undrained strengths in the liquefied materials are gradually reduced to the residual values. During this reduction in strength, the loss in resistance to lateral deformation along the layer of clayey silt is transformed to lateral load on the piles. The piles continue to deflect until an equilibrium state is achieved.

3.1 Flow Deformation Analysis

The weak layer of clayey silt in the dam, shown cross-hatched in Fig. 2 is located between EL200 and EL205. After liquefaction the lateral pressures on the lead pile are shown in Fig. 4.

The distribution of moments along the lead pile on the downstream side of the remediated zone is shown in Fig. 5. The peak moment is 11,000 kip in, just above the weak layer.

The distributions of maximum shears and moments in piles along the remediated section are shown in Fig. 6 and Fig. 7, respectively. It is evident that the lead row of piles takes by far the largest moments and that these piles present the

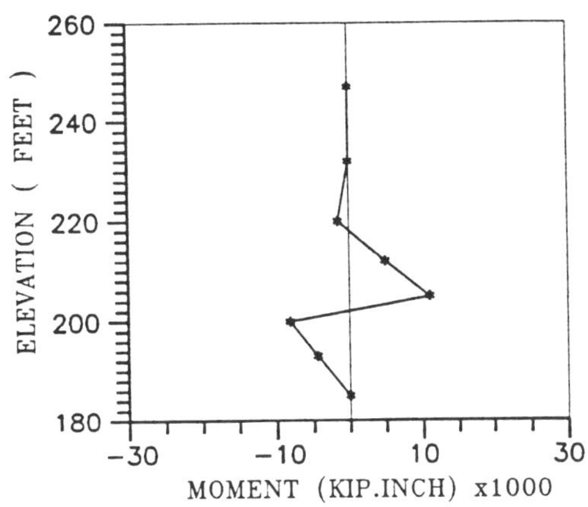

Figure 5. Distribution of moments along lead piles after liquefaction.

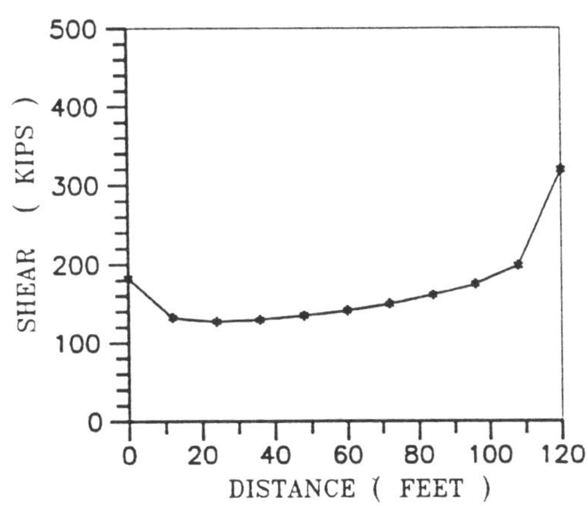

Figure 6. Distribution of maximum shears in piles along remediated section.

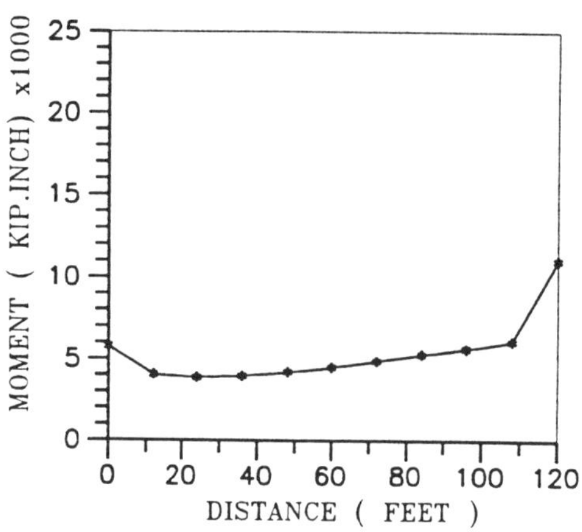

Figure 7. Distribution of maximum moments in piles along remediated section.

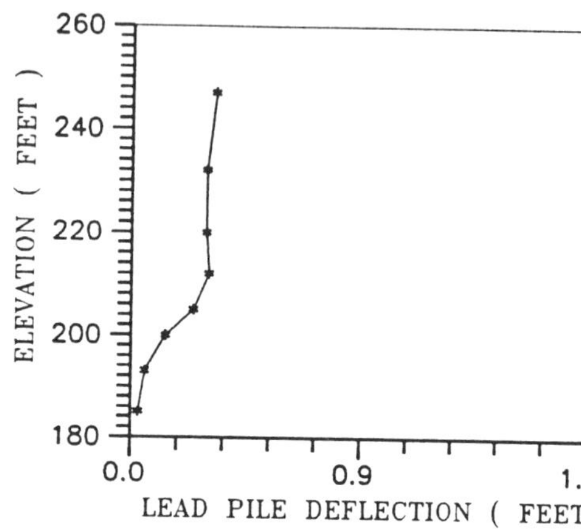

Figure 8. Deflection pattern in the leading row of piles.

major challenge to effective design. The lead piles play such a dominant role because in the absence of a pile cap the only load transfer mechanism between the lead piles and the interior piles is the deflection of the lead piles towards the interior piles and the transmission of the associated increases in lateral pressures from one pile to another.

The deflection of piles in the leading row are shown in Fig. 8. The peak deflection is of the order of 3 inches.

3.2 Seismic Analysis by TARA-3

The dynamic response of the remediated dam after triggering of residual strength by liquefaction was evaluated using the program TARA-3 (Finn et al., 1986). The input motions were the first 20s of the S00E acceleration component of the 1940 El Centro earthquake scaled to 0.2g.

The time history of dynamic moments in the leading row of piles at the top of the weak layer is shown in Fig. 9. The dynamic moments are plotted starting from the post-liquefaction moment. This plot therefore, gives the total moment at any instant during shaking. The peak moment occurring at t = 2.2s in the record is 17,000 kip in and the residual moment in the pile after the earthquake is about 9,000 kip in.

The distribution of maximum dynamic moments among the rows of piles is shown in Fig. 10 and the distribution of total moments in Fig. 11. Pile location is measured with respect to the upstream end of the remediated zone at 0 ft.

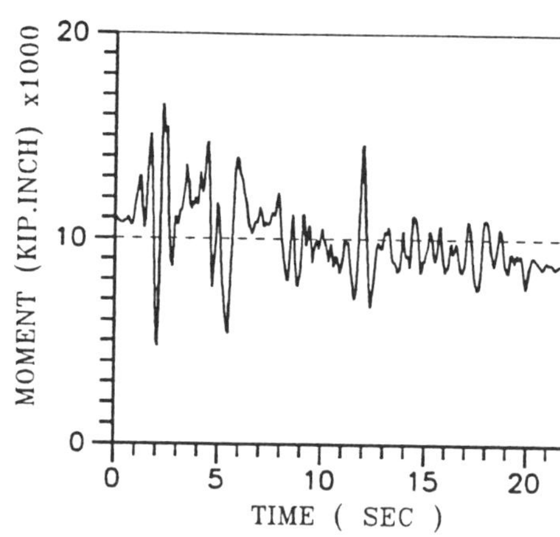

Figure 9. Time history of total moment in the lead piles at top of weak layer.

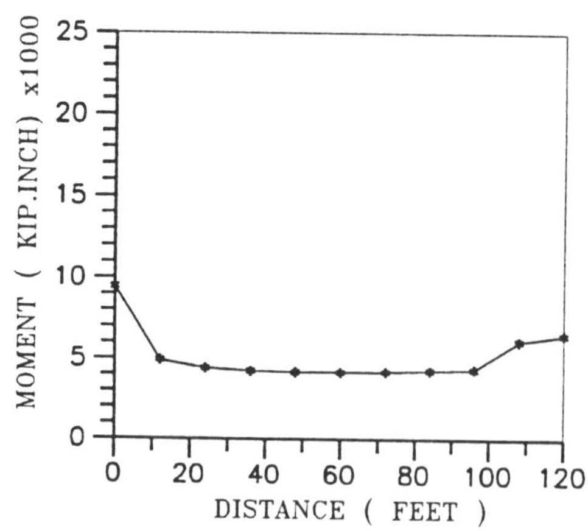

Figure 10. Distribution of maximum dynamic moments in piles along remediated section.

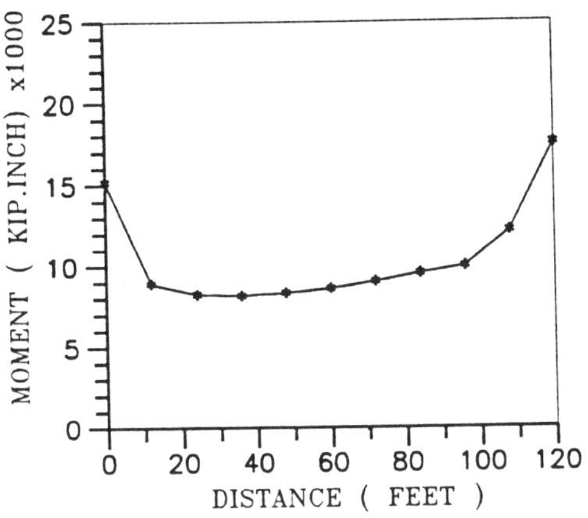

Figure 11. Distribution of combined maximum static and dynamic moments in piles along remediated section.

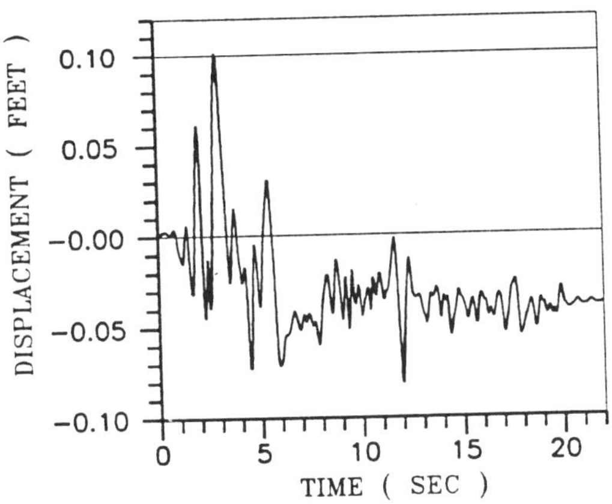

Figure 12. Time history of deformations at the top of the lead piles.

The time history of dynamic deformations at the top of the lead pile is shown in Fig. 12. The combined peak static and dynamic deflections of the lead pile upstream is of the order of 4 inches.

4 CONCLUSIONS

The analyses show that potential flow deformations in Sardis Dam resulting from liquefaction can be restrained by piles. The behaviour of the structure after liquefaction with the piles in place is a coupled problem. Any solution must track the mutual interaction between soil and piles as the soil softens due to increasing porewater pressures and especially when residual strength is triggered. The loads that are transferred to the piles and the interactive response depend on the processes of stress transfer from soil to piles. These loads cannot be calculated independently. Neither can the subsequent response be modelled accurately if the residual strengths are included in the structure from the beginning instead of being triggered during the earthquake.

It is also necessary to treat the dynamic response of the structure in a fully coupled manner to estimate reliably the dynamic stresses and displacements in the restraining piles.

TARA-3 and TARA-3FL analyses with the focus on deformation estimation are a valuable adjunct to the conventional factor of safety approach used to assess the extent of remediation required in structures with some potential for liquefaction. An estimate of the deformation field associated with a planned degree of remediation will allow a cost-effective judgement to be made about the most effective locations for remediation and the scope of remediation required to satisfy performance criteria.

The piles must resist not only the static bending moments due to flow deformations but also the peak dynamic moments. The selection of the design moment is the key factor controlling design of the piles.

The leading row of piles takes about two times the peak moment of the second row. This is a potentially dangerous situation. Designers should take into account the possibility of progressive failure of the rows of piles.

These preliminary studies indicated that it would be feasible to restrain potential deformations in the dam by nailing the upstream slope to the stable lower strata using piles. Further studies are now underway using more refined analyses to determine the optimum pile size and spacing.

5 REFERENCES

Finn, W.D. Liam. (1990) Seismic Response of Embankment Dams. Invited State-of-the-Art Paper, Dam Engineering, Vol. I, January, pp. 59-75.

Finn, W.D. Liam. (1988) Dynamic Analysis in Geotechnical Engineering. Earthquake Eng. and Soil Dynamics II - Recent Advances in Ground Motion Evaluation, Geotechnical

Special Publication No. 20, ASCE, August, pp. 523-591.

Finn, W.D. Liam, Ledbetter, R.H., Fleming Jr., R.L., Templeton, A.E., Forrest, T.W. and Stacy, S.T. (1991) Dam on Liquefiable Foundation: Safety Assessment and Remediation. Proc., 17th Int. Congress on Large Dams, Vienna, pp. 531-553, June .

Finn, W.D. Liam and Yogendrakumar, M. (1989) TARA-3FL - Program for Analysis of Liquefaction Induced Flow Deformations. Dept. of Civil Engineering, University of British Columbia, Vancouver, B.C., Canada.

Finn, W.D. Liam, Yogendrakumar, M, Yoshida, N. and Yoshida, H. (1986). TARA-3: A Program for Nonlinear Static and Dynamic Effective Stress Analysis. Soil Dynamics Group, University of British Columbia, Vancouver, B.C..

Newmark, N.M. (1965) Effects of Earthquakes on Dams and Embankments, 5th Rankine Lecture, Geotechnique, Vol. 15, No. 2, June, pp. 139-160.

Impact of Computational Mechanics on Engineering Problems, Pulmano & Murti (eds)
© 1993 Balkema, Rotterdam, ISBN 90 5410 324 8

Author index